Studies in Computational Intelligence

Volume 1001

The series “Studies in Computational Intelligence” (SCI) publishes new developments and advances in the various areas of computational intelligence—quickly and with a high quality. The intent is to cover the theory, applications, and design methods of computational intelligence, as embedded in the fields of engineering, computer science, physics and life sciences, as well as the methodologies behind them. The series contains monographs, lecture notes and edited volumes in computational intelligence spanning the areas of neural networks, connectionist systems, genetic algorithms, evolutionary computation, artificial intelligence, cellular automata, self-organizing systems, soft computing, fuzzy systems, and hybrid intelligent systems. Of particular value to both the contributors and the readership are the short publication timeframe and the world-wide distribution, which enable both wide and rapid dissemination of research output.

Indexed by SCOPUS, DBLP, WTI Frankfurt eG, zbMATH, SCImago.

All books published in the series are submitted for consideration in Web of Science.

More information about this series at https://link.springer.com/bookseries/7092

Mohamed Lahby · Al-Sakib Khan Pathan ·
Yassine Maleh · Wael Mohamed Shaher Yafooz
Editors

Combating Fake News with Computational Intelligence Techniques

Springer

Editors
Mohamed Lahby
Hassan II University
Casablanca, Morocco

Yassine Maleh
Sultan Moulay Slimane University
Khouribga, Morocco

Al-Sakib Khan Pathan
United International University
Dhaka, Bangladesh

Wael Mohamed Shaher Yafooz
Taibah University
Madinah, Saudi Arabia

ISSN 1860-949X ISSN 1860-9503 (electronic)
Studies in Computational Intelligence
ISBN 978-3-030-90086-1 ISBN 978-3-030-90087-8 (eBook)
https://doi.org/10.1007/978-3-030-90087-8

This Springer imprint is published by the registered company Springer Nature Switzerland AG
The registered company address is: Gewerbestrasse 11, 6330 Cham, Switzerland

Preface

With the advancement of electronics and information technologies, social media and virtual space have become a significant part of human life today. People can share and exchange information and ideas globally just with a single click. A piece of information or news can impact the people and enterprises. Such an influence can be positive or negative. Online news has made the hustling lives of the present society much convenient. Traditionally, news articles would be published by credible sources such as newspapers, TV channels, and magazines. Relevant authorities could regulate these sources. However, falsified news can often spread swiftly through social media networks such as Facebook, Twitter, Instagram, and YouTube because these platforms are easily accessible by the general public nowadays.

Detecting fake news is a challenge among the Web mining community, especially on social media. Despite the efforts of major social media platforms to fight against disinformation, things like miracle cures and conspiracy theories often surface on Facebook and Twitter. Artificial intelligence can be a bulwark against the diversity of fake news on the Internet and social networks. In the recent days, researchers have shown great interest in tackling such issues through novel methods, approaches, datasets, and techniques.

This book presents the latest cutting-edge research, theoretical methods, and novel applications in computational intelligence techniques to combat fake news. The discussion of new models, practical solutions, and technological advances related to detecting and analyzing falsified news is based on computational intelligence models and techniques. These help decision-makers, managers, professionals, and researchers design new paradigms with the unique opportunities associated with computational intelligence techniques. Furthermore, the book helps readers understand the computational intelligence techniques to combat fake news systematically and forthrightly.

This book contains a total of 21 chapters classified into four main parts. The first part presents a state of the art related to the use of machine learning techniques and deep learning techniques for combating fake news. The second part focuses on machine learning techniques for combating fake news. The third part presents frameworks and valuable case studies that can be used in the context of fake news.

Finally, the last part provides some solutions based on computational Intelligence techniques to deal with fake news in the period of COVID-19 pandemic.

We want to take this opportunity and express our sincere thanks to the contributors to this volume and the reviewers for their great efforts in reviewing and providing interesting feedback to the authors of the chapters. The editors would like to thank Dr. Thomas Ditzinger (Springer, Editorial Director, Interdisciplinary Applied Sciences), Prof. Janusz Kacprzyk (Series Editor in Chief), and Mr. Ramamoorthy Rajangam (Springer Project Coordinator), for the editorial assistance and support to produce this important scientific work. Without this collective effort, this book would not have been possible to be completed.

Casablanca, Morocco — Prof. Mohamed Lahby
Dhaka, Bangladesh — Prof. Al-Sakib Khan Pathan
Khouribga, Morocco — Prof. Yassine Maleh
Madinah, Saudi Arabia — Prof. Wael Mohamed Shaher Yafooz

Contents

State-of-the-Art

Online Fake News Detection Using Machine Learning Techniques: A Systematic Mapping Study 3
Mohamed Lahby, Said Aqil, Wael M. S. Yafooz, and Youness Abakarim

Using Artificial Intelligence Against the Phenomenon of Fake News: A Systematic Literature Review 39
Mustafa A. Al-Asadi and Sakir Tasdemir

Fake News Detection in Internet Using Deep Learning: A Review 55
Israel Barrutia-Barreto, Renzo Seminario-Córdova, and Brian Chero-Arana

Machine Learning Techniques and Fake News

Early Detection of Fake News from Social Media Networks Using Computational Intelligence Approaches 71
Roseline Oluwaseun Ogundokun, Micheal Olaolu Arowolo, Sanjay Misra, and Idowu Dauda Oladipo

Fandet Semantic Model: An OWL Ontology for Context-Based Fake News Detection on Social Media 91
Anoud Bani-Hani, Oluwasegun Adedugbe, Elhadj Benkhelifa, and Munir Majdalawieh

Fake News Detection Using Machine Learning and Natural Language Processing 127
Mansi Patel, Jeel Padiya, and Mangal Singh

Fake News Detection Using Ensemble Learning and Machine Learning Algorithms 149
Sanaa Elyassami, Safa Alseiari, Maryam ALZaabi, Anwar Hashem, and Nouf Aljahoori

Evaluation of Machine Learning Methods for Fake News Detection 163
Dimitrios Papakostas, George Stavropoulos, and Dimitrios Katsaros

Credibility and Reliability News Evaluation Based on Artificial Intelligent Service with Feature Segmentation Searching and Dynamic Clustering 185
Ming-Shen Jian

Deep Learning with Self-Attention Mechanism for Fake News Detection 205
Ivana Cvitanović and Marina Bagić Babac

Modeling and Solving the Fake News Detection Scheduling Problem 231
Said Aqil and Mohamed Lahby

Case Studies and Frameworks

The Multiplier Effect on the Dissemination of False Speeches on Social Networks: Experiment during the Silly Season in Spain 245
Cristóbal Fernández-Muñoz, Ángel Luis Rubio-Moraga, and David Álvarez-Rivas

Detecting News Influence in a Country: One Step Forward Towards Understanding Fake News 259
Cristian Pop and Alexandru Popa

Factors Affecting the Intention of Using Fintech Services in the Context of Combating of Fake News 277
Lam Oanh Ha, Van Chien Nguyen, Do Dinh Thuy Tien, and Bui Thi Bich Ngoc

Crowd Sourcing and Blockchain-Based Incentive Mechanism to Combat Fake News 299
Munaza Farooq, Aqsa Ashraf Makhdomi, and Iqra Altaf Gillani

Framework for Fake News Classification Using Vectorization and Machine Learning 327
Yogita Dubey, Pushkar Wankhede, Amey Borkar, Tanvi Borkar, and Prachi Palsodkar

Fact Checking: An Automatic End to End Fact Checking System 345
Sajjad Ahmed, Knut Hinkelmann, and Flavio Corradini

Fake News and COVID-19 Pandemic

False Information in a Post Covid-19 World 369
Mohiuddin Ahmed, Chris Martin, Tristram Walker, and James Van Rooyen

Applying Fuzzy Logic and Neural Network in Sentiment Analysis for Fake News Detection: Case of Covid-19 ... 387
Bahra Mohamed, Hmami Haytam, and Fennan Abdelhadi

Analyzing Deep Learning Optimizers for COVID-19 Fake News Detection ... 401
Ayan Chakraborty and Anupam Biswas

Detecting Fake News on COVID-19 Vaccine from YouTube Videos Using Advanced Machine Learning Approaches ... 421
Wael M. S. Yafooz, Abdel-Hamid Mohamed Emara, and Mohamed Lahby

About the Editors

Mohamed Lahby is working as Assistant Professor of Computer Science at the Higher Normal School (ENS) University Hassan II of Casablanca, Morocco. He obtained his Ph.D. in Computer Science from the Faculty of Sciences and Technology of Mohammedia, University Hassan II of Casablanca, in 2013. His research interests are wireless communication and network, mobility management, QoS/QoE, Internet of things, smart cities, optimization, and machine learning. He has published more than 35 papers (chapters, international journals, and conferences), 3 edited books, and 2 authored books. He has served and continues to serve on executive and technical program committees of numerous international conferences such as IEEE PIMRC, ICC, NTMS, IWCMC, WINCOM, and ISNCC. He also serves as a referee of many prestigious Elsevier journals: *Ad Hoc Networks*, *Applied Computing and Informatics*, and *International Journal of Disaster Risk Reduction*. He organized and participated in more than 40 conferences and workshops. He is Chair of many international workshops and special sessions such as MLNGSN'19, CSPSC'19, MLNGSN'20, AI2SC '20, WCTCP'20, and CIOT'2021.

Al-Sakib Khan Pathan is a Professor at Computer Science and Engineering department, United International University, Bangladesh and Adjunct Professor at Independent University, Bangladesh. He received Ph.D. degree in Computer Engineering in 2009 from Kyung Hee University, South Korea, and B.Sc. degree in Computer Science and Information Technology from Islamic University of Technology (IUT), Bangladesh, in 2003. In his academic career so far, he worked as Faculty Member at the CSE Department of Southeast University, Bangladesh, during 2015–2020; Computer Science Department, International Islamic University Malaysia (IIUM), Malaysia, during 2010–2015; BRACU, Bangladesh, during 2009–2010; and NSU, Bangladesh, during 2004–2005. He was Guest Lecturer for the STEP project at the Department of Technical and Vocational Education, Islamic University of Technology, Bangladesh, in 2018. He also worked as Researcher at Networking Lab, Kyung Hee University, South Korea, from September 2005 to August 2009 where he completed his MS leading to Ph.D. His research interests include wireless sensor networks, network security, cloud computing, and e-services technologies. Currently, he is also working on some multidisciplinary issues. He is a recipient of several awards/best paper awards and has several notable publications in these areas. So far, he has delivered over 20 keynotes and invited speeches at various international conferences and events. He was awarded the IEEE Outstanding Leadership Award for his role in IEEE GreenCom'13 Conference. He is currently serving as Editor in Chief of *International Journal of Computers and Applications*, Taylor & Francis, UK; Editor of *Ad Hoc and Sensor Wireless Networks*, Old City Publishing, *International Journal of Sensor Networks*, Inderscience Publishers, and *Malaysian Journal of Computer Science*; Associate Editor of *Connection Science*, Taylor & Francis, UK, and *International Journal of Computational Science and Engineering*, Inderscience; Area Editor of *International Journal of Communication Networks and Information Security*; Guest Editor of many special issues of top-ranked journals; and Editor/Author of 21 books. One of his books has been included twice in Intel Corporation's Recommended Reading List for Developers, 2nd half of 2013 and 1st half of 2014; 3 books were included in IEEE Communications Society's (IEEE ComSoc) Best Readings in Communications and Information Systems Security, 2013, 2 other books were indexed with all the titles (chapters) in Elsevier's acclaimed abstract and citation database, Scopus, in February 2015, and a seventh book is translated to simplified Chinese language from English version. Also, 2 of his journal papers and 1 conference paper were included under different categories in IEEE Communications Society's (IEEE ComSoc) Best Readings Topics on Communications and Information Systems Security, 2013. He also serves as a referee of many prestigious journals. He received some awards for his reviewing activities like: one of the most active reviewers of IAJIT several times, Elsevier Outstanding Reviewer for Computer Networks, Ad Hoc Networks, FGCS, and JNCA in multiple years. He is Senior Member of the Institute of Electrical and Electronics Engineers (IEEE), USA.

Yassine Maleh is Associate Professor at the National School of Applied Sciences at Sultan Moulay Slimane University, Morocco. He received his Ph.D. degree in Computer Science from Hassan 1st University, Morocco. He is Cybersecurity and Information Technology Researcher and Practitioner with industry and academic experience. He worked for the National Ports Agency in Morocco as IT Manager from 2012 to 2019. He is Senior Member of IEEE and Member of the International Association of Engineers IAENG and the Machine Intelligence Research Labs. He has made contributions in the fields of information security and privacy, Internet of things security, wireless and constrained networks security. His research interests include information security and privacy, Internet of things, networks security, information system, and IT governance. He has published over 50 papers (chapters, international journals, and conferences/workshops), 7 edited books, and 3 authored books. He is Editor in Chief of the *International Journal of Smart Security Technologies* (IJSST). He serves as Associate Editor for IEEE Access (2019 Impact Factor 4.098), the *International Journal of Digital Crime and Forensics* (IJDCF), and the *International Journal of Information Security and Privacy* (IJISP). He was also Guest Editor of a special issue on Recent Advances on Cyber Security and Privacy for Cloud-of-Things of the *International Journal of Digital Crime and Forensics* (IJDCF), Volume 10, Issue 3, July–September 2019. He has served and continues to serve on executive and technical program committees and as a reviewer of numerous international conferences and journals such as Elsevier *Ad Hoc Networks*, *IEEE Network Magazine*, *IEEE Sensor Journal*, *ICT Express*, and Springer *Cluster Computing*. He was Publicity Chair of BCCA 2019 and General Chair of the MLBDACP 19 symposium.

Wael Mohamed Shaher Yafooz is Associate Professor in the Computer Science Department, Taibah University, Saudi Arabia. He was Associate Professor in the Information Technology Department at Al-Madinah International University (MEDIU), Malaysia. He was Dean of the Faculty of Computer and Information Technology. He received his bachelor degree in the area of computer science from Egypt in 2002 while a Master of Science in Computer Science from the University of MARA Technology (UiTM) in 2010 as well as a Ph.D. in Computer Science in 2014 from UiTM. He was awarded many Gold and Silver Medals for his contribution to a local and international expo of innovation and invention in the area of computer science. Besides, he was awarded the Excellent Research Award from UiTM. He served as Member of various committees in many international conferences. Additionally, he chaired IEEE international conference on smart computing and electronic enterprise 2018. Moreover, he supervised many students at the master and Ph.D. level. Furthermore, he delivered and conducted many workshops in the research area and practical courses in data management and visualization. He was invited as a speaker in many international conferences held in Bangladesh, Thailand, India, China, and Russia. His research interest includes big data, data mining, machine learning, and data management.

State-of-the-Art

Online Fake News Detection Using Machine Learning Techniques: A Systematic Mapping Study

Mohamed Lahby, Said Aqil, Wael M. S. Yafooz, and Youness Abakarim

Abstract This last decade, the amount of data exchanged on the Internet and more specifically on social media networks is growing exponentially. Fake News phenomenon has become a major problem threatening the credibility of these social networks. Machine Learning (ML) techniques represent a promising solution to deal with this issue. For that, several solutions and algorithms using Machine Learning have been proposed in literature in the recent time for detecting fake news generated by different digital media platforms. This chapter aims to conduct a systematic mapping study to analyze and synthesize studies concerning the utilization of machine learning techniques for detecting fake news. Therefore, a total number of 76 relevant papers published on this subject between 1 January 2010 and 30 June 2021 were carefully selected. The selected articles were classified and analyzed according to eight criteria: channel and year of publication, research type, study domain, study platform, study context, study category, feature, and machine learning techniques used to handle categorical data. The results showed that most of the selected papers use both features text/content and linguistic to design machine learning models. Furthermore, SVM technique, and Deep Neural Network (DNN) technique were the most binary classification algorithms used to combat fake news.

Keywords Fake news · Social networks · Machine learning · Classification · SMS

M. Lahby (✉) · Y. Abakarim
University Hassan II, Casablanca, Morocco
e-mail: lahby@ieee.org

S. Aqil
Faculty of sciences and Technology of Mohammedia, University Hassan II, Casablanca, Morocco

W. M. S. Yafooz
Department of Computer Science, College of Computer Science and Engineering, Taibah University, Madinah, Saudi Arabia
e-mail: Wyafooz@taibahu.edu.sa

M. Lahby et al. (eds.), *Combating Fake News with Computational Intelligence Techniques*, Studies in Computational Intelligence 1001,
https://doi.org/10.1007/978-3-030-90087-8_1

1 Introduction

The recent fast growth and evolution of wireless technologies such as 3G, 4G and 5G, have shifted the behavior of users concerning the utilization of the Internet [1]. At the same time, the development and deployment of several streaming servers have given the users the privilege to use different multimedia applications. Among these services, social networks such as Facebook, YouTube, Twitter, Facebook, Youtube, Instagram and Tik-ToK have a good standing and they represent the most used applications by the people, everywhere and for all times so far. Moreover, the traffic generated by digital media platforms (social networks and digital newspapers) represent more than half of the web traffic worldwide [2].

Furthermore, the use of these digital media platforms has greatly influenced o ur habits on the Internet and in our daily life. As a result, we must radically be wary of the contents posted and shared via social networks. Indeed, with widespread dissemination of information via digital media platforms, users are required to judge the credibility of the information by eliminating fake news. According to [3], there are several kinds and forms of fake news, such as misinformation, disinformation, rumors, satire news, Clickbait, fake reviews, opinion spam, fake advertisements and conspiracy theories.

Moreover, this phenomenon is becoming a worldwide issue, it is considered as one of the greatest threats to many disciplines such as political, education, science, sport, economics, health, E-commerce, etc. For instance, in relation with political domain, several face news have widely spreaded on Twitter, during the 2016 US Presidential primaries and general election campaign. In E-commerce domain, fake reviews significantly affect online consumers, online vendors, and retailers [4]. More recently, the term fake news has grown in importance in health domain due to the Covid-19 pandemic. Indeed, during this global health crisis, several fake news has been spread globally from discovering the symptoms, effect areas, and source of the pandemics. In addition, the Covid19 vaccines began in Feb 2021 was the title of many rumors, making many people worried and scared about the vaccine.

On the other hand, Machine Learning techniques [5] have shown promise to be powerful tools in a huge variety of different environments: computer vision [6], speech recognition [7], language processing [8], human-computer interaction [9], drug discovery [10], image analysis [11], recommender systems [12], bank failure prediction [13], fraud detection [14], loan credit approval [15], robotics [16], cyber-security [17] and others domains. In this context, in the recent time, several machine learning models have been developed to deal with widespread dissemination of information generated by digital media platforms.

In this chapter we conduct a systematic mapping study (SMS) to analyze and synthesize studies concerning the utilization of machine learning techniques for detecting fake news. We notice that a SMS is a method that consists of searching in the literature for all articles published in a given field, in order to carry out a statistical study based on research questions [18].

This chapter is organized as follows. In Sect. 2, we present an overview of the existing survey and reviews in literature related to fake news detection based on machine learning techniques. In Sect. 3, we describe the research methodology used in our study to conduct this survey. In Sect. 4, we report the results and we discuss the findings of this SMS. We present the implications for researchers in Sect. 5. Finally, we conclude our work and look into our future research in Sect. 6.

2 Related Work

Recently, several guidelines have been proposed to structure and understand a literature study related to the last research published in many different areas and disciplines [19]. Systematic Literature Review (SLR) and Systematic Mapping Study (SMS) are among the most commonly used approaches to conduct any survey at any research field [19].

The SLR process allows a deep analysis and synthesis of a particular topic. While the SMS process permits only superficial overview of a particular research field by counting and classifying of different contributions published in this field. It is important to point out that both SLRs and SMSs have based on rigorous research questions [19] for performing such studies. Hence, SLRs are more driven by specific research questions whilst research questions in SMS are of a higher-level.

In this section, our goal consist to better understand our research field which is focused on fake news detection based on Machine Learning. For this reason, we outline and we analyze the previous reviews associated with our topic.

On one hand, several surveys and reviews have been published in the literature, in order to summarize and understand the existing techniques for fake detection [20–24].

In the reference, Bondielli et al. [20] surveyed the different approaches to automatic detection of fake news in the recent literature. The authors Dwivedi et al. [21], presented a literature survey on various fake news detection methods. Zhang et al. [22], presented an overview of the existing datasets and fake news detection approaches. Karishma et al. [23], highlighted the existing methods applicable to both identification and mitigation for fake news detection. The authors also highlighted the significant advances in each method and their advantages and limitations. In the [24], Islam et al. provided a state-of-the-art review on the existing deep learning (DL) techniques that are used for misinformation detection on online social networks.

However, these surveys suffer from two major limitations. First, they did not use any an appropriate approach or methodology (SMS or SLR) for conducting these studies. Secondly, they did not cover all aspects of fake reviews, such as all recent machine learning algorithms and all existing datasets.

On the other hand, amongst different review based on SLR that are recently performed we would like to point out the three existing references [25–27]. In [25], Habib et al. presented a systematic literature review of the types of false information (rumors, fake news, misinformation and hoax), and described how these particularities influence detection across online content. The outcomes of this SLS also provided an overview of existing machine learning method and deep learning techniques for each type of false information.

The authors Dylan et al. [26] conducted a systematic literature review of the existing techniques that are used to identify fake news. Based on this review, these techniques can be classified into five approaches: (1) language approach, (2) topic-agnostic approach, (3) machine learning approach, (4) knowledge-based approach, and (5) hybrid approach. In the last SLR presented in [27], the authors provided the percentage of published papers related to the field of fake news and rumor detection. This statistical study concerns different models, different datasets, and experimental setups for content and behavior analysis of fake news circulating online.

In this chapter, we perform an SMS related to the application of machine learning techniques for detecting fake news in digital media. Our motivation to conduct this SMS was primarily due to three factors. The first one, there are only three SLRs performed previously for fake news detection. The second factor is that the SMS approach can ensure significant benefits over SLR technique. Finally, to the best of our knowledge, there are only five references, namely, Refs. [28–31], which have introduced the SMS methodology for conducting literature review on fake news detection. In what follows, we describe each of these references in order to identify its advantages and disadvantages.

The authors [28] conducted a systematic mapping study of recent published articles that deal with the misinformation spread on social media. The selected studies were analyzed and classified according to three criteria: areas of misinformation, research type and publication channels. The outcome of this SMS indicated that most studies focus on providing solutions to detect misinformation spread on social media platforms. In [29], the authors have presented a systematic mapping study related to fake news detection based machine learning techniques. The objective of the proposed SMS is analyzing the current state the research on machine learning techniques used to combat fake news phenomenon and identifying the main challenges and methodological gaps to motivate future research. In [30], de Souza et al. performed a systematic mapping study of papers published to combat fake news in social media. A total number of 87 papers were selected out of 1333 candidates in order to identify and classify the existing computational solutions for the automatic detection of fake news in social networks. Finally, in [31], Caio et al. carried out a systematic mapping study of recent articles published for combating fake news in the Big Data context. A total number of 35 articles were selected and analyzed in order to identify and analyze intelligent computing techniques used to deal with this phenomenon.

However, the major limitations of these few existing review articles are related to the quality of the articles selected and the proposed research questions to conduct these different systematic mapping studies. Therefore, in this chapter, our proposed SMS is based on 76 relevant papers published in the period from 1 January 2010 and 30 June 2021. The main purpose of this SMS is having a clear vision about the recent advances in fake news detection based machine learning techniques. Particularly, we aim to give readers the opportunity to understand the followings keys: (1) the most frequently used machine learning algorithms to detect fake news, (2) the social media platforms that represent the major source of fake news, (3) the domain fields that have been targeted the literature, (4) where the literature has been published, (5) which kind of fake news have been covered in in the literature and (6) which kind of features have been used for fake news detection.

3 Research Methodology

In this section, we describe the research methodology used to conduct this SMS on fake news detection. This methodology is based on the guidelines presented in [19]. The basic idea of this guidelines is shown in Fig. 1. According to this figure, the SMS process is composed of five steps: (1) defining the Research Questions (RQ); (2) formulating the search string; (3) Selecting studies; (4) Extracting Data from selected paper and (5) analysis and classification of Data. Thereafter, we describe the role of each step and how it can be applied in the context of fake news detection.

3.1 Defining Research Questions

This step, allows the preparation of research questions, in order to obtain a comprehensive overview of published works about the use of machine learning techniques for detecting fake news in digital media. For that, we define eight RQ that are presented in Table 1. For each research question, we explain the rationale behind its adoption for this study.

3.2 Formulating Search Strings

Based on the research questions obtained in the previous step, in this step we identify different keywords which can be used for formulating the search strings and facilitating the literature search. For that, in this SMS study, we use a specialized model developed by Kitchenham and Charters [31], called PICO. According to this model, each question be separated in terms of four key components: Population (P), Intervention (I), Comparison (C) and Outcome (O). Table 2 describes the PICO elements in the context of fake news detection and Table 3 defines the search terms extracted from each PICO element. After obtaining the set of keywords that can be used to for-

Fig. 1 The systematic mapping study process

mulate the search strings, we classify these keywords into seven groups as shown in Table 4. We notice that each group contains similar keywords. The similarity between the keywords is ensured by one of these two properties: have the same synonyms or have the same semantic meaning. After that, we apply the following two rules to formulate our search string:

- Rule 1 : we use the Boolean OR to concatenate terms in the same group.
- Rule 2 : we use the Boolean AND to join groups of terms.

The definitive search string obtained is (False information OR Fake news OR Fake information OR Misinformation OR Disinformation OR Rumors) AND (Collect OR Analyze OR Learn OR Classify OR Predict OR Detect OR Classification OR Detection) AND (Data OR Behavior) AND (Approach OR Technique OR Method OR Algorithm) AND (Application OR Tool OR Framework OR Solution) AND (Optimization OR Performance) AND (Accuracy OR Satisfaction).

3.3 Selecting Papers from Digital Databases

This step represents the core of the SMS process, because it permits to extract the relevant documents related to fake news detection from different Digital Databases (DD). The most DD commonly used to publish any research at any flied are IEEE Xplore, ACM Digital Library, ScienceDirect, and Springer Link. We report that, before executing our obtained search string, we should adapted it according to the search roles on each digital library [19]. In order to specify one or more pieces of information to find related documents, we have adopted advanced search options. As result, we have obtained several ways to search for papers within all digital libraries, we can search our query in title, in author, in abstract, in full text, etc. In our study, for all digital libraries we apply our query within full text option. All searches were restricted to the studies published between 01 January 2010 and 30 June 2021.

After the execution of the search strings used for each database, several amount of documents is being provided as search results. In general, the four digital libraries do not provide the same research results. In order to select the most relevant papers to answer the RQs, we apply inclusion and exclusion criteria to the candidate papers retrieved by executing the search strings. During this process, we included some articles and we excluded other based on titles and abstracts, as well as full-text reading and quality assessment. In this chapter, the quality assessment is mainly associated to the citation number of each candidate paper. Table 5 presents different inclusion and exclusion criteria applied for each selected paper.

Table 1 Definition and rationale of different research questions

Id	Research question	Rationale
RQ1	In which years, sources, and publication channels papers were published?	To indicate where studies concerning this research area can be found and whether there are specific publication channels. It also determines when efforts in this filed of research have been made
RQ2	Which research types are adopted in selected papers?	To highlight the different types of research published in the literature regarding the use of Machine Learning in the context of fake news detection
RQ3	Which domain fields are targeted in selected papers?	Various domains are infected by fake news phenomenon such as politics, education, health, sports, tourism, etc. Our goal was to identify in which domain fields Machine Learning techniques were used and researchers were interested
RQ4	Which social media platforms are the major source of fake news to the end users.	To determine in which social media platforms Machine Learning techniques were carried out and researchers were interested
RQ5	Which contexts are targeted in selected papers.	To identify in which context different studies were carried out and published in literature regarding the application of ML in fake news
RQ6	What kinds of fake news are targeted in selected papers?	To determine the different forms of fake news used in the selected papers
RQ7	Which types of features are exploited for fake news detection.	To explore the types of features used in the selected papers
RQ8	Which ML models, tasks, and techniques are used to detect fake news	To provide an overview of different ML techniques used in the selected papers for dealing with fake news detection

3.4 Data Extraction Strategy

This step allows to extract the pertinent information from the selected papers retrieved in the previous step, in order to answer to the different RQs defined in Table 1. To achieve this tack, we use the following template shown in Table 6. Each data extraction field is characterized by three keys which are: data item, a value and a research question to which they refer. In the following, we describe each data item and we provide some values related to each data item as example.

Publication Channel presents any source of communication used to publish each selected paper such as scientific journal, conference, workshop, or symposium.

Publication Source indicates the name of the journal or the academic events (conference, workshop, or symposium) that have used to publish each selected paper.

Research type refers to a classification of the selected articles based on the nature of the contribution made in each article. According to Petersen et al. in [19] and Wieringa et al. in [33], research is classified into five categories: (i) Validation Research (VR), (ii) Solution proposal (SP), (iii) Evaluation Research (ER), (iv) Philosophical Papers (PP), and (v) Opinion Papers (OP). Validation Research (VR), this category presents a novel technique, approach, or strategy that has not been implemented in practice, but whose effectiveness has been evaluated In-depth through experiments, simulations, prototyping, etc. Solution Proposal (SP), contains studies that propose a novel solution or an improvement of an existing solution and argues for its relevance without a validation. Evaluation Research (ER), this category contains studies that empirically evaluate a technique, approach, or strategy in practice. Philosophical Papers (PP), this category contains studies which provide a new vision or concept of looking at things or a new conceptual framework. Finally, Opinion Papers (OP), this category contains studies that give an opinion about what is good or wrong related to something, how we should do something.

Study Domain refers to the domain of the application of the study such as Politics, Sports, Health, Financial markets, Tourism, Education and science. A study that does not specify the application domain is considered as Generic.

Study Platform refers to social media platforms of the study such as Facebook, Twitter, WhatsApp, Skype, You tube, and Tik Tok. A study that does not specify the the platform is considered as Generic.

Study Category refers to fake news form used in each selected paper such as fake news, fake review, misinformation, false information, disinformation, rumors, satire news, and opinion spam. A study that does not specify the category is considered as Generic.

Study Context refers to the context in which the study was performed. According to the reference [34], Kitchman and al. have introduced four contexts: Academic, Organization, Industrial, and Government.

Feature refers to the most relevant data that can be used to design machine learing models. Among these features the following can be cited: linguistic, text/content, user, network, propagation, temporal, structural and visual [39].

ML Model refers to the model of learning that is applied to extract knowledge from data in order to detect fake news. In this chapter, we consider four categories of ML models: Supervised Learning (SL) [35], Unsupervised Learning (UL) [36], Semi-Supervised Learning (SSL) [37], and Reinforcement Learning (RL) [38]. Supervised learning is one of the main categories of machine learning. In supervised machine learning, input data also known as training examples, comes with a label, and the goal of learning is to predict the label for new, unforeseen examples. Unsupervised learning is a machine learning type that learns from data that has not been labeled. The goal of unsupervised learning is to detect patterns in the data. Semi-supervised learning is a machine learning technique that falls between supervised and unsupervised learning. The algorithms of this category include some labeled data with a large amount of unlabeled data in order to create a new model. Finally, Reinforcement learning is a goal oriented learning that is based on interaction with the environment. The algorithms of this category attempt to discover an association between the goal

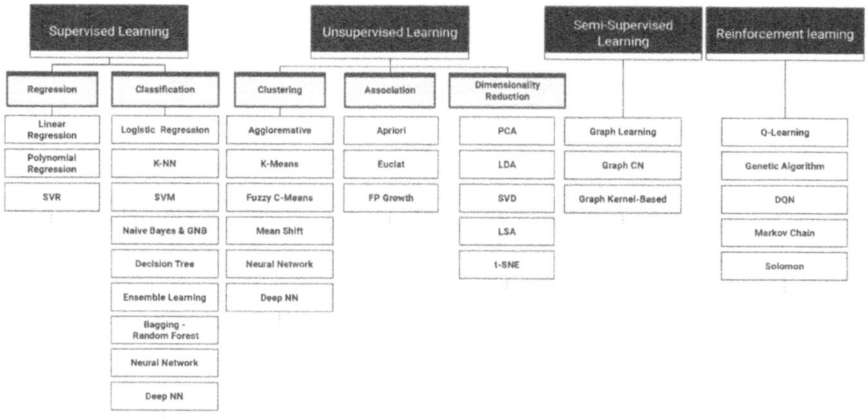

Fig. 2 The taxonomy of machine learning algorithms

and the sequence of events that leads to a successful outcome. The main characteristic of reinforcement learning is the use of trial and error mechanism in order to achieve a goal or to maximize the total reward.

ML techniques refer to the algorithms that are used to perform the learning task for solving any problem. These techniques can be classified into several ML Tasks according to each Machine learning Model. For instance, for Supervised Learning models, we have classification task and regression task. While for Unsupervised Learning models we have both tasks: association and clustering. Finally, for Semi Supervised Learning Models, we use the same ML tasks of Supervised learning Models, because the SSL models are considered as variants of the SL models. The taxonomy of machine learning algorithms mostly used in practice is shown in Fig. 2.

3.5 Analysis and Classification

This step, presents the main findings of our systematic mapping, because we analysis and we classify the results of our study. In the first time, the information for each item extracted in the previous step was tabulated and visually illustrated. In the second time, we discussed the results of our systematic mapping related to the questions presented in Table 1. We notice that all results achieved in this step are presented in the next section "Results and Discussion".

Table 2 Pico definition for our SMS study

Element	Description
Population	This element refers to fake news detection studies
Intervention	The aim of this SMS is to provide an overview related to the benefits when machine learning algorithms be applied to deal with fake news detection. For that we envisage two interventions: the collected Data by the each solution proposed in published works and behaviors analyzed by each solution
Comparison	We compare different machine learning algorithms used to detect fake news in digital medial
Outcome	In the context of this study, the outcomes represents the factors that will be used to compare the interventions in order to minimize or to avoid the impact of fake news phenomenon

Table 3 The extracted keywords from the PICO definition

Element	Description
Population	False information, Fake news, Fake information, misinformation, disinformation, rumors
Intervention	Collect, data, behavior, analyze, learn, classify, predict,detect
Comparison	Approach, technique, method, algorithm, benchmark, application, tool, framework, solution
Outcome	Classification, detection, optimization, performance, accuracy, satisfaction

4 Results and Discussion

In this section we provide the findings related to this systematic map. In the first time, we introduce an overview of the result of the selection process. In the second time we report the results concerning each research question.

4.1 Overview of the Selected Studies

After the execution of our search string described previously on the four digital libraries, we have retrieved 6757 candidate papers between 01 January 2010 and 30 June 2021. Afterward, we have applied the inclusion and exclusion criteria to filter the candidate papers. The outcome of this process leading to the identification of 76 relevant articles regarding the use of machine learning techniques in context of fake news in digital social media. Table 7 presents the number of selected papers according to each step of the selection process. In fact, we have updated the selected papers by withdrawing the duplicate papers as well as the papers reporting the same study. Besides, we have considered only all papers written in English and accessible in full-text. Then, we have excluded also papers that have added to each data base after

Table 4 Classification of different keywords according to their similarity

Terms	G1	G2	G3	G4	G5	G6	G7
False information	*						
Fake news	*						
Fake information	*						
Misinformation	*						
Disinformation	*						
Rumors	*						
Collect		*					
Analyze		*					
Learn		*					
Classify		*					
Predict		*					
Detect		*					
Data			*				
Behavior			*				
Approach				*			
Technique				*			
Method				*			
Algorithm				*			
Benchmark					*		
Application					*		
Tool					*		
Framework					*		
Solution					*		
Classification		*					
Detection		*					
Optimization						*	
Performance						*	
Accuracy							*
Satisfaction							*

30 June 2021. Finally, we have applied a full-text review for many papers in order to decide about their inclusion or exclusion in our study. It is clear that the number of selected papers proposed for conducting this SMS is very reasonable and reflects the importance and the relevance of this research study. Moreover, the number of 76 selected papers will permit us to ensure a credible and a relevant overview about Fake news detection based on machine learning techniques. The list of 76 selected papers starts from the reference [40], and ends with the reference [114]. For more information the sms file is available via this reference [115].

Table 5 Inclusion and exclusion criteria

Category	Criteria
Inclusion	Studies presenting methods and techniques to develop intelligent solution for combating fake news in digital media
	Studies presenting methods and techniques to analyze the content of social medial in the context of fake news
	Studies presenting an overview or a comparison between different Machine learning techniques used in the literature for detecting fake news
	Studies published between 2010 and 2021
Exclusion	Studies not accessible in full-text
	Studies not presented in English
	Books and gray literature
	Studies that are duplicates of other studies

Table 6 Data extraction template

Data	item Value	RQ
Authors	Set of names of the authors	RQ1
Title	Title of the paper	RQ1
Publication channel	Kind of publication channel	RQ1
Publication source	Name of publication source	RQ1
Year of publication	Calendar year	RQ1
Research type	Which research strategy was followed	RQ2
Study domain	In which domain fields the research was applied	RQ3
Study platform	In which social media platforms the research was applied	RQ4
Study context	In which context the research was conducted	RQ5
Study category	Which forms of fake news studied in the selected papers.	RQ6
Feature	Which kinds of feature were used by selected papers	RQ7
ML model	Which machine learning approach was adopted	RQ8
ML task	Which data mining tasks were used by selected paper	RQ8
ML technique	Which data mining techniques were used by selected papers	RQ8

4.2 *RQ1: In Which Years, Sources, and Publication Channels Papers Were Published?*

Figure 3 presents the variation of relevant studies published over the period between 2010 and the first half of 2021. Based on this figure, we remark that the annual number of publications related to fake news increases significantly every year. Moreover, in the last five years between 2017 and 2021, we have seen a high rate of publication

Table 7 The number of retained papers after each step of selection process

Data Base	Returned studies	Year filter (2010–2021)	Title/Abstract review	Full-text review	Add manually	Retained studies
IEEE	1322	1148	59	17	3	20
Springer	1523	1326	53	17	4	21
ACM	4202	3370	77	22	5	27
SCDirect	998	913	24	7	1	8

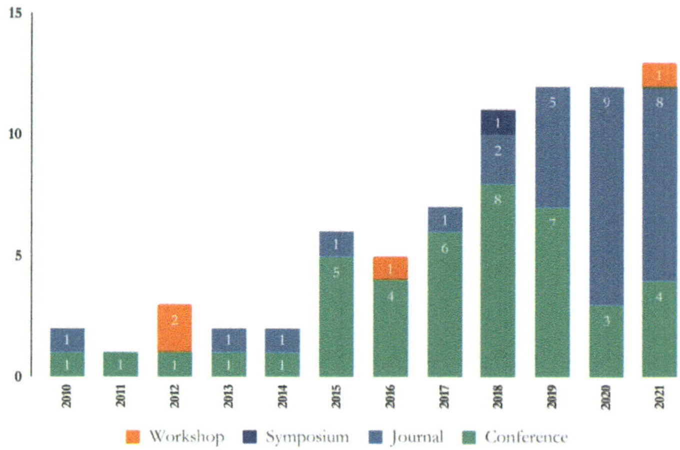

Fig. 3 The distribution of selected papers during the years 2010–2021

with 72.37% of the total selected studies. In addition, 13 papers were published in the first half of 2021, which is a significant number if it is compared to other years. As result, it is very clear that researchers are becoming more and more interested in this research filed. This growth in terms of publications is justified by the trend of Machine learning usage in different domains over the last decade.

Figure 4 shows that the period between 2010 and 2019 the number of papers published in conferences is more important than the other channels. In reality, the number of articles published in conferences is always very large compared to the number of articles published in journals. However, related to our SMS study, the most of the selected studies have high quality. As result, for the period between 2020 and 2021, we remark that the number of papers published in journals is more important than the other channels. Figure 4 shows that 56.3% of selected papers were published in conferences, while 38% were published in journals, 4.2% were published in workshop and 1.4% were published in symposium. The high percentage of conference papers is justified by the fact that the academic authors and the researchers often submit their first papers for conferences. As we have already mentioned above, our

Fig. 4 The percentage of each publication channel in selected papers

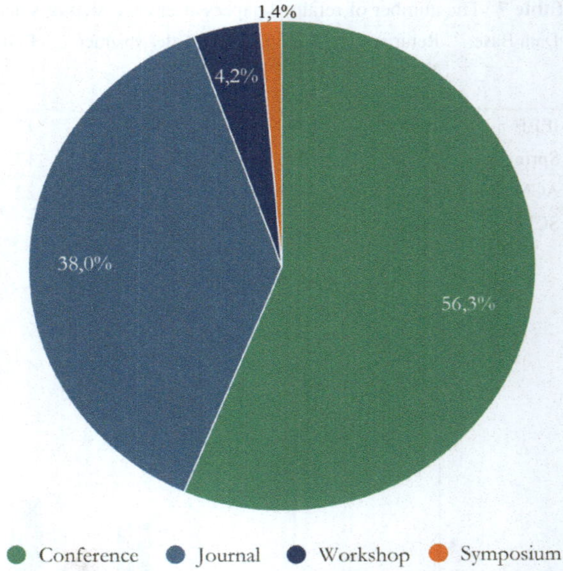

goal is ensuring a credible and a relevant SMS study related to fake news detection. For that, the most of the selected studies have been published in publication sources with a high-ranking. From 76 selected papers, 76.2% have published in prestigious conference or prestigious journal.

Table 8 presents the most frequent publication sources concerning our selected studies. Thus, the most frequent publication sources are journals Q1 with a percentage of 32.89 and 3.95% represent journals Q2. While 26.32% of selected papers were published in conferences A* and 9.21% in conferences A.

4.3 *RQ2: Which Research Types Are Adopted in Selected Papers?*

Figure 5 presents the percentage of each research type in selected papers during the years 2010–2021. From this figure, we observe that that ER (Evaluation Research) is the most adopted research type with a value of 57,8%. The second most adopted research type is SP (Solution Proposal) with a percentage of 25,3%. Moreover, 14,1% of selected papers were adopted VR (Validation Research) strategy. Finally, PP (philosophical paper) is the less adopted research type with a value of 2,8%. Based on these values, it's clear that ER type is dominant over other search types, because several papers have proposed final solutions that are evaluated and experimented in practice. Also, from these values we can deduce that the majority of proposed

Table 8 Most frequent publication sources in the selected papers

Name	Type	Libraries	Ranking	Number
IEEE ACCESS	Journal	IEEE	Q1	4
IEEE Transactions on Computational Social Systems	Journal	IEEE	Q1	2
International Joint Conference on Neural Networks (IJCNN)	Conference	IEEE	B	2
ACM Transactions on Knowledge Discovery from Data	Journal	ACM	Q1	2
International World Wide Web Conference	Conference	ACM	A*	6
International on Conference on Information and Knowledge Management	Conference	ACM	A	4
Expert Systems with Applications	Journal	SCDirect	Q1	5
Neural Computing and Applications	Journal	Springer	Q1	1
Social Network Analysis and Mining	Journal	Springer	Q1	1
International Conference on Computational Science	Conference	Springer	A	1
International Conference on Advanced Data Mining and Applications	Conference	Springer	B	1

solutions are not implemented or experimented in real context. Finally, there are only a few PP works that present a new conceptual framework solution without any implementation.

Figure 6 presents the percentage of research types published in journals and conferences during the years 2010–2021. We observe that the ER papers are published with equitable manner in both conferences and journals, because these works are already experimented and validated. While both SP and VR papers are often published in conferences because these works are firstly reported as primary results. Finally, we observe that there are only two PP works which are published in journals during the years 2010–2021.

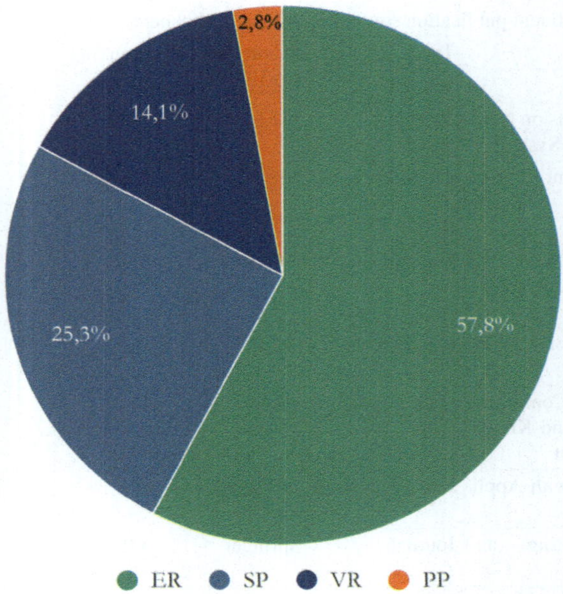

Fig. 5 The percentage of each research type in selected papers between 2010 and 2021

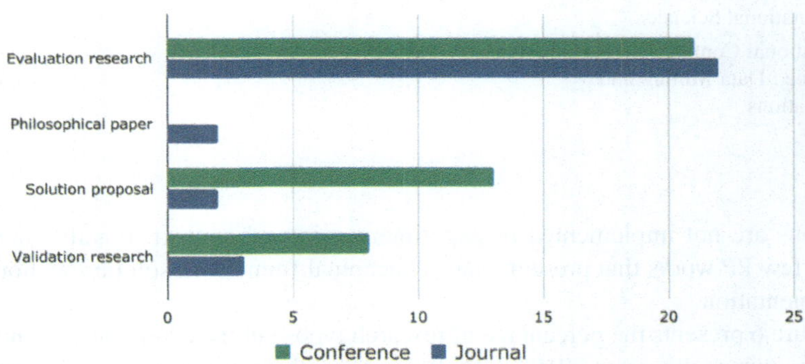

Fig. 6 The number of research types in journals and conferences

4.4 *RQ3: Which Domain Fields Are Targeted in Selected Papers?*

Figure 7 presents the percentage of selected papers for both domains generic or specific during the years 2010–2021. Based on this figure, we observe that the percentage of selected papers that focus on a generic domain is 59.2%. While the percentage of selected papers that focus on specific domain is 40.8%. It's clear that higher value is

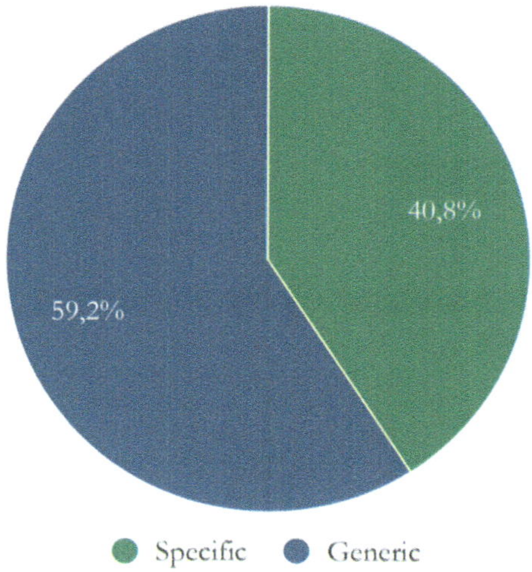

Fig. 7 The percentage of selected papers for both domains generic and specific

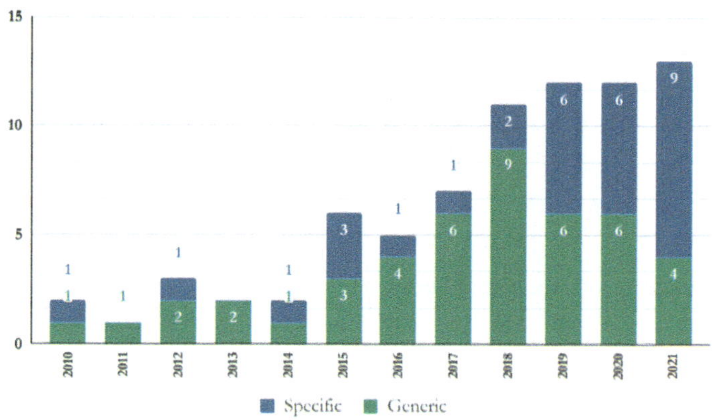

Fig. 8 Distribution of selected papers for generic and specific domains

provided by the generic domain. Therefore, we confirm that researchers are focusing more and more on generic domains.

Figure 8 presents the distribution of selected papers for generic and specific domains in the period between 2010 and 2021. We observe that between 2010 and 2018, the total number of papers that focus on a generic field is geater than thos that are specific. Moreover, between 2019 and 2021, we observe that the total number of selected papers based on specific solution is greather than thos are generic. Therefore, we note that researchers are focusing more and more on specific solution during two last years.

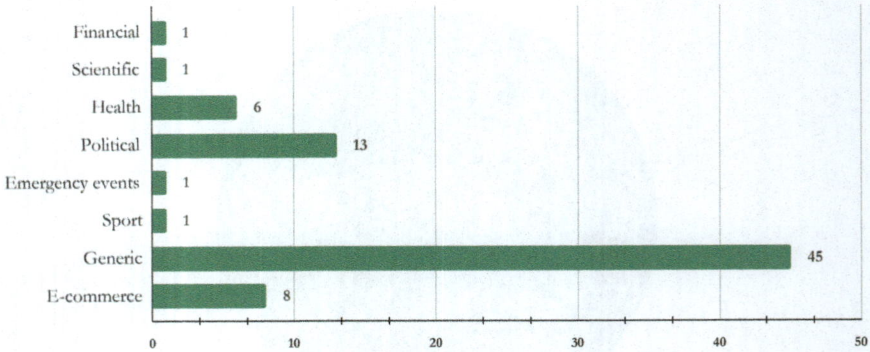

Fig. 9 The number of selected papers for both domains generic or specific

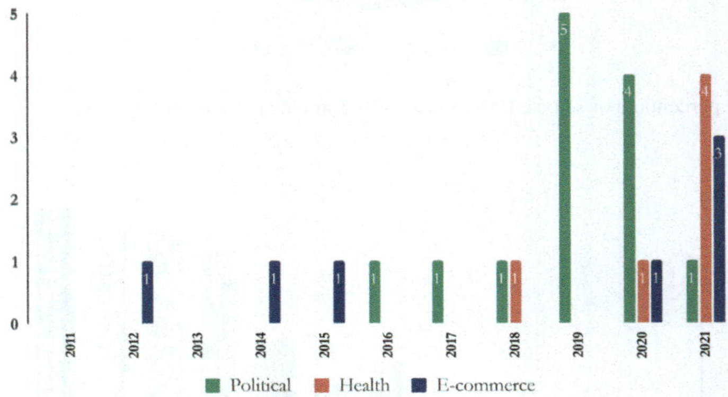

Fig. 10 The evolution of three domains in the selected papers

Figure 9 presents the number of papers in specific domains adopted in selected papers. It can be observed from this figure that the most targeted domains affected by fake news are political domain with 13 papers (17.1%), E-commerce domain with 8 papers (10.5%) and health domain with 6 papers (7.9%). The dominance of political and health domains can be explained by the fact that the majority of social media users are interested to have more information about the political situation and health especially with this COVID-19 pandemic. While the dominance of E-commerce domain can be explained the high use of this service in recent years.

Figure 10 presents the evolution of three domains E-commerce, political and health in the selected papers between 2010 and 2021. From this figure, we observe that between 2010 and 2015 the most targeted domain affected by fake news is E-commerce. Also we remark that between 2016 and 2020 the most targeted domain is political domain. Finally, from 2021 the most targeted domain is health due to COVID-19 pandemic.

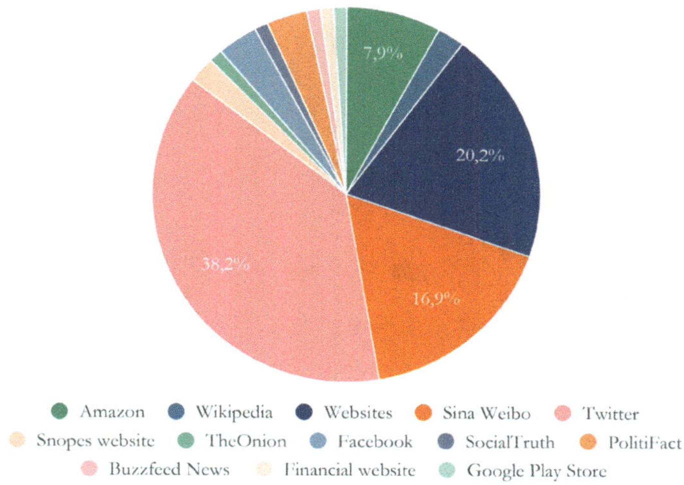

Fig. 11 The percentage of each social media platform used in selected papers

4.5 RQ4: Which Social Media Platforms Are the Major Source of Fake News to the End Users?

Figure 11 presents the percentage of each social media platform used in selected papers during the years 2010–2021. As can be shown in this figure, the digital media that spreads fake news is twitter with the value of 38.2%, succeeded by websites with 20.2%, Sina Weibo with 16.9% and Amazon with 7.9%. Moreover, the percentage of utilization concerning the other platforms such as Facebook, Google Play Store, Wikipedia, etc., does not exceed 3.5%. Therefore, we confirm that researchers are focusing more and more on twitter platform. The dominance of this social platform can be explained by the fact that the majority of users of twitter are politicians, show business stars, institutions and gouvernements.

Figure 12 presents the evolution usage of four platforms Twitter, Websites, Sina Weibo and Amazon. It can be observed from this figure that between 2010 and 2015 there is no use of websites in the selected papers. Moreover, between 2016 and 2021 we remark the use of websites in the selected papers. In addition, we observe that twitter platform is most used every year during 2010 and 2021. Finally, we observe that Sina weibo is also used every year. The growing number concerning the use of websites can be explained by the fact that researchs have used others platforms.

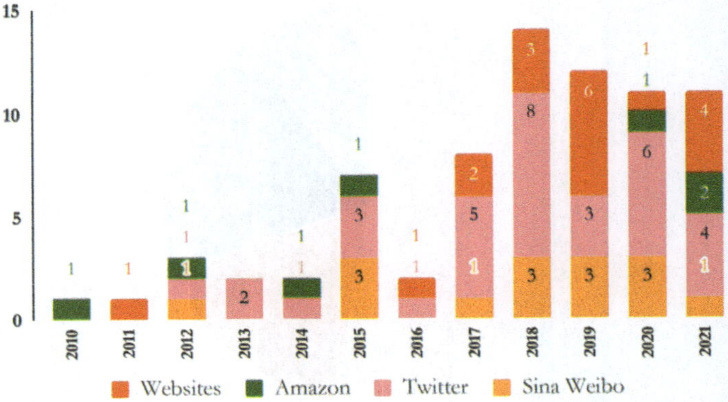

Fig. 12 The evolution usage of the most used platforms during 2010 and 2021

Fig. 13 The percentage of each study context in selected papers between 2010 and 2021

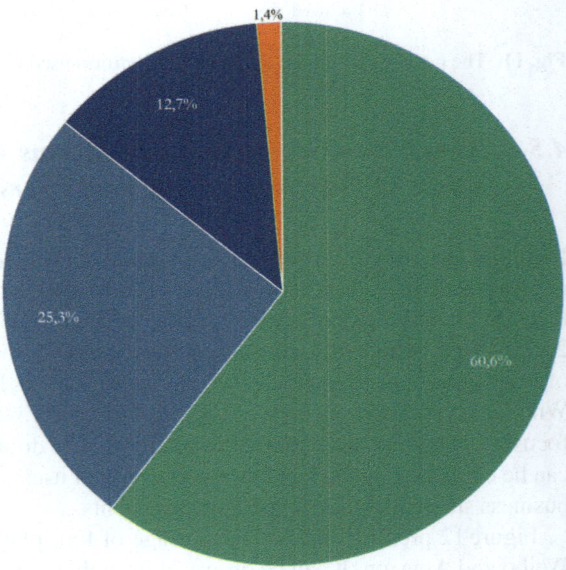

4.6 RQ5: Which Contexts Are Targeted in Selected Papers?

Figure 13 presents the percentage of each study context in selected papers over the period between 2010 and 2021. Based on this figure, we remark that 60.6% of selected papers were conducted in an academic context, 25.3% were adopted by government institutes, 12.7% were funded by organizations, and only 1.4% of selected papers were conducted in an industrial context. We deduce that the majority of selected studies are done in academic context.

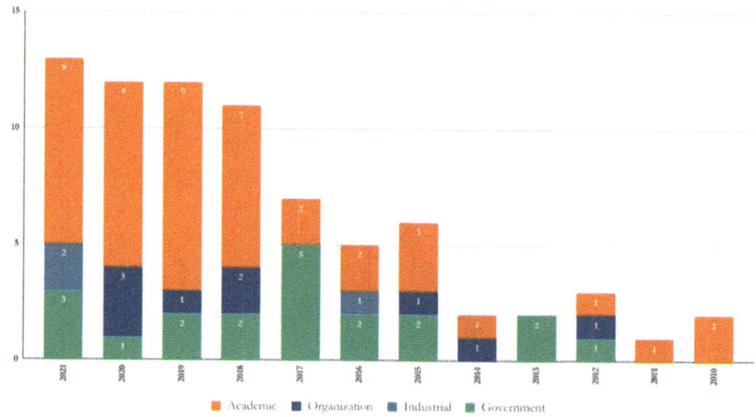

Fig. 14 Distribution of selected papers between 2010 and 2021 for each research context

Figure 14 presents the distribution of selected studies during the years 2010–2021, according to each research context. Based on this figure, we observe that over the period 2015–2021 there are at least two studies per year that were conducted in a government context and at least on study per year that were conducted in a an organization context. Moreover, during 2021, there are two studies that were conducted in industrial context during 2021.

The growing number of studies funded by governments can be explained by the fact each country's institutions often fall victim to fake news. Indeed, fake news issue can have a real impact on several sectors such as economy, security, health, education, tourism, etc. As result, gouvernements have begun to invest heavily in research for combating this phenomenon. Finally, regarding the number of works funded by organizations, it can be explained by the fact organizations are becoming increasingly aware of the consequences of fake news in many fields such as economic, political, and social.

4.7 RQ6: What Kinds of Fake News Are Targeted in Selected Papers?

Figure 15 presents the percentage of each kind of fake news in selected papers. Based on this figure, we observe that the most selected papers deal with fake news issue with the value of 43.4%, succeeded by rumors with 28.9%, and fake review with 10.5%. Moreover, the other kind of fake news are less used in the selected papers, misinformation, false information, hoax and spam are used with the values of 6.6%, 5.3%, 2.6% and 2.6% respectively. Therefore, we confirm that researchers are focusing more and more on fake news.

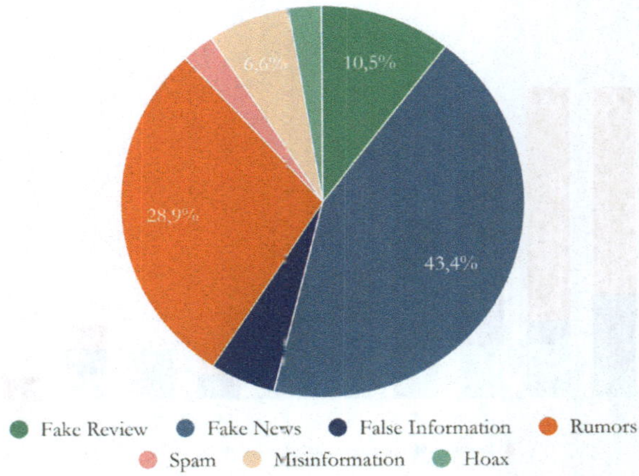

Fig. 15 The percentage of each kind of fake news in selected papers

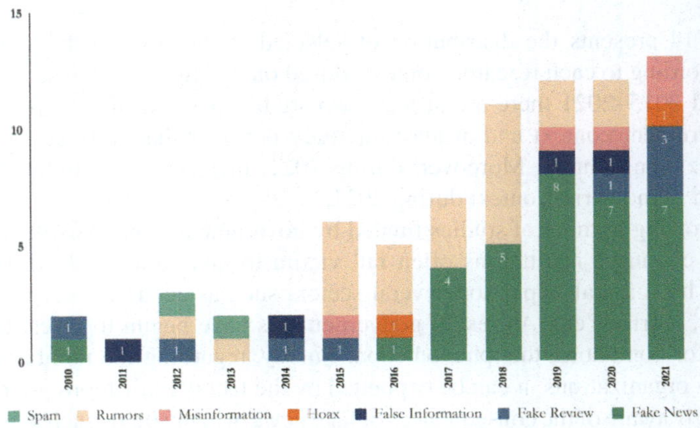

Fig. 16 The distribution of selected papers according to each type of fake news

Figure 16 presents the distribution of selected papers according to each type of fake news during the years 2010–2021. Based on this figure, we observe that rumors type is dominant during the period 2015–2016. While fake news type is dominant during the period 2017–2021.

4.8 *RQ7: Which Types of Features Are Exploited for Fake News Detection?*

Figure 17 presents the percentage of each feature used in selected papers to design machine-learning models. We remark that text/content is the most feature used with

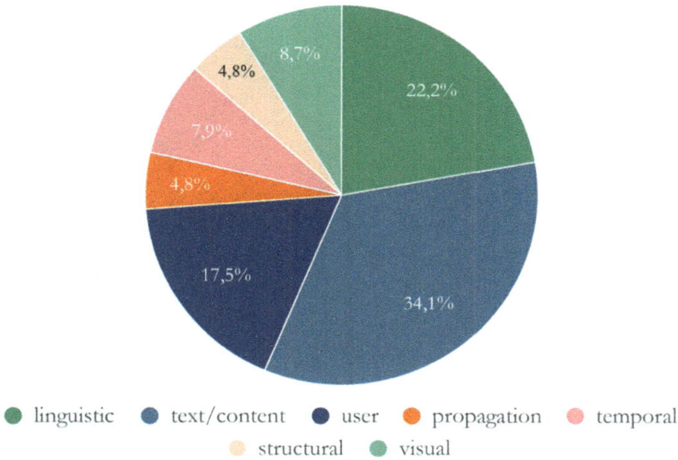

Fig. 17 The percentage of each feature used in selected papers

a percentage of 34.1%, succeeded by linguistic feature with 22.2%, user feature with 17.5%, visual feature with 8.7%, temporal feature with 7.9%, propagation feature with 4.8%, and structural feature with 4.8%. But it must be emphasized that, in the same work, several kinds of features can be used simultaneous to design machine-learning models.

Text/content feature and linguistic feature are the most used because the nature of fake news is the text. Also user feature is more used because this feature provides some information such as name, image, gender, account status, number of friends, location, etc. There is also the tendency to use visual feature for online fake news detection, because this feature give important information related to resolution of image, image visual features and image statistical features. Finally, both propagation and structural features are not very used to design machine-learning models.

4.9 RQ8: Which Machine Learning Models, Data Mining Tasks, and Techniques Are Used to Deal with Fake News?

Figure 18 presents the percentage utilization of each machine learning model in selected papers during the years 2010–2021. From this figure, we observe that the most selected papers have used the supervised learning models with a value of 85.5%. While 7.9% of selected papers have used semi supervised learning, 3.9% have used unsupervised learning. Finally, 2.6% of selected papers have used reinforcement learning models.

Figure 19 presents the percentage of adopted data analysis tasks in selected papers. It can be observed from this figure that the higher value is provided by classification

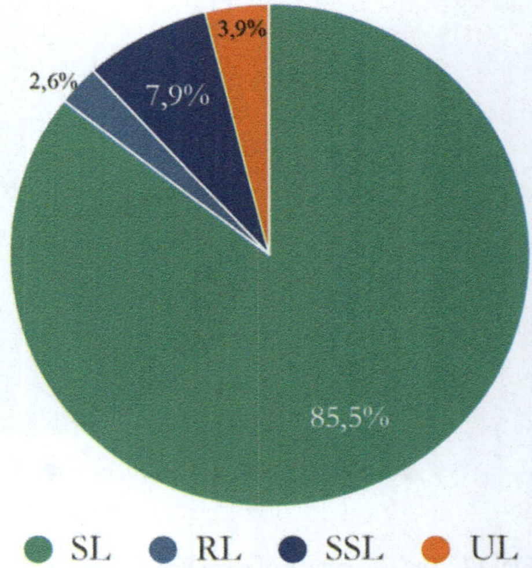

Fig. 18 The percentage utilization of each machine learning model

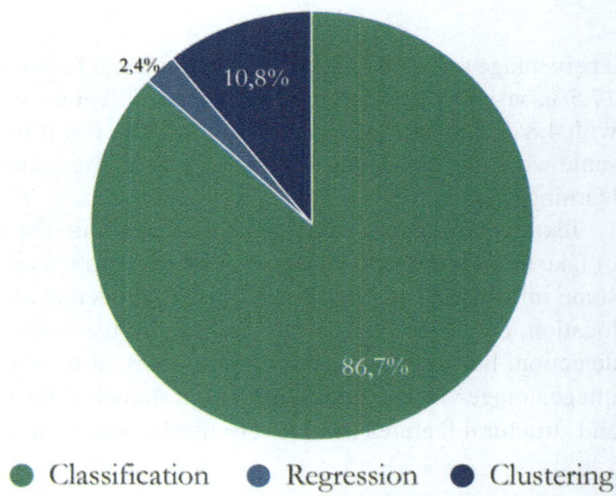

Fig. 19 The percentage of adopted data analysis tasks in selected papers

task with a percentage of 86.7%, succeeded by clustering task with a percentage of 10.8% and regression task in the last range with a percentage of 2.4%.

The dominance of classification task can be explained by the fact that most selected papers have used the supervised learning models. Moreover, this dominance confirm the nature of fake news detection which is classification task. Although regression task is supervised learning task, this task is less used in the context of fake news detection.

Table 9 presents machine learning techniques that are applied at least three time in the selected studies. It can be observed from this table that DNN is the most frequently technique used in 33 papers, followed by SVM technique that was used

Table 9 Frequent machine learning techniques used in selected papers

Techniques	Number of papers
DNN (Deep Neural Network)	33
SVM (Support Vector Machine)	26
DT (Decision Tree)	13
RF (Random Forest)	12
LR (Logistic Regression)	10
KNN (K-Nearest Neighbour)	7
NBC (Naïve Bayesian)	5
K-means	4
ANN (artificial neural network)	4
Ensemble learning model	3

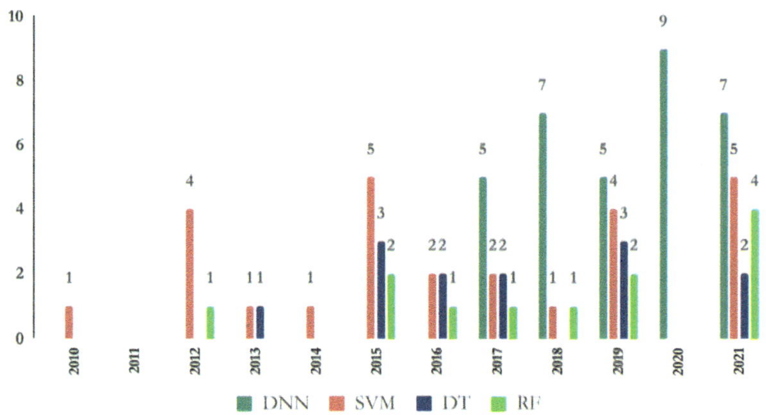

Fig. 20 The evolution of frequent machine learning techniques usage in selected papers

in 26 studies, DT technique that was used in 13 studies, RF technique that was used in 12 studies and LR technique that was used in 10 studies. While, the other kind of machine learning techniques are less used in the selected papers, KNN, NBC, K-means, ANN and Ensemble Learning Model were used in 7 papers, 5 papers, 4 papers, 4 papers and 3 papers respectively. Finally, we point out that there are other techniques that are not mentioned in Table such as LSV, Linear Regression, Active Learning, Solomon, Gradient Boost, Association Rule, etc. These techniques were used only one time in the selected papers.

Figure 20 presents the evolution of frequent machine learning techniques usage in selected papers between 2010 and 2021. It can be observed from this figure that SVM technique is the only technique that was used in all years from 2010 until 2021. Moreover, this technique is the most dominant algorithm in selected studies between 2010 and 2016. Also Fig. 20 shows that from 2017, the researchers have used Deep Neural Network (DNN) in the selected papers. In addition, from the same year, the DNN became the popular method that was used for fake news detection. Therefore, we confirm that researchers are focusing more and more on Deep Learning techniques for combating fake news.

We conclude that the utilization of supervised learning models and precisely the classification task is intuitive due to the nature of fake news issue. Indeed, for fake news phenomenon, we need only to classify the news are fake or not, we need to classify information it's true or false, etc. In addition, the classification techniques are known by their simplicity because they attempt to learn from the training dataset in order to find the appropriate model. Finally, we conclude that both unsupervised learning models and reinforcement learning models are the less used for combating fake news issue.

5 Implications for Researchers

In this section we provide some implications and recommendations that are critical for researchers based on the results obtained from the previous section. These implications and recommendations are categorized according to the RQs, as follows:

5.1 RQ1

Despite the large number of papers retrieved from different data bases, also were a significant number of discarded papers during different levels of the selection process. This is mainly due to the fact that we have used only four digital libraries: IEEE, Springer, ACM and SCDirect. However, we did not consider all the relevant publishers such as the Hindawi, Taylor& CRC, and MDPI. As result, it is possible that relevant studies were not considered in our selection process. So, researchers must extend their researches into these publishers. Moreover, due to high quality of most of the selected studies, many excluded papers are published in prestigious conference (B or C) or prestigious journal (Q2 or Q3). It is therefore recommended that researchers analyze and synthesize their works in order to ensure an acceptable level of scientific credibility. Finally, more than 50% of selected papers have published in the last 3 years. So, researchers must focus their research works from from 2018 until today.

5.2 RQ2 and RQ3

The majority of selected papers provide only empirical study without any implementation in practice. In this context, we recommend that researchers may work for implementing these solutions. Moreover, the majority of selected papers were focused on a generic domain. It is therefore recommended that researchers focus more on specific domains. Finally, the specific domains that most used in selected papers are political, health and E-commerce. We encourage researchers to focus on other specific domain such as sport, financial and scientific.

5.3 RQ4

For the majority of selected studies, we found that most used platforms are twitter, Sina Weibo, amazon and Websites. Furthermore, there is a lack of information about the websites used in this SMS. For instance, we don't know the link of each website, the number of users, the nature of content. A result the nature of collected data from the websites used are not the same. So, it is recommended for researchers to analyze more each website used in this SMS. Finally, many other popular social media that such as facebook, Instagram, Tik Tok and Youtube are less used. It is therefore recommended that researchers investigate more effort to use these social media.

5.4 RQ5

The majority of selected studies are done in academic context except that recently, some selected studies are done in governments and organizations. It is therefore recommended that researchers should continue to collaborate with governments and organizations in order to get funding for their research projects. Finally, there is a lack in terms of collaboration with industry partners. In this context, we encourage researchers to collaborate with industry for the future works.

5.5 RQ6 and RQ7

For the majority of selected studies, we found that the authors focus only on three kind of false information: fake news, rumors and fake review. So, it is strongly recommended for researchers to explain why three forms of false information are more used than other kinds such as spam, misinformation, and hoax. Furthermore, there is a lack in terms of two kinds of fake news which are fake images and fake videos. So, researchers must make more effort to study and to implement new solutions based on fake images and fake videos.

5.6 RQ8

The majority of selected studies, have based on supervised learning models. This choice is intuitive due to the nature of fake news which is classification task. However, the use of unsupervised learning models and reinforcement learning models are not justified. It is therefore recommended that researchers investigate more effort in order to clarify all reasons behind this choice. Furthermore, deep learning algorithms have been widely applied to deal with fake news detection in many works. However, these algorithms are often criticized for being used as mysterious "black box" and we necessitate the interpretability of this mysterious black box deep learning models. So, it is strongly recommended for researchers to take into consideration the interpretability of different deep learning techniques and also to explain how each input of the each model influence the performance. Finally, there is a lack in terms of algorithms that based on Natural Language Processing (NLP) or text mining in the context of fake news. So, it is recommended for researchers to conduct more research on fake news by using NLP techniques or text mining techniques.

6 Conclusion

In this chapter we have conducted a systematic mapping study regarding the application of ML techniques for combating fake news. A total of 76 relevant papers published between between 1 January 2010 and 30 June 2021 were retrieved from four relevant digital databases in order in order to provide different responses related to different research questions presented in Table 1. To achieve this task, the selected papers were analyzed by year and publication source, research by year, sources and publication channel, research type, study domain, platform, study context and category, kind of collected feature, and finally machine learning models adopted, tasks applied, and techniques used.

The obtained results show that the number of published papers in the field of fake news detection has increased significantly over the last decade, and particularly in the last two years. These papers were published in prestigious conferences (A*, A and B) and prestigious journals (Q1 and Q2). The majority of the selected articles were adopted evaluation search methodology. Political, Health and E-commerce are the most targeted domain fields. Twitter, Sina Weibo and Amazon were used massively than other social media platforms. The majority of selected studies were done in academic context. Fake news, rumors and fake review are the most kind of false information in the selected studies. Text/content feature and linguistic feature are the most feature used to design machine learning models. The majority of selected articles have applied supervised learning models and more specifically classification tasks. Both SVM and DNN are the most used machine learning techniques.

In order to improve the quality of this study, some recommendations can be addressed promptly, for researchers. The first recommendation for researchers consist to focus their works on other specific domain such as security, education, sport, financial, and scientific domain and other popular social media that such as facebook, Instagram, Tik Tok and Youtube. The second recommendation is about context study, we are seeing that there is a lack in terms of collaboration with industry partners. So, we recommend researchers to collaborate with industry for the future works in order to obtain result in real context with real data. The third recommendation concerns the use of different features, it is recommended that future works consider this orientation of finding the suitable combination between different features. The fourth recommendation concerns the use of Deep Neural Network (DNN) to classify fake news. Therefore, researchers must make more effort to deal with the interpretability of different deep learning techniques used. The last recommendation is about the choice of unsupervised learning models and reinforcement learning models which are not justified. It is therefore recommended that researchers investigate more effort in order to clarify all reasons behind this choice.

Finally, for our future works, we intend to conduct a systematic mapping study on fake news based Deep learning techniques. Furthermore, we intend to the realize a systematic literature review (SLR) that will make an in depth analysis of all selected papers published recently on the topic of fake news detection.

References

1. Lahby, M., Essouiri, A., & Sekkaki, A. (2019). A novel modeling approach for vertical handover based on dynamic k-partite graph in heterogeneous networks. *Digital Communications and Networks, 5*(4), 297–307.
2. Dolega, L., Rowe, F., & Branagan, E. (2021). Going digital? The impact of social media marketing on retail website traffic, orders and sales. *Journal of Retailing and Consumer Services, 60*, 102–501.
3. Di Domenico, G., Sit, J., Ishizaka, A., & Nunan, D. (2021). Fake news, social media and marketing: A systematic review. *Journal of Business Research, 124*, 329–341.
4. Wu, Y., Ngai, E. W., Wu, P., & Wu, C. (2020). Fake online reviews: Literature review, synthesis, and directions for future research. *Decision Support Systems, 132*, 113280.
5. Pramod, A., Naicker, H. S., & Tyagi, A. K. (2021). Machine learning and deep learning: Open issues and future research directions for the next 10 years. In *Computational analysis and deep learning for medical care: Principles, methods, and applications* (pp. 463–490).
6. Shekhar, H., Seal, S., Kedia, S., & Guha, A. (2020). Survey on applications of machine learning in the field of computer vision. In Emerging technology in modelling and graphics (pp. 667-678). Springer.
7. Malik, M., Malik, M. K., Mehmood, K., & Makhdoom, I. (2021). Automatic speech recognition: A survey. *Multimedia Tools and Applications, 80*(6), 9411–9457.
8. Chowdhary, K. (2020). Natural language processing. In *Fundamentals of artificial intelligence* (pp. 603–649). Springer.
9. Yang, Y., & Sun, J. (2020). Machine learning and human-computer interaction technologies in media and cognition course. In *International Conference on Human-Computer Interaction* (pp. 690-697). Springer, Cham.
10. Réda, C., Kaufmann, E., & Delahaye-Duriez, A. (2020). Machine learning applications in drug development. *Computational and Structural Biotechnology Journal, 18*, 241–252.
11. Ker, J., Wang, L., Rao, J., & Lim, T. (2017). Deep learning applications in medical image analysis. *IEEE Access, 6*, 9375–9389.

12. Najafabadi, M. K., Mohamed, A. H., & Mahrin, M. N. R. (2019). A survey on data mining techniques in recommender systems. *Soft Computing, 23*(2), 627–654.
13. Abakarim, Y., Lahby, M., & Attioui, A. (2020). Bank failure prediction: A deep learning approach. In *Proceedings of the 13th International Conference on Intelligent Systems: Theories and Applications* (pp. 1–7).
14. Abakarim, Y., Lahby, M., & Attioui, A. (2018). An efficient real time model for credit card fraud detection based on deep learning. In *Proceedings of the 12th International Conference on Intelligent Systems: Theories and Applications* (pp. 1–7).
15. Abakarim, Y., Lahby, M., & Attioui, A. (2018). Towards an efficient real-time approach to loan credit approval using deep learning. In *2018 9th International Symposium on Signal, Image, Video and Communications (ISIVC)* (pp. 306–313). IEEE.
16. Asada, M., et al. (2009). Cognitive developmental robotics: A survey. *IEEE Transactions on Autonomous Mental Development, 1*(1), 12–34.
17. Buczak, A. L., & Guven, E. (2015). A survey of data mining and machine learning methods for cyber security intrusion detection. *IEEE Communications Surveys and Tutorials, 18*(2), 1153–1176.
18. Kitchenham, B. (2004). Procedures for performing systematic reviews. *Keele, UK, Keele University, 33*(2004), 1–26.
19. Petersen, K., Vakkalanka, S., & Kuzniarz, L. (2015). Guidelines for conducting systematic mapping studies in software engineering: An update. *Information and Software Technology, 64*, 1–18.
20. Bondielli, A., & Marcelloni, F. (2019). A survey on fake news and Rumohr detection techniques. *Information Sciences, 497*, 38–55.
21. Dwivedi, S. M., & Wankhade, S. B. (2020). Survey on fake news detection techniques. In *International Conference on Image Processing and Capsule Networks* (pp. 342–348).
22. Zhang, X., & Ghorbani, A. A. (2020). An overview of online fake news: Characterization, detection, and discussion. *Information Processing and Management, 57*(2), Article 102025.
23. Sharma, K., Qian, F., Jiang, H., Ruchansky, N., Zhang, M., & Liu, Y. (2019). Combating fake news: A survey on identification and mitigation techniques. *ACM Transactions on Intelligent Systems and Technology (TIST), 10*(3), 1–42.
24. Islam, M. R., Liu, S., Wang, X., & Xu, G. (2020). Deep learning for misinformation detection on online social networks: a survey and new perspectives. *Social Network Analysis and Mining, 10*(1), 1–20.
25. Habib, A., Asghar, M. Z., Khan, A., Habib, A., & Khan, A. (2019). False information detection in online content and its role in decision making: A systematic literature review. *Social Network Analysis and Mining, 9*(1), 1–20.
26. de Beer, D., & Matthee, M. (2020). Approaches to identify fake news: A systematic literature review. In Springer International (Ed.), *Integrated science in digital age 2020, Tatiana Antipova* (pp. 13–22). Cham: Publishing.
27. Meel, P., & Vishwakarma, D. K. (2020). Fake news, rumor, information pollution in social media and web: A contemporary survey of state-of-the-arts, challenges and opportunities. *Expert Systems with Applications, 153*, 112986.
28. Almaliki, M. (2019). Online misinformation spread: A systematic literature map. In *Proceedings of the 2019 3rd International Conference on Information System and Data Mining* (pp. 171–178).
29. Choraś, M., et al. (2020). Advanced Machine Learning techniques for fake news (online disinformation) detection: A systematic mapping study. *Applied Soft Computing, 107050*.
30. De Souza, J. V., Gomes, J., Jr., de Souza Filho, F. M., de Oliveira Julio, A. M., & de Souza, J. F. (2020). A systematic mapping on automatic classification of fake news in social media. *Social Network Analysis and Mining, 10*(1), 1–21.
31. Meneses Silva, C. V., Silva Fontes, R., & Colaço Júnior, M. (2021). Intelligent fake news detection: A systematic mapping. *Journal of Applied Security Research, 16*(2), 168–189.
32. Kitchenham, B. A. (2012). Systematic review in software engineering: Where we are and where we should be going. In *Proceedings of the 2nd International Workshop on Evidential Assessment of Software Technologies* (pp. 1–2).

33. Wieringa, R., Maiden, N., Mead, N., & Rolland, C. (2006). Requirements engineering paper classification and evaluation criteria: A proposal and a discussion. *Requirements Engineering, 11*(1), 102–107.
34. Keele, S. (2007). Guidelines for performing systematic literature reviews in software engineering (Vol. 5). Technical report, Ver. 2.3 EBSE Technical Report. EBSE.
35. Marsland, S. (2011). *Machine learning: An algorithmic perspective*. Chapman and Hall/CRC.
36. Han, J., Pei, J., & Kamber, M. (2011). *Data mining: Concepts and techniques*. Elsevier.
37. Van Engelen, J. E., & Hoos, H. H. (2020). A survey on semi-supervised learning. *Machine Learning, 109*(2), 373–440.
38. Sutton, R. S., & Barto, A. G. (2018). Reinforcement learning: An introduction. MIT press.
39. Meel, P., & Vishwakarma, D. K. (2020). Fake news, rumor, information pollution in social media and web: A contemporary survey of state-of-the-arts, challenges and opportunities. *Expert Systems with Applications, 153*, 112–986.
40. Lai, C. L., Xu, K. Q., Lau, R. Y., Li, Y., & Jing, L. (2010, November). Toward a language modeling approach for consumer review spam detection. In *2010 IEEE 7th International Conference on E-Business Engineering* (pp. 1–8). IEEE.
41. Lavergne, T., Urvoy, T., & Yvon, F. (2011). Filtering artificial texts with statistical machine learning techniques. *Language Resources and Evaluation, 45*(1), 25–43.
42. Yin, X., & Tan, W. (2011). Semi-supervised truth discovery. In *Proceedings of the 20th International Conference on World Wide Web* (pp. 217–226).
43. Yang, F., Liu, Y., Yu, X., & Yang, M. (2012). Automatic detection of rumor on sina weibo. In *Proceedings of the ACM SIGKDD Workshop on Mining Data Semantics* (pp. 1–7).
44. Mukherjee, A., Liu, B., & Glance, N. (2012, April). Spotting fake reviewer groups in consumer reviews. In *Proceedings of the 21st International Conference on World Wide Web* (pp. 191–200).
45. Gupta, A., & Kumaraguru, P. (2012, April). Credibility ranking of tweets during high impact events. In *Proceedings of the 1st Workshop on Privacy and Security in Online Social Media* (pp. 2–8).
46. Kwon, S., Cha, M., Jung, K., Chen, W., & Wang, Y. (2013). Prominent features of rumor propagation in online social media. In *2013 IEEE 13th International Conference on Data Mining* (pp. 1103-1108). IEEE.
47. Martinez-Romo, J., & Araujo, L. (2013). Detecting malicious tweets in trending topics using a statistical analysis of language. *Expert Systems with Applications, 40*(8), 2992–3000.
48. Nabeshima, K., Mizuno, J., Okazaki, N., & Inui, K. (2014, August). Mining false information on twitter for a major disaster situation. In *International Conference on Active Media Technology* (pp. 96–109). Springer.
49. Ball, L., & Elworthy, J. (2014). Fake or real? The computational detection of online deceptive text. *Journal of Marketing Analytics, 2*(3), 187–201.
50. Wu, K., Yang, S., & Zhu, K. Q. (2015). False rumors detection on sina weibo by propagation structures. In *2015 IEEE 31st International Conference on Data Engineering* (pp. 651–662). IEEE.
51. Liang, G., He, W., Xu, C., Chen, L., & Zeng, J. (2015). Rumor identification in microblogging systems based on users' behavior. *IEEE Transactions on Computational Social Systems, 2*(3), 99–108.
52. Zhao, Z., Resnick, P., & Mei, Q. (2015, May). Enquiring minds: Early detection of rumors in social media from enquiry posts. In *Proceedings of the 24th International Conference on World Wide Web* (pp. 1395–1405).
53. Ma, J., Gao, W., Wei, Z., Lu, Y., & Wong, K. F. (2015, October). Detect rumors using time series of social context information on microblogging websites. In *Proceedings of the 24th ACM International on Conference on Information and Knowledge Management* (pp. 1751–1754).
54. Antoniadis, S., Litou, I., & Kalogeraki, V. (2015). A model for identifying misinformation in online social networks. In *OTM Confederated International Conferences On the Move to Meaningful Internet Systems* (pp. 473–482). Springer.

55. Chang, T., Hsu, P. Y., Cheng, M. S., Chung, C. Y., & Chung, Y. L. (2015). Detecting fake review with rumor model—Case study in hotel review. In *International Conference on Intelligent Science and Big Data Engineering* (pp. 181–192). Springer.
56. Kumar, S., West, R., & Leskovec, J. (2016). Disinformation on the web: Impact, characteristics, and detection of wikipedia hoaxes. In *Proceedings of the 25th International Conference on World Wide Web* (pp. 591–602).
57. Sampson, J., Morstatter, F., Wu, L., & Liu, H. (2016). Leveraging the implicit structure within social media for emergent rumor detection. In *Proceedings of the 25th ACM International on Conference on Information and Knowledge Management* (pp. 2377–2382).
58. Chang, C., Zhang, Y., Szabo, C., & Sheng, Q. Z. (2016, December). Extreme user and political rumor detection on twitter. In *International Conference on Advanced Data Mining and Applications* (pp. 751–763). Springer.
59. Yang, H., Zhong, J., Ha, D., & Oh, H. (2016). Rumor propagation detection system in social network services. In *International Conference on Computational Social Networks* (pp. 86-98). Springer, Cham.
60. Elkasrawi, S., Dengel, A., Abdelsamad, A., & Bukhari, S. S. (2016). What you see is what you get? Automatic image verification for online news content. In *2016 12th IAPR Workshop on Document Analysis Systems (DAS)* (pp. 114–119). IEEE.
61. Buntain, C., & Golbeck, J. (2017). Automatically identifying fake news in popular twitter threads. In *2017 IEEE International Conference on Smart Cloud (SmartCloud)* (pp. 208–215). IEEE.
62. Ruchansky, N., Seo, S., & Liu, Y. (2017). Csi: A hybrid deep model for fake news detection. In *Proceedings of the 2017 ACM on Conference on Information and Knowledge Management* (pp. 797–806).
63. Jin, Z., Cao, J., Guo, H., Zhang, Y., & Luo, J. (2017). Multimodal fusion with recurrent neural networks for rumor detection on microblogs. In *Proceedings of the 25th ACM International Conference on Multimedia* (pp. 795–816).
64. Vosoughi, S., Mohsenvand, M. N., & Roy, D. (2017). Rumor gauge: Predicting the veracity of rumors on Twitter. *ACM Transactions on Knowledge Discovery From Data (TKDD), 11*(4), 1–36.
65. Duong, C. T., Nguyen, Q. V. H., Wang, S., & Stantic, B. (2017). Provenance-based rumor detection. In *Australasian Database Conference* (pp. 125–137). Springer, Cham.
66. Singhania, S., Fernandez, N., & Rao, S. (2017). 3han: A deep neural network for fake news detection. In *International Conference on Neural Information Processing* (pp. 572–581). Springer.
67. Dandekar, A., Zen, R. A., & Bressan, S. (2017). Generating fake but realistic headlines using deep neural networks. In *International Conference on Database and Expert Systems Applications* (pp. 427–440). Springer, Cham.
68. Poddar, L., Hsu, W., Lee, M. L., & Subramaniyam, S. (2018). Predicting stances in twitter conversations for detecting veracity of rumors: A neural approach. In *2018 IEEE 30th International Conference on Tools with Artificial Intelligence (ICTAI)* (pp. 65–72). IEEE.
69. Zubiaga, A., Aker, A., Bontcheva, K., Liakata, M., & Procter, R. (2018). Detection and resolution of rumours in social media: A survey. *ACM Computing Surveys (CSUR), 51*(2), 1–36.
70. Wang, Y., et al. (2018). Eann: Event adversarial neural networks for multi-modal fake news detection. In *Proceedings of the 24th ACM sigkdd International Conference on Knowledge Discovery and Data Mining* (pp. 849–857).
71. Wu, L., & Liu, H. (2018). Tracing fake-news footprints: Characterizing social media messages by how they propagate. In *Proceedings of the 11th ACM International Conference on Web Search and Data Mining* (pp. 637–645).
72. Ajao, O., Bhowmik, D., & Zargari, S. (2018). Fake news identification on twitter with hybrid cnn and rnn models. In *Proceedings of the 9th International Conference on Social Media and Society* (pp. 226–230).

73. Ma, J., Gao, W., & Wong, K. F. (2018). Detect rumor and stance jointly by neural multi-task learning. In *Companion Proceedings of the Web Conference 2018* (pp. 585–593).
74. Guo, H., Cao, J., Zhang, Y., Guo, J., & Li, J. (2018). Rumor detection with hierarchical social attention network. In *Proceedings of the 27th ACM International Conference on Information and Knowledge Management* (pp. 943–951).
75. Al Asaad, B., & Erascu, M. (2018). A tool for fake news detection. In *2018 20th International Symposium on Symbolic and Numeric Algorithms for Scientific Computing (SYNASC)* (pp. 379–386). IEEE.
76. Dong, M., Yao, L., Wang, X., Benatallah, B., Sheng, Q. Z., & Huang, H. (2018). Dual: A deep unified attention model with latent relation representations for fake news detection. In *International Conference on Web Information Systems Engineering* (pp. 199–209). Springer.
77. Chen, T., Li, X., Yin, H., & Zhang, J. (2018). Call attention to rumors: Deep attention based recurrent neural networks for early rumor detection. In *Pacific-Asia Conference on Knowledge Discovery and Data Mining* (pp. 40–52). Springer.
78. Sicilia, R., Giudice, S. L., Pei, Y., Pechenizkiy, M., & Soda, P. (2018). Twitter rumour detection in the health domain. *Expert Systems with Applications, 110,* 33–40.
79. Dong, M., Yao, L., Wang, X., Benatallah, B., Zhang, X., & Sheng, Q. Z. (2019). Dual-stream self-attentive random forest for false information detection. In *2019 International Joint Conference on Neural Networks (IJCNN)* (pp. 1–8). IEEE.
80. Katsaros, D., Stavropoulos, G., & Papakostas, D. (2019). Which machine learning paradigm for fake news detection?. In *2019 IEEE/WIC/ACM International Conference on Web Intelligence (WI)* (pp. 383–387). IEEE.
81. Benamira, A., Devillers, B., Lesot, E., Ray, A. K., Saadi, M., & Malliaros, F. D. (2019). Semi-supervised learning and graph neural networks for fake news detection. In *2019 IEEE/ACM International Conference on Advances in Social Networks Analysis and Mining (ASONAM)* (pp. 568–569). IEEE.
82. Al-Sarem, M., Boulila, W., Al-Harby, M., Qadir, J., & Alsaeedi, A. (2019). Deep learning-based rumor detection on microblogging platforms: A systematic review. *IEEE Access, 7,* 152788–152812.
83. Horne, B. D., Nørregaard, J., & Adali, S. (2019). Robust fake news detection over time and attack. *ACM Transactions on Intelligent Systems and Technology (TIST), 11*(1), 1–23.
84. Shu, K., Cui, L., Wang, S., Lee, D., & Liu, H. (2019). defend: Explainable fake news detection. In *Proceedings of the 25th ACM SIGKDD International Conference on Knowledge Discovery and Data Mining* (pp. 395–405).
85. Vicario, M. D., Quattrociocchi, W., Scala, A., & Zollo, F. (2019). Polarization and fake news: Early warning of potential misinformation targets. *ACM Transactions on the Web (TWEB), 13*(2), 1–22.
86. Karduni, A., et al. (2019). Vulnerable to misinformation? Verifi!. In *Proceedings of the 24th International Conference on Intelligent User Interfaces* (pp. 312–323).
87. Khattar, D., Goud, J. S., Gupta, M., & Varma, V. (2019, May). Mvae: Multimodal variational autoencoder for fake news detection. In *The World Wide Web Conference* (pp. 2915–2921).
88. Geng, Y., Lin, Z., Fu, P., & Wang, W. (2019). Rumor detection on social media: A multi-view model using self-attention mechanism. In *International Conference on Computational Science* (pp. 339–352). Springer, Cham.
89. Rajabi, Z., Shehu, A., & Purohit, H. (2019). User behavior modelling for fake information mitigation on social web. In *International Conference on Social Computing, Behavioral-Cultural Modeling and Prediction and Behavior Representation in Modeling and Simulation* (pp. 234–244). Springer, Cham.
90. Gravanis, G., Vakali, A., Diamantaras, K., & Karadais, P. (2019). Behind the cues: A benchmarking study for fake news detection. *Expert Systems with Applications, 128,* 201–213.
91. Hassan, F. M., & Lee, M. (2020). Political fake statement detection via multistage feature-assisted neural modeling. In *2020 IEEE International Conference on Intelligence and Security Informatics (ISI)* (pp. 1–6). IEEE.

92. Ksieniewicz, P., et al. (2020). Fake news detection from data streams. In *2020 International Joint Conference on Neural Networks (IJCNN)* (pp. 1–8). IEEE.
93. Ghanem, B., Rosso, P., & Rangel, F. (2020). An emotional analysis of false information in social media and news articles. *ACM Transactions on Internet Technology (TOIT), 20*(2), 1–18.
94. Rosenfeld, N., Szanto, A., & Parkes, D. C. (2020, April). A kernel of truth: Determining rumor veracity on twitter by diffusion pattern alone. In *Proceedings of The Web Conference 2020* (pp. 1018–1028).
95. Braşoveanu, A. M., & Andonie, R. (2020). Integrating machine learning techniques in semantic fake news detection. *Neural Processing Letters*, 1–18.
96. Jahanbakhsh-Nagadeh, Z., Feizi-Derakhshi, M. R., & Sharifi, A. (2020). A semi-supervised model for Persian rumor verification based on content information. *Multimedia Tools and Applications*, 1–29.
97. Li, Q., Hu, Q., Lu, Y., Yang, Y., & Cheng, J. (2020). Multi-level word features based on CNN for fake news detection in cultural communication. *Personal and Ubiquitous Computing, 24*(2), 259–272.
98. Hajek, P., Barushka, A., & Munk, M. (2020). Fake consumer review detection using deep neural networks integrating word embeddings and emotion mining. *Neural Computing and Applications, 32*(23), 17259–17274.
99. Choudhary, A., & Arora, A. (2021). Linguistic feature based learning model for fake news detection and classification. *Expert Systems with Applications, 169*, 114171.
100. Huang, Y. F., & Chen, P. H. (2020). Fake news detection using an ensemble learning model based on self-adaptive harmony search algorithms. *Expert Systems with Applications, 159*, 113584.
101. Zeng, J., Zhang, Y., & Ma, X. (2021). Fake news detection for epidemic emergencies via deep correlations between text and images. *Sustainable Cities and Society, 66*, 102652.
102. Zhao, Y., Da, J., & Yan, J. (2021). Detecting health misinformation in online health communities: Incorporating behavioral features into machine learning based approaches. *Information Processing and Management, 58*(1), 102390.
103. Al-Rakhami, M. S., & Al-Amri, A. M. (2020). Lies kill, facts save: Detecting COVID-19 misinformation in twitter. *IEEE Access, 8*, 155961–155970.
104. Zhi, X., Xue, L., Zhi, W., Li, Z., Zhao, B., Wang, Y., & Shen, Z. (2021). Financial Fake News Detection with Multi fact CNN-LSTM Model. In *2021 IEEE 4th International Conference on Electronics Technology (ICET)* (pp. 1338–1341). IEEE.
105. Kaliyar, R. K., Fitwe, K., Rajarajeswari, P., & Goswami, A. (2021). Classification of Hoax/Non-Hoax news articles on social media using an effective deep neural network. In *2021 5th International Conference on Computing Methodologies and Communication (ICCMC)* (pp. 935–941). IEEE.
106. Li, D., Guo, H., Wang, Z., & Zheng, Z. (2021). Unsupervised fake news detection based on autoencoder. *IEEE Access, 9*, 29356–29365.
107. Abdelminaam, D. S., Ismail, F. H., Taha, M., Taha, A., Houssein, E. H., & Nabil, A. (2021). Coaid-deep: An optimized intelligent framework for automated detecting COVID-19 misleading information on twitter. *IEEE Access, 9*, 27840–27867.
108. Rathore, P., Soni, J., Prabakar, N., Palaniswami, M., & Santi, P. (2021). Identifying groups of fake reviewers using a semisupervised approach. *IEEE Transactions on Computational Social Systems*.
109. Yao, J., Zheng, Y., & Jiang, H. (2021). An ensemble model for fake online review detection based on data resampling, feature pruning, and parameter optimization. *IEEE Access, 9*, 16914–16927.
110. Yin, C., Cuan, H., Zhu, Y., & Yin, Z. (2021). Improved fake reviews detection model based on vertical ensemble tri-training and active learning. *ACM Transactions on Intelligent Systems and Technology (TIST), 12*(4), 1–19.
111. Thirumuruganathan, S., Simpson, M., & Lakshmanan, L. V. (2021). To intervene or not to intervene: Cost based intervention for combating fake news. In *Proceedings of the 2021 International Conference on Management of Data* (pp. 2300–2309).

112. Patwa, P., et al. (2021). Fighting an infodemic: Covid-19 fake news dataset. In *International Workshop on Combating On line Ho st ile Posts in Regional Languages during Emergency Situation* (pp. 21–29). Springer, Cham.
113. Kasseropoulos, D. P., & Tjortjis, C. (2021). An approach utilizing linguistic features for fake news detection. In *IFIP International Conference on Artificial Intelligence Applications and Innovations* (pp. 646–658). Springer, Cham.
114. Sahoo, S. R., & Gupta, B. B. (2021). Multiple features based approach for automatic fake news detection on social networks using deep learning. *Applied Soft Computing, 100*, 106983.
115. https://github.com/mlahby/mlahby/raw/cf165d8ef6da2a74033e0f875644738b7b77bda5/SMS_Fake_News.xlsx

Using Artificial Intelligence Against the Phenomenon of Fake News: A Systematic Literature Review

Mustafa A. Al-Asadi and Sakir Tasdemir

Abstract Social networks like Facebook and Twitter have become an important way for people to connect and share their thoughts. The most important feature of social networks is the rapid sharing of information. In this context, users often share fake news without even knowing it. Fake news affects people's daily lives and its consequences can range from mere disturbing to misleading societies or even countries. The aim of this study was to provide a literature review that investigates how artificial intelligence tools are used in detecting fake news on social media and how successful they are in different fields. The study was developed using the methodology presented by Keela (2007), which is a formal methodology in computer science. The results of the study show that artificial intelligence tools such as machine learning and deep learning are widely used to develop systems for detecting fake news in various fields such as politics, sports, business, etc. and that these two tools have proven to be effective in classifying fake news. This study is intended to guide researchers as well as people involved in this field. It is believed that this study will help fill a gap in this field by presenting the main tools used for this purpose and shed light on further research. It is also hoped that this study will be a guide for researchers and individuals interested in the detection of fake news.

Keywords Artificial intelligence · Machine learning · Deep learning · Fake news detection · Social networks

1 Introduction

Nowadays, social networks such as Facebook, Twitter, and YouTube have become one of the most important communication tools for people due to their easy accessibility and the trend of sharing information and discussing events. Therefore, almost everyone has a few social media accounts today. As a result of this use of social

---

M. A. Al-Asadi (✉) · S. Tasdemir
Faculty of Technology, Department of Computer Engineering, Selçuk University, Konya, Turkey
e-mail: masadi@lisansustu.selcuk.edu.tr

M. Lahby et al. (eds.), *Combating Fake News with Computational Intelligence Techniques*, Studies in Computational Intelligence 1001,
https://doi.org/10.1007/978-3-030-90087-8_2

networks by society, many news publications have spread that are not verified or validated, giving rise to the term fake news.

Paskin [50] defines fake news as certain news articles that originate from either mainstream or social media and have no factual basis [50]. The main purpose of fake news is to influence the public on various political, sports, economic and other issues. Fake news can spread very quickly through social media. Therefore, social networks are increasingly using digital tools to detect fake news and train the public to recognize fake news [18].

In the political arena, there have been several cases of fake news on social media that have had global repercussions. A very popular case was in the United States elections in 2016. In the study presented by Allcott and Gentzkow [6], posts on social media surrounding the US presidential elections were analyzed. In this panorama, 115 fake news related to Donald Trump were counted. 41 fake news were shared on Facebook about the other presidential candidate, Hillary Clinton. Another relevant case in the financial sector was that of United Airlines, which caused the company's share price to fall by uploading a fake article on the Internet. The impact of this fake news was severe and its effect lasted for a while before recovering [16]. Another case that also had some impact was the news that was spread on the different platforms about chlorine dioxide. According to the content of the fake news, this chemical was the cure for Covid-19, as a result of which many people decided to buy and consume it, without knowing the consequences that this chemical can have if taken without medical supervision [30].

The cases described above are just a few examples of fake news from recent years. This has drawn the attention of researchers to the development of tools aimed at detecting fake news on social networks and the Internet in general. In this sense, Artificial Intelligence (AI) and particularly machine learning (ML) and deep learning (DL) techniques have been used to develop predictive systems capable of classifying fake news on social networks. As their main component, these systems use data-based learning model, which is trained on tweets, posts or web missions downloaded from social networks, microblogs or websites.

Recently, machine learning algorithms (e.g., decision trees, random forests, Naïve Bayes, support vector machines), have been successfully used [21, 42]. Similarly, deep learning models (e.g. convolutional networks, memory, recurrent networks) have been used independently or together with other neural networks [39]. Many models for the classification of fake news in social networks have emerged.

The aim of this study is to investigate which are the main AI models proposed in the literature to detect fake news on social networks. For this purpose, a literature review based on a formal methodology is proposed, such as the methodology of Keele [36], which is widely used in computer science for the development of this type of studies. The results of this study will allow describing the experience of researchers and the level of accuracy achieved by automatic fake news detection systems implemented with machine learning and deep learning models.

The remainder of this article is organized as follows. Section 2 reports the related work. Section 3 explains the methodology used in this paper, focusing on the search strategy developed. Section 4 presents the results, i.e. it discusses in detail each of the research questions (RQs) formulated based on the analysis of the primary studies found. In addition, these findings are briefly discussed. Finally, Sect. 5 describes the conclusions.

2 Related Work

Related work falls broadly into the following categories: (1) Exploratory analysis of Fake News features, (2) Detection based on traditional machine learning algorithms, (3) Deep learning-based detection, (4) Advanced language model based detection, and (5) Literature reviews.

Some authors have conducted systematic literature reviews to determine a conceptual approach to solving the problem of detecting fake news on social networks using artificial intelligence tools. While these literature reviews focus on different aspects of the problem [7, 17–19, 23, 25, 46, 47], they did not provide any information about the datasets used, the scope of application and the software tools used for the previous work.

This work differs from the previous one in that it exclusively analyses the proposed methods for detecting Fake News in social media. Moreover, we adopted a systematic approach to find the works in an unbiased way. To the best of our knowledge, there is no systematic survey that analyzes the existing proposed solutions in the literature that can support automatic detection of Fake News in social media. This work provides a quantitative overview of a number of features of the found works, as well as an overview of what has already been done in this area and where these works have been published, to help new researchers make their first contact with this topic.

3 Methodology

A systematic review of the literature consists of identifying, evaluating, and interpreting the most relevant studies on a given topic. To conduct the systematic review proposed in this study, the methodology proposed by Keele [36] was used, which is a formal methodology for carrying out this type of work in computer science.

3.1 Research Questions

Fake news has attracted a lot of interest from AI researchers. Fake news has therefore become a global problem that needs to be addressed. In this research, the following main question was asked:

RQ: How have AI learning models contributed to the creation of classifiers for automatic detection of fake news on social networks?

To answer this question, the following sub-questions were asked:

SRQ1: What artificial intelligence tools are used to detect fake news on social networks and how accurate are they?
SRQ2: What software and data tools have been used to build predictive models to detect fake news on social media and in what areas has false news been detected on social networks?

3.2 Search Strategy

The search conducted in this study considered works available in the literature that have addressed the automatic detection of fake news in social networks from different perspectives. The search strategy was divided into five steps: Defining the sources of the information, formulating the search equation, selecting the primary studies found, quality assessment, and creating a flowchart for the search process.

3.2.1 Definition of Information Sources

The first step of the search strategy was to identify and determine the sources of information that would be used to carry out the systematic review. To identify relevant research articles, we used different digital libraries, such as Google Scholar (https://scholar.google.com), ACM Digital Library (www.acm.org), IEEE Xplore (https://ieeexplore.ieee.org), Springer Link (https://link.springer.com), and Science Direct (www.sciencedirect.com).

3.2.2 Formulation of the Search Equation

A search string was defined to search for the primary studies, described as follows: ((“machine learning*”) OR (“deep learning*”) OR (“artificial intelligence*”) AND (“fake news*”)). This string was applied only to the titles of documents. The search string was adapted to be applied in each of the selected libraries.

3.2.3 Inclusion and Exclusion Criteria

The inclusion criteria defined for this study aimed to include only those proposals that analyzed and used AI mechanisms to detect false news on social networks. In addition, four exclusion criteria were defined that aimed to exclude studies that did not contribute to the research. These exclusion criteria were as follows: duplicate documents, documents written in a language other than English, inaccessible documents, and documents published before 2017. Articles found using the search term that did not meet these criteria were not analyzed.

3.2.4 Quality Assessment

In addition to the exclusion criteria, the quality of all included papers was assessed based on the research presented in them. Papers in which researchers discussed the use of machine learning and deep learning to detect fake or false news were considered high quality for inclusion in this literature review. We also created a quality standard based on four aspects that affect the quality of the study, which were collected to provide a comprehensive measure of the quality of the study:

1. Is the process of data analysis appropriate?
2. Does the study include the details of the data sets and their citations?
3. Has the data been divided into training and tested?
4. Was the result of accuracy or other analysis used to measure the quality of the models?

3.2.5 Flowchart of Search Process

Searching the above string, we first found 118 papers between 2017 and Mid-August 2021. The selection of studies was then done in two steps; (i) reading the title, abstract and conclusions of each article, and (ii) reading the articles in full. After excluding non-target articles, 49 papers were finally selected. Figure 1 shows a flowchart of the search process: the identification of articles, the screening and selection process, and the number of articles included.

4 Results and Discussion

Based on the information and findings described in the 40 articles examined. This section presents the results of the main search formulated in this paper. These results are summarized as follows:

SRQ1: What artificial intelligence tools are used to detect fake news on social networks and how accurate are they?

To classify fake news on social networks, both machine learning and deep learning algorithms have been used. In the first block of Table 1, 18 studies are listed in which only the use of at least one of the machine learning algorithms was proposed to train automatic classifiers of fake news in social networks. In block 2 of the same table, it can be seen that 19 other studies are limited to the exclusive use of deep learning models (neural networks) to train fake news classifiers. Finally, In Block 3 of the

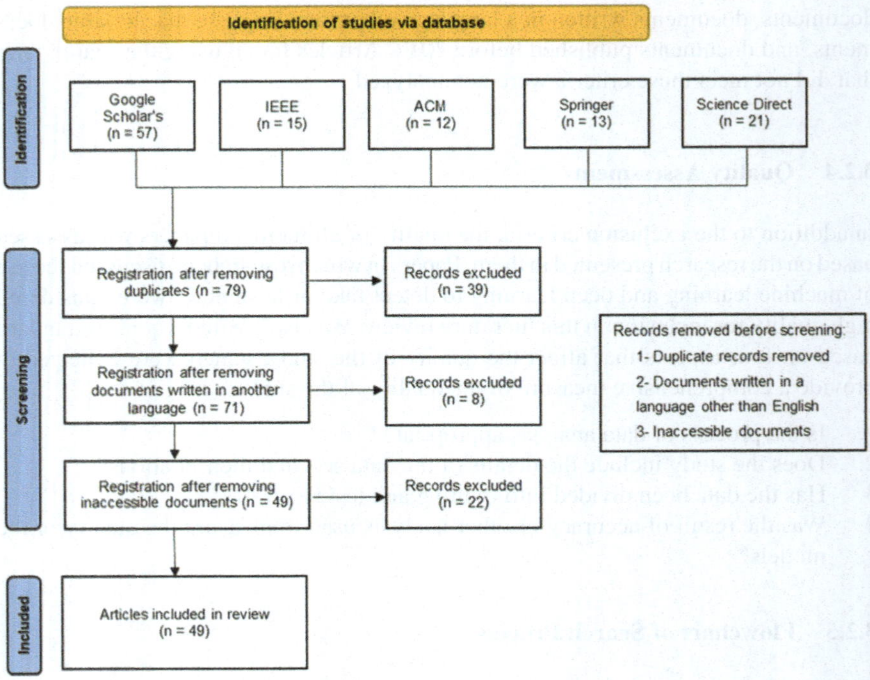

Fig. 1 Flowchart of the selection process

same table, it was observed that 13 other studies use both machine learning models and deep learning models. Besides, other hybrid models are less common. These are hybrid models mainly developed to improve accuracy when machine learning and deep learning algorithms are applied independently (such as S33 and S34).

Table 1 shows that the use of applied AI to detect fake news has been extensive in recent years. Most of the analyzed proposals were published between 2019 and 2020 (35 studies), in addition to three studies analyzed in the current year 2021. It is noticeable that most of the studies were developed by researchers from India (15 studies). This is because this country has been working on a strong campaign to spread fake news since 2019. It is followed by the United States of America with 5 studies. Therefore, science plays an important role in solving this problem, which is destabilizing from a political point of view.

Regarding the accuracy of the artificial intelligence tools used to detect fake news, both machine learning and deep learning algorithms have been used, as mentioned earlier. Table 2 shows the most popular algorithms used for this purpose. From Table 2, it can be seen that support vector machine algorithms have been used most frequently (9 studies), followed by random forest (8 studies) and logistic regression (7 studies). At the same time, algorithms based on Bayes theory were used, decision tree with 5 studies respectively. Other algorithms, although less frequently used,

Table 1 Selected studies aimed at uncovering fake news using machine and deep learning

Group	S	Country	year	Source	Reference	T
Group 1 Machine learning	S1	Canada	2021	Sciencedirect	[26]	
	S2	India	2020	IEEE	[24]	17
	S3	Tunisia	2020	Springer	[27]	
	S4	India	2020	GoogleScholar	[57]	
	S5	India	2019	GoogleScholar	[42]	
	S6	Brazil	2019	ACM	[54]	
	S7	Malaysia	2019	GoogleScholar	[48]	
	S8	Pakistan	2019	IEEE	[34]	
	S9	Canada	2019	Springer	[31]	
	S10	India	2019	IEEE	[33]	
	S11	India	2019	IEEE	[32]	
	S12	India	2019	GoogleScholar	[40]	
	S13	UAE	2018	Sciencedirect	[5]	
	S14	Colombia	2018	Springer	[2]	
	S15	India	2017	IEEE	[21]	
	S16	USA	2017	GoogleScholar	[59]	
	S17	Canada	2017	Springer	[3]	
Group 2 Deep learning	S18	India	2021	Springer	[62]	19
	S19	Turkey	2021	GoogleScholar	[38]	
	S20	UK	2020	ACM	[13]	
	S21	USA	2020	GoogleScholar	[55]	
	S22	Spain	2020	IEEE	[37]	
	S23	UK	2019	IEEE	[28]	
	S24	Slovakia	2019	IEEE	[39]	
	S25	Singapore	2019	IEEE	[44]	
	S26	Algeria	2019	IEEE	[8]	
	S27	UK	2019	GoogleScholar	[49]	
	S28	Jordon	2019	ACM	[1]	
	S29	Korea	2019	Sciencedirect	[43]	
	S30	Jordon	2019	IEEE	[53]	
	S31	Korea	2019	IEEE	[4]	
	S32	India	2019	IEEE	[61]	
	S33	Portugal	2019	ACM	[14]	
	S34	Belgium	2018	GoogleScholar	[20]	
	S35	USA	2018	GoogleScholar	[60]	
	S36	Egypt	2018	IEEE	[22]	

(continued)

Table 1 (continued)

Group	S	Country	year	Source	Reference	T
Group 3 Machine & Deep learning	S37	USA	2020	ACM	[58]	13
	S38	India	2020	Sciencedirect	[51]	
	S39	India	2020	GoogleScholar	[41]	
	S40	India	2019	IEEE	[29]	
	S41	India	2019	IEEE	[52]	
	S42	India	2019	IEEE	[12]	
	S43	India	2019	Springer	[56]	
	S44	Bangladesh	2019	IEEE	[45]	
	S45	Cyprus	2019	IEEE	[35]	
	S46	Romania	2019	Springer	[15]	
	S47	India	2019	Springer	[11]	
	S48	Thailand	2018	IEEE	[9]	
	S49	USA	2017	GoogleScholar	[10]	

include extra trees classifier, k-nearest neighbours algorithm, gradient enhancement, stochastic gradient descent, XGBoost, linear discriminant analysis, and Quadratic discriminant analysis. However, the algorithms that achieved the best accuracy in detecting fake news were: random forest with 99.30%, decision tree with 99.29% and Bayes Theorem with 98.70%.

Machine learning algorithms are not the only ones being used to try to develop AI systems to detect fake news on social networks. Very popular and efficient models based on neural networks have also been used to develop a more efficient classifier than the machine learning based models. The second group of Table 2 lists 19 proposals in which neural networks were used to detect fake news. In addition, 13 other studies listed in the third group of the same table used these deep learning models in conjunction with machine learning models. As with the approach described above, the main reason for using both models was to compare and determine which of the two AI learning approaches was best suited for the fake news datasets used to train the models.

Although there is a variety of models within Deep Learning that aim to process information using artificial neural networks, our research found that only three of the existing neural network models were primarily used to create classifiers for fake news on social networks. These neural network models, summarized in Table 3, include the following: Memory Networks (14 studies), Convolutional Networks (13 studies), and Recurrent Networks (RNN and GRU) (9 studies). The results in terms of accuracy show that the best-fit neural network model was a generic neural network without specification with 99.90% (S48), followed by convolutional networks with 99.00% (S20). Finally, the BERT network model with an accuracy of 98.41% (S18). However, as shown in S39, by developing hybrid models, it is possible to achieve more competent accuracy values than with models that operate independently. For example, a

Table 2 Use of machine learning models in the analyzed studies

N	RF	DT	ET	NB	BN	SVM	KNN	GB	LR	SGD	XG Boost	LDA, QDA
S1	98.45	99.29	97.59									
S2				65.00		69.00			72.30		76.20	
S3	97.96	97.93		98.19		97.61	97.12		98.07			98.21 97.38
S4	59.00			60.00					65.00			
S5	90.70	82.70				75.50	79.20					
S6											88.00	
S7									77.00			
S8	65.00	64.00		68.00		68.00			69.00			
S9									40.00			
S10								86.00				
S11						39.60						
S12						59.00						
S13	99.30			98.70	94.4				99.4			
S14	88.10											
S15	64.80	67.60				73.60		65.77		65.70		
S16						87.00						
S17						92.00						
$\sum$	8	5	1	5	1	9	2	2	7	1	2	2

RF = random forests, DT = decision tree, ET = extra trees classifier, NB = naive Bayes, BN = Bayesian network, SVM = supporting vector machine, KNN = k-nearest neighbors, GB = gradient enhancement, LR = logistic regression, SGD = stochastic gradient descent, XGBoost = extreme gradient boosting, LDA = linear discriminant analysis, QDA = Quadratic discriminant analysis

model called bidirectional CNN + LSTM achieved an accuracy of 88.78%. While the accuracy of a convolutional network (CNN) is 73.29% and that of a memory network (LSTM) is 80.62%. This opens a number of opportunities to further explore hybrid deep learning models (e.g., S30, S42).

It is important to note that the accuracy achieved by machine learning and deep learning algorithms depends directly on the dataset used to train the models. Moreover, it also depends on the feature extraction method applied to the data. In most studies, different methods were used, such as Term frequency (TF) (e.g., S4, S8), Term Frequency-Inverse Document Frequency (TF-IDF) (e.g., S2, S8, S15, S25, S38, S41, S44), global vectors (GloVe) (e.g., S25, S45), Bag-Of-Words (e.g., S4, S12), N-Grams (e.g., S4, S17) and the CountVectorizer method (e.g., S25, S45).

Table 3 Use of deep learning models in the analyzed studies

N	CNN	RNN	DNN	GNN	MLP	Generic ANN	GDL	GRU	LSTM	BERT
S18	93.00								93.00	98.41
S19	93.70									
S20	99.00									
S21								97.00		
S22	93.00								91.00	
S23							92.70			
S24	97.50				89.8				91.80	
S25	92.40								95.30	
S26	96.00									
S27				92.70						
S28									72.20	
S29	52.80								41.70	
S30	87.47									
S31										96.25
S32								91.90	94.30	
S33								83.38		
S34				94.40						
S35			94.21							
S36		21.50						21.70	21.66	
S37										72.10
S38					93.00				93.30	
S39	73.29								80.62	
S40			91.00							
S41						49.90				
S42									88.61	
S43					72.00					
S44								74.00	78.00	
S45	60.00				58.00					
S46	29.00							54.90	32.40	
S47					88.36					
S48						99.90				
S49	97.00	91.00			89.00			89.00	93.00	
∑	13	2	2	2	6	2	1	7	14	3

CNN = convolutional neural network, RNN = recurrent network, GRU = gated recurrent unit, GDL = geometric deep learning, DNN = deep neural network, LSTM = long short-term memory, MLP = multilayer perceptron, GNN = graph neural network

SRQ2: What software and data tools were used to build predictive models to detect fake news on social media, and in what areas was fake news detected on social networks?

The development tools used to create classifiers for fake news on social networks were mainly using the Python programming tool (29 studies). Python is currently the most popular and widely used tool for developing AI systems. In particular, this programming language is used for the development of text mining systems that integrate automatic classification models. The use of Python libraries for machine learning has also been demonstrated, such as sci-kit learn (14 studies; S1, S2, S3, S4, S8, S11, S13, S15, S21, S38, S39, S42, S44, S45), Keras (8 studies; S19, S21, S22, S24, S25, S42, S44, S46) and TensorFlow (7 studies; S18, S24, S26, S32, S34, S42, S49). Complementarily, the natural language processing library, NLTK, Natural Language Toolkit, was also used for its English acronym (12 studies; S4, S8, S7, S15, S21, S24, S28, S29, S32, S35, 45, 47) has also been used. In other studies, the use of Google's natural language processing tool was also evidenced (3 studies; S6, S9, S30). Other Python libraries have also been used such as NetworkX to study the structure of complex networks (e.g., S3, and the Gensim framework to make the conversion of natural language texts to the Vector Space Model as easy and natural as possible (e.g., S21). Similary, some studies did not describe aspects of the development tools in detail, such as studies S7, S10, S12, S16, S20, S27, S29, S31, S33, S36, S41, S43, S48, making it impossible to replicate the experiments in some cases.

Regarding the data used by the analyzed proposals, they mainly used publicly available datasets in repositories such as Kaggle (11 studies; S5, S9, S12, S14, S19, S24, S26, S39, S41, S47, S49), GitHub (2 studies; S2, S30) and PolitiFact (8 studies; S4, S7, S22, S23, S34, S36, S39, S45). It also appeared that some studies used other data sources to a lesser extent, such as Twitter (6 studies; S3, S18, S21, S34, S44, S48), The New York Times and Washington Post (S22), Reuter.com or BuzzFeed (S2, S6, S23, S27, S34).

In general, text mining models were applied to predefined datasets. In this way, researchers optimized the time required for collecting and preprocessing the data, which is the input for the text mining process, and for training the classifier by applying supervised learning algorithms that provide the outputs. Likewise, some of the data was used to implement the testing phase of the model. However, in some proposals, the researchers created their own datasets to develop their predictive models to classify fake news. Information from news websites and blogs served as the source. In summary, the datasets used were the following: NYT (S25, S37), LIAR (S1, S36, S45 and S46), News (S40), Fake News (S9, S26), Fake News Net (S38), Fake Real News (S14) and FNC-1 (S28, S30, S33, and S43).

Regarding the areas in which fake news was exposed in social networks, the proposals studied focused on four areas: Politics (32 studies; S3, S4, S6, S8, S9, S10, S12, S15, S16, S17, S19, S21, S22, S23, S24, S25, S26, S27, S28, S29, S30, S31, S33, S34, S37, S38, S39, S41, S42, S45, S46, S47), business and economy (3 studies; S29, S31, S37), society, sport and culture (2 studies; S8, S37), science, technology and health (3 studies; S8, S18, S37), and entertainment (2 studies; S8, S29). In addition, there were 11 studies in which the scope of the fake news analyzed was not specified (S2, S5, S7, S11, S14, S20, S32, S35, S40, S43, S44, S49). It has been shown that the largest proportion of false news is found in the political sphere, i.e. information about leaders and their actions. For this reason, government at the national level have an interest in spreading fake news, many of which destabilize their governments and cause chaos among the population.

5 Conclusion

Artificial intelligence is a suitable tool for use in social networks, and several applications have successfully used AI tools to improve the fight against fake news. The results show that AI learning models are widely used to create automatic systems for detecting fake news, with both high and low accuracy rates. Among the analyzed studies, 17 studies mainly rely on machine learning algorithms based on the following methods: Naive Bayes, k-nearest neighbours, logistic regression, supporting vector machine, linear discriminant analysis, decision tree, and random forests. The latter had the best level of accuracy at 99.30%. The remaining 32 studies used neural networks (NR). The most common of these networks were: convolutional networks, and recurrent networks, and long short-term memory (LSTM). Convolutional networks were the most commonly used networks with an accuracy of 99% and a generic neural network of which no details were given with 99.90%. This suggests that classifiers using machine learning were more accurate in classifying fake news from tweets. However, many models combine both machine learning and deep learning to optimize the process at different stages to achieve better accuracy in detecting fake news.

Although a wide range of predictive models have been developed that aim to classify fake news from tweets, it is still a challenge to integrate these models into the Twitter social network to warn users against sharing content. The proposed models have only identified the main features of tweets that contain fake news. Thus, there is a gap in the detection of fake news compatible not only with the Twitter social network, but also with any other of the widely used social networks such as Facebook or the blogs available on the Internet.

The spread of fake news on social networks or on the Internet harms society in general. Therefore, both human wisdom and digital tools must be used in this process. We hope that some of these measures will remain in place and that digital media platform operators and the public will take responsibility and work together to detect and combat fake news.

We believe that this study will help fill a gap in this area by introducing the main tools used for this purpose and shedding light on further research. In addition, this study is intended to be a guide for researchers and individuals interested in detecting fake news.

6 Declarations

6.1 Competing Interests

Mustafa A. AL-ASADI certifies that the submission is original work and is not under review at any other publication. There is no financial interest to report.

6.2 Funding

Not applicable.

References

1. Abedalla, A., Al-Sadi, A., Abdullah, M. (2019). A closer look at fake news detection: A deep learning perspective. In *Proceedings of the 2019 3rd International Conference on Advances in Artificial Intelligence* (pp. 24–28).
2. Agudelo, G. E. R., Parra, O. J. S., Velandia, J. B. (2018). Raising a model for fake news detection using machine learning in Python. In *Conference on e-Business, e-Services and e-Society* (pp. 596–604). Springer.
3. Ahmed, H., Traore, I., Saad, S. (2017). Detection of online fake news using n-gram analysis and machine learning techniques. In *International Conference on Intelligent, Secure, and Dependable Systems in Distributed and Cloud Environments.* (pp. 127–138). Springer.
4. Ahn, Y.-C., Jeong, C.-S. (2019). Natural language contents evaluation system for detecting fake news using deep learning. In *2019 16th International Joint Conference on Computer Science and Software Engineering (JCSSE)* (pp. 289–292). IEEE.
5. Aldwairi, M., & Alwahedi, A. (2018). Detecting fake news in social media networks. *Procedia Computer Science, 141*, 215–222.
6. Allcott, H., & Gentzkow, M. (2017). Social media and fake news in the 2016 election. *Journal of Economic Perspectives, 31*(2), 211–236.
7. Almaliki, M. (2019) Online misinformation spread: A systematic literature map. In *Proceedings of the 2019 3rd International Conference on Information System and Data Mining* (pp. 171–178)
8. Amine, B. M., Drif, A., Giordano, S. (2019). Merging deep learning model for fake news detection. In *2019 International Conference on Advanced Electrical Engineering (ICAEE)* (pp. 1–4). IEEE.
9. Aphiwongsophon, S., Chongstitvatana, P. (2018). Detecting fake news with machine learning method. In *2018 15th International Conference on Electrical Engineering/Electronics, Computer, Telecommunications and Information Technology (ECTI-CON)* (pp. 528–531). IEEE.
10. Bajaj, S. (2017). The pope has a new baby! fake news detection using deep learning. CS 224N:1–8.
11. Bali, A. P, S., Fernandes, M., Choubey, S., Goel, M. (2019). Comparative performance of machine learning algorithms for fake news detection. In *International conference on advances in computing and data sciences* (pp. 420–430). Springer.

12. Barua, R., Maity, R., Minj, D., Barua, T., Layek, A. K. (2019) F-NAD: An application for fake news article detection using machine learning techniques. In *2019 IEEE Bombay Section Signature Conference (IBSSC)* (pp. 1–6). IEEE.
13. Bogale Gereme, F., Zhu, W. (2020). Fighting fake news using deep learning: Pre-trained word embeddings and the embedding layer investigated. In *2020 The 3rd International Conference on Computational Intelligence and Intelligent Systems* (pp. 24–29).
14. Borges, L., Martins, B., & Calado, P. (2019). Combining similarity features and deep representation learning for stance detection in the context of checking fake news. *Journal of Data and Information Quality (JDIQ), 11*(3), 1–26.
15. Braşoveanu, A. M., Andonie, R. (2019). Semantic fake news detection: A machine learning perspective. In *International Work-Conference on Artificial Neural Networks* (pp. 656–667). Springer.
16. Carvalho, C., Klagge, N., & Moench, E. (2011). The persistent effects of a false news shock. *Journal of Empirical Finance, 18*(4), 597–615.
17. Choraś, M., Demestichas, K., Giełczyk, A., Herrero, Á., Ksieniewicz, P., Remoundou, K., Urda, D., Woźniak, M. (2020). Advanced machine learning techniques for fake news (online disinformation) detection: A systematic mapping study. *Applied Soft Computing* 107050.
18. De Beer, D., Matthee, M. (2020). Approaches to identify fake news: A systematic literature review. In *International Conference on Integrated Science* (pp. 13–22). Springer.
19. de Souza, J. V., Gomes, J., Jr., de Souza Filho, F. M., de Oliveira Julio, A. M., & de Souza, J. F. (2020). A systematic mapping on automatic classification of fake news in social media. *Social Network Analysis and Mining, 10*(1), 1–21.
20. Deligiannis, N., Huu, T., Nguyen, D. M., Luo, X. (2018). Deep learning for geolocating social media users and detecting fake news. In *NATO Workshop.*
21. Gilda, S. (2017). Evaluating machine learning algorithms for fake news detection. In *2017 IEEE 15th Student Conference on Research and Development (SCOReD)* (pp. 110–115).
22. Girgis, S., Amer, E., Gadallah, M. (2018). Deep learning algorithms for detecting fake news in online text. In *2018 13th International Conference on Computer Engineering and Systems (ICCES)* (pp. 93–97). IEEE.
23. Goksu, M., Cavus, N. (2019). Fake news detection on social networks with artificial intelligence tools: systematic literature review. In *International Conference on Theory and Application of Soft Computing, Computing with Words and Perceptions* (pp. 47–53). Springer.
24. Goswami, S., Hudnurkar, M., & Ambekar, S. (2020). Fake news and hate speech detection with machine learning and NLP. *PalArch's Journal of Archaeology of Egypt/Egyptology, 17*(6), 4309–4322.
25. Habib, A., Asghar, M. Z., Khan, A., Habib, A., & Khan, A. (2019). False information detection in online content and its role in decision making: A systematic literature review. *Social Network Analysis and Mining, 9*(1), 1–20.
26. Hakak, S., Alazab, M., Khan, S., Gadekallu, T. R., Maddikunta, P. K. R., & Khan, W. Z. (2021). An ensemble machine learning approach through effective feature extraction to classify fake news. *Future Generation Computer Systems, 117*, 47–58.
27. Hamdi, T., Slimi, H., Bounhas, I., Slimani, Y. (2020). A hybrid approach for fake news detection in twitter based on user features and graph embedding. In *International Conference on Distributed Computing and Internet Technology* (pp. 266–280). Springer.
28. Han, W., Mehta, V. (2019) Fake news detection in social networks using machine learning and deep learning: Performance evaluation. In *2019 IEEE International Conference on Industrial Internet (ICII)* (pp. 375–380). IEEE.
29. Hiramath, C.K., Deshpande, G. (2019) Fake news detection using deep learning techniques. In *2019 1st International Conference on Advances in Information Technology (ICAIT)* (pp. 411–415). IEEE.
30. Huang, B., Carley, K.M. (2020). Disinformation and misinformation on twitter during the novel coronavirus outbreak. arXiv preprint. arXiv:200604278.
31. Ibrishimova, M. D., Li, K. F. (2019). A machine learning approach to fake news detection using knowledge verification and natural language processing. In *International Conference on Intelligent Networking and Collaborative Systems* (pp. 223–234). Springer.

32. Jain, A., Shakya, A., Khatter, H., Gupta, A. K. (2019). A smart system for fake news detection using machine learning. In *2019 International Conference on Issues and Challenges in Intelligent Computing Techniques (ICICT)* (pp. 1–4). IEEE.
33. Kaliyar, R. K., Goswami, A., Narang, P. (2019). Multiclass fake news detection using ensemble machine learning. In *2019 IEEE 9th International Conference on Advanced Computing (IACC)* (pp. 103–107). IEEE.
34. Kareem, I., Awan, S. M. (2019) Pakistani media fake news classification using machine learning classifiers. In *2019 International Conference on Innovative Computing (ICIC)* (pp. 1–6). IEEE.
35. Katsaros, D., Stavropoulos, G., Papakostas, D. (2019). Which machine learning paradigm for fake news detection? In *2019 IEEE/WIC/ACM International Conference on Web Intelligence (WI)* (pp. 383–387). IEEE.
36. Keele, S. (2007). Guidelines for performing systematic literature reviews in software engineering. Citeseer.
37. Kong, S. H., Tan, L. M., Gan, K. H., Samsudin, N. H. (2020). Fake news detection using deep learning. In *2020 IEEE 10th Symposium on Computer Applications & Industrial Electronics (ISCAIE)* (pp. 102–107). IEEE.
38. Korkmaz, T., Çetinkaya, A., Aydin, H., & Barişkan, M. A. (2021). Analysis of whether news on the Internet is real or fake by using deep learning methods and the TF-IDF algorithm. *International Advanced Researches and Engineering Journal, 5*(1), 31–41.
39. Krešňáková, V. M., Sarnovský, M., Butka, P. (2019). Deep learning methods for fake news detection. In *2019 IEEE 19th International Symposium on Computational Intelligence and Informatics and 7th IEEE International Conference on Recent Achievements in Mechatronics, Automation, Computer Sciences and Robotics (CINTI-MACRo)* (pp. 000143–000148). IEEE.
40. Kumar, A., Singh, S., Kaur, G. (2019). Fake news detection of Indian and United States election data using machine learning algorithm.
41. Kumar, S., Asthana, R., Upadhyay, S., Upreti, N., & Akbar, M. (2020). Fake news detection using deep learning models: A novel approach. *Transactions on Emerging Telecommunications Technologies, 31*(2), e3767.
42. Lakshmanarao, A., Swathi, Y., & Kiran, T. S. R. (2019). An effecient fake news detection system using machine learning. *International Journal of Innovative Technology and Exploring Engineering, 8*(10), 3125–3129.
43. Lee, D.-H., Kim, Y.-R., Kim, H.-J., Park, S.-M., & Yang, Y.-J. (2019). Fake news detection using deep learning. *JIPS, 15*(5), 1119–1130.
44. Liu, H. (2019). A location independent machine learning approach for early fake news detection. In *2019 IEEE International Conference on Big Data (Big Data)* (pp. 4740–4746). IEEE.
45. Mahir, E.M., Akhter, S., Huq, M. R. (2019). Detecting fake news using machine learning and deep learning algorithms. In *2019 7th International Conference on Smart Computing & Communications (ICSCC)* (pp. 1–5). IEEE.
46. Meel, P., & Vishwakarma, D. K. (2020). Fake news, rumor, information pollution in social media and web: A contemporary survey of state-of-the-arts, challenges and opportunities. *Expert Systems with Applications, 153*, 112986.
47. Meneses Silva, C. V., Silva Fontes, R., & Colaço Júnior, M. (2021). Intelligent fake news detection: A systematic mapping. *Journal of Applied Security Research, 16*(2), 168–189.
48. Mokhtar, M. S., Jusoh, Y. Y., Admodisastro, N., Pa, N., & Amruddin, A. Y. (2019). Fakebuster: Fake news detection system using logistic regression technique in machine learning. *International Journal of Engineering and Advanced Technology, 9*(1), 2407–2410.
49. Monti, F., Frasca, F., Eynard, D., Mannion, D., Bronstein, M.M. (2019). Fake news detection on social media using geometric deep learning. arXiv preprint. arXiv:190206673.
50. Paskin, D. (2018). Real or fake news: Who knows? *The Journal of Social Media in Society, 7*(2), 252–273.
51. Pereira, N., Dabreo, S., Rodrigues, L., & Thomas, P. M. (2020). Comparative analysis of fake news detection using machine learning and deep learning techniques. *International Journal of Emerging Technologies and Innovative Research, 7*(4), 1379–1385.

52. Poddar, K., Umadevi, K. (2019). Comparison of various machine learning models for accurate detection of fake news. In *2019 Innovations in Power and Advanced Computing Technologies (i-PACT)* (pp. 1–5). IEEE.
53. Qawasmeh, E., Tawalbeh, M., Abdullah, M. (2019) Automatic identification of fake news using deep learning. In *2019 Sixth International Conference on Social Networks Analysis, Management and Security (SNAMS)* (pp. 383–388). IEEE.
54. Reis, J. C., Correia, A., Murai, F., Veloso, A., Benevenuto, F. (2019). Explainable machine learning for fake news detection. In *Proceedings of the 10th ACM Conference on Web Science* (pp. 17–26).
55. Sabeeh, V., Zohdy, M., Mollah, A., Al Bashaireh, R. (2020). Fake news detection on social media using deep learning and semantic knowledge sources. *International Journal of Computer Science and Information Security (IJCSIS) 18*(2).
56. Saikh, T., Anand, A., Ekbal, A., Bhattacharyya, P. (2019). A novel approach towards fake news detection: deep learning augmented with textual entailment features. In *International Conference on Applications of Natural Language to Information Systems* (pp. 345–358). Springer.
57. Sharma, U., Saran, S., Patil, S. M. (2020). Fake news detection using machine learning algorithms. *International Journal of Creative Research Thoughts (IJCRT), 8*(6).
58. Singh, R., Chun, S. A., Atluri, V. (2020). Developing machine learning models to automate news classification. In *The 21st Annual International Conference on Digital Government Research* (pp. 354–355).
59. Singh, V., Dasgupta, R., Sonagra, D., Raman, K., Ghosh, I. (2017). Automated fake news detection using linguistic analysis and machine learning. In *International Conference on Social Computing, Behavioral-Cultural Modeling, & Prediction and Behavior Representation in Modeling and Simulation (SBP-BriMS)* (pp. 1–3).
60. Thota, A., Tilak, P., Ahluwalia, S., & Lohia, N. (2018). Fake news detection: A deep learning approach. *SMU Data Science Review, 1*(3), 10.
61. Verma, A., Mittal, V., Dawn, S. (2019). FIND: Fake information and news detections using deep learning. In *2019 Twelfth International Conference on Contemporary Computing (IC3)* (pp. 1–7). IEEE.
62. Wani, A., Joshi, I., Khandve, S., Wagh, V., Joshi, R. (2021). Evaluating deep learning approaches for covid19 fake news detection. arXiv preprint. arXiv:210104012.

Fake News Detection in Internet Using Deep Learning: A Review

Israel Barrutia-Barreto, Renzo Seminario-Córdova, and Brian Chero-Arana

Abstract The main objective of this research was to explore, from a reflexivity approach, the current state of Deep learning techniques for automatic detection of fake news on the Internet, analyzing the most important Deep learning algorithms and studies on their effectiveness in detecting distrustful information. The research methodology employed was bibliographic, documentary and descriptive. The information was collected from several scientific articles provided by indexed journals and web platforms, using keywords such as "fake news", "Deep learning" and "neural networks" for the compilation. As a result of this research, it was concluded that Deep learning techniques present a better performance than conventional methods and will be of great importance in the future of war against fake news due to their potential in automatic detection.

Keywords Fake news · Disinformation · Deepfakes · Artificial intelligence · Deep learning · Neural networks · Social networks

1 Introduction

The spread of false information is not a recent concern; rather, it is a problem as old as the existence of mankind itself. Although such spread has been taking place since at least the times of Ancient Greece, it was only after the nineteenth century that these started to spread with greater force, assisted by emerging technologies such as the printing press and later the radio [41]. This action took on a new meaning in recent decades with the emergence of mass media and the Internet itself, for the global interconnection resulting from the massification of these technologies in the world population greatly facilitated the spread of misleading information [13].

In essence, the term "fake news" refers to the production and dissemination of false information disguised as real news, which is spread for profit or with the intention of causing public harm [32]. The spread of fake news can be done deliberately or

I. Barrutia-Barreto (✉) · R. Seminario-Córdova · B. Chero-Arana
Innova Scientific, Lima, Perú

M. Lahby et al. (eds.), *Combating Fake News with Computational Intelligence Techniques*, Studies in Computational Intelligence 1001,
https://doi.org/10.1007/978-3-030-90087-8_3

intentionally for political, economic or knowledge purposes [27]. In recent years, it has even been discussed to replace the term "fake news" by "disinformation", taking into account that the first one does not adequately encompass the concept of disinformation on the Internet, being this a wide and complex topic [31].

When reliable information is shared through platforms such as social networks, it is often overshadowed by the large amount of false or misleading information produced by several users, who take advantage of the anonymity and the lack of veracity controls on these platforms in order to spread these hoaxes [13]. These factors have made them one of the main means of diffusion of fake news on the Internet, which are often used as information sources in an irresponsible way [32]. The ease with which anyone can now publish a news story or mere statement online and make it go viral has led to fake news being considered a threat to commerce, journalism and democracy around the world [23].

This situation sparked the interest of researchers in studying fake news in depth in order to find ways to identify them and prevent them from spreading. In the field of computer science, scientists sought to apply artificial intelligence to address this problem. For that matter, they started applying techniques such as natural language processing or digital forensics in order to develop ways to automatically identify fake news [27]. One of these methods is known as Deep learning, a branch within machine learning which includes algorithms capable of analyzing information at multiple levels of representation, obtaining higher level information from lower level information [45].

Deep learning techniques have gained relevance lately for being considered an effective tool for automatic detection of fake news and have been the subject of several studies in recent years. In this context, the aim of this book chapter is to explore, from a reflexivity approach, the evolution of Deep learning algorithms in recent years, their presence within fake news and their potential usefulness in detecting them and becoming an important tool to curb disinformation and media manipulation in the near future.

2 Methodology

Regarding the employed research methodology, after the selection of the topic, a general review of the bibliography related to fake news and the use of Deep learning for its automatic detection was carried out. With this information, it was possible to identify the most important topics and thus define a main outline in order to elaborate this chapter. In this way, the chapter was focused on providing definitions and exploring research conducted on the use of Deep learning techniques in fake news detection.

Scientific articles were collected from scientific databases such as Google Scholar. Several relevant articles were found using keywords such as "fake news", "Deep learning", "applications" and "methods". This bibliographic analysis focused on collecting recent information published not earlier than 2019 for specific topics such as research, while there were no restrictions for information on definitions needed according to the defined topics.

The literature review retrieved a large number of articles when the keywords were introduced in the database search engine. From this amount, a total of approximately 70 articles were initially collected after being considered relevant by briefly reviewing their title and abstract. From this resulting bibliography, the articles were reviewed in their entirety to finally select a total of 51 bibliographic references.

3 Fake News: Why Are They Used?

Historically, fake stories were conceived in order to increase newspaper sales or to generate fear or anger in the population [16]. However, in modern times there are countless reasons for causing disinformation among the general population, the main motivations being political, social or financial [16, 23]. The term "fake news", used to refer to false information, even gained popularity following statements by former U.S. President Donald Trump regarding the large amount of false information that circulated during the 2016 presidential election [9].

Nowadays, there is a great number of disinformation campaigns, targeting issues such as climate change, vaccines, health, food, nutrition, curing diseases, generic drugs, nuclear energy, immigration impact, etc. [41]. This strong presence of false information has led to the prediction that, by 2022, false information will have spread to such an extent that the Western public will consume on average more fake news than real ones [13]. This is particularly harmful, as fake news tend to show disapproval or incite hatred towards the news shared, and seek to implant these negative ideas in the reader [40].

Furthermore, it is known that fake news are 70% more likely to achieve a greater overall reach than real ones [32]. One of the several causes may be a change in Facebook's algorithms that took place in 2016. These new algorithms began highlighting posts from friends and family over those from informational sites, causing a 25% drop in traffic to the latter ones [26]. Considering that in countries such as the United States, where 66% of adults get their information from social networks, and Spain, where 70% of the population has the Internet as their main source of information, this disregard for factual information is problematic [41].

In the midst of the current health crisis caused by COVID-19, also named the first global pandemic of social networks, fake news had a new moment to shine. Since the beginning, a large amount of fake news of all kinds has been spread about the current situation, often even endorsed and shared by politicians, celebrities and influencers [7]. As many of these news stories portrayed the pandemic as a conspiracy and underestimated the magnitude of the situation, this spreading of false information has had adverse effects on the effectiveness of virus containment strategies, for they affected people's perceptions and responses toward the virus [42].

Although the interest in detecting fake news is not new, there has been a growing interest lately in improving such detection through a series of research and novel proposals aimed at combating the misinformation that plagues social media [46]. Companies such as Google, Facebook and Twitter have been trying for years to take action in order to curb the problem, as their platforms have been the main means of diffusion. However, the measures taken did not contribute to solving the problem, as users in general kept encountering a large number of websites plagued by false information [3].

4 Deep Learning and Fake News

4.1 Fake News Detection

Given the complexity that the spread of false information can have in terms of range and variety, it is not an easy task to detect them. The first proposals to combat it involved human intervention to verify the veracity of the information, validating the quality of the publications by means of tools in social networks [10]. As another alternative for an immediate solution, the importance of reinforcing the so-called digital literacy, or in other words, that the public develops the critical capacity to differentiate on its own between real and false information, was highlighted [11]. However, as time went by, more sophisticated and effective proposals emerged.

Most of the proposals developed addressed the problem of false information detection as a classification problem. In other words, researchers sought to develop ways of labeling texts as true or false based on an analysis of their content [5]. For this purpose, several studies focused their research on detecting fake news through the use of artificial intelligence, specifically supervised learning methods.

This type of methods or algorithms need to be previously trained by means of a reliable database, which will be used as a reference to later learn how to analyze news, user profiles, electronic messages, social context, etc., and classify them appropriately [44]. In order to achieve a good automatic detection of fake news, an algorithm capable of identifying the meaning in the words or sentences used in the analyzed text at a deep level is required, so that they can be interpreted properly and classified correctly [39].

Machine learning models have been widely studied for use in automatic detection of fake news as the best available option [21]. However, over time one of its recent branches, Deep learning, started to gain more relevance in the field as it was further researched. This is due to the presence of more than one hidden layer between the input and output in deep learning techniques, which allows them to outperform conventional machine learning techniques in several fields [25].

It should be noted that the use of Deep learning is a double-edged sword, as it can also be used to assist in the dissemination of fake news [1]. One of the best known negative applications of Deep learning is the development of deepfakes, which are false photos or videos manipulated with artificial intelligence in a way that results indistinguishable for the human eye [24]. However, despite the negative applications it may have, the benefits of its use far outweigh them.

4.2 Deep Learning Applications

Researchers working with Deep learning have tried to put it to good use by studying these algorithms for use in fake news detection. This trend has been ongoing for several years, although without much success until recently. These initial experiments applied several popular deep learning algorithms such as CNN, LSTM, and RNN, but they also had a tendency to fail for they were carried out on a small scale and with unrealistic news. something that was improved later with more complete and improved databases [12].

Before implementing machine learning or deep learning algorithms, it is necessary that the text to be analyzed goes through a pre-processing stage in order to remove unnecessary information and elements, and leave only the relevant parts for the algorithm [39]. Although such pre-processing may vary depending on the algorithm to be used [12], their objective is to retrieve the main terms and characteristics of a text. A general outline can be seen in Fig. 1 [15].

Once the words have been separated, they need to be stored in multidimensional vectors in order to be used with deep learning techniques in a process that is known as Word embedding [28]. Through said process, words are mapped and stored in real number vectors of discrete or distributed representation for further use [19]. This pre-processing step can be done automatically by using algorithms such as Word2Vec and Google News [17, 29].

It is also necessary to set up an initial database, manually collecting information about publications from various users online [33] or by using specific databases assembled by third parties and available on the Internet [29]. For a database to be

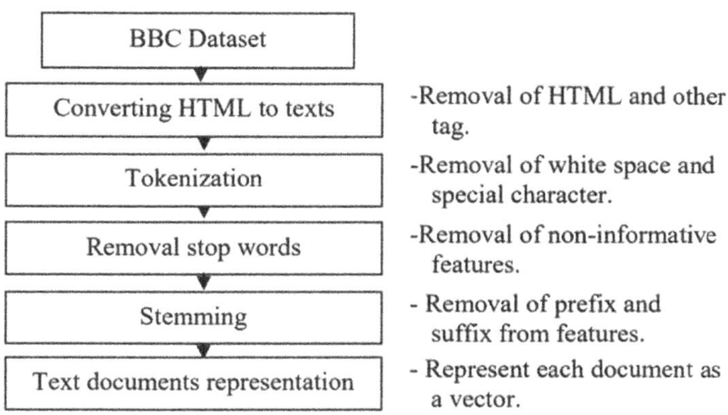

Fig. 1 Text pre-processing steps. *Source* Kadhim [15]

used as a reference, it must first go through a filtering process, analyzing existing information on users and publications and marking it as false or legitimate [33]. Deep learning techniques use these provided and classified references as a starting point in order to analyze further incoming news and classify them in the same way.

5 Deep Learning Algorithms

Within machine learning, the algorithms involved use the so-called neural networks in order to perform their tasks. These techniques are intended to resemble the human nervous system and the structure of the brain and perform tasks by emulating human reasoning [36]. Neural networks are also a vital part of the deep learning algorithms, but at a greater depth than in machine learning, as these improved techniques have a greater number of layers and parameters that allow them to analyze large amounts of information and extract important data from it [35].

Among the main deep learning algorithms used in the detection of fake news there are:

Convolutional Neural Network (CNN). These are a type of neural networks used for data processing, which get their name from the mathematical operation known as convolution [17]. This type of algorithms are based on the human visual cortex [36], which allowed them to gain popularity in the area of natural language processing [14]. These neural networks require fewer parameters than conventional ones, allowing them to solve tasks of higher complexity [2].

Recurrent Neural Network (RNN). This is another type of neural network focused on recognizing patterns in sequences of data, such as text, video, speech, language, etc., and using them to classify, group and make predictions about them [20]. These algorithms have layers that form a feedback loop, i.e., the output of one layer becomes the input of the next one. This allows the network to store information about previous states and use it to influence the current output [36].

Long-Short Term Memory (LSTM). Within the RNNs, there is an important type of neural networks focused on processing sequential values, known as long-short term memory [17]. Thus, these networks can perform tasks that require memory or state awareness by constructing context-based models [36]. This technique stands out among the other RNN techniques for improving aspects such as training with long sequences and retaining memory [20].

Geometric Deep learning. In recent times, there has been a growing interest in research on geometric deep learning, a branch that encompasses a set of deep learning techniques with a non-Euclidean approach [22]. Since non-Euclidean data tends to have a large volume, as is the case of information in social networks, conventional deep learning techniques are not enough when dealing with these cases. This led to the research of new techniques specialized in working with this type of information [6].

6 Current Research

Although machine learning techniques have already been investigated before for their application in fake news detection with decent results, the implementation of deep learning techniques in this field is still a recent topic that has generated its own research lately due to their exceptional performance in detection tasks.

We have the investigation of Kumar et al. [18], who experimented with identifying fake news from the PolitiFact database using 7 variations of both CNN and LSTM. Nasir et al. [23] continued to explore the combination of CNN and RNN for fake news detection, while Monti et al. [22] sought to implement geometric methods with deep learning. On the other hand, there is research such as that of Pathwar and Gill [28], who, in the context of the current COVID-19 pandemic, conducted experiments on fake news detection with various deep learning models and two data mining methods.

On the other hand, although it is part of the problem in terms of creating deepfakes, deep learning can also be part of the solution. For example, there is research such as the one conducted by Tariq et al. [37], who developed a method based on a convolutional LSTM for analyzing deepfake videos. To complement the above, there are analyses such as the one performed by Almars [4], which focused on making a compilation of the most recent proposals for automatic detection of deepfakes, both photos and videos. Table 1 explores in more detail the recent investigations regarding the use of Deep learning for fake news and deepfakes detection and the most important results obtained.

7 Future Research

Deep learning algorithms are considered by many to be the future in terms of automatic detection of fake news; however, there are still some gaps to be filled in the near future regarding these methods [30]. Some of the main problems encountered are the issues they present when having to process massive databases and the large amount of time required to receive the necessary training [14]. This is not to mention the limited amount of quality databases available for training purposes [45].

Current experimental studies on fake news detection through deep learning focus their future research primarily on testing their methods with larger databases to keep testing their degree of effectiveness and accuracy [28]. In addition, their aim is to increase the capacity of these algorithms as a result of constant research on all available algorithms and even the fusion of two or more in hybrid methods [23, 33]. This would allow them to detect the information in the texts at a higher level of detail and perform a more accurate analysis.

Another important problem arises when trying to generalize these fake news detection models to be used with more databases or social networks. When performing experiments with deep learning algorithms, specific databases are used for the training stage, usually about a particular social network. This could mean that only

Table 1 Recent research on deep learning in fake news and deepfakes detection

Field	Investigation	Results	Reference
Fake News	Experiment with a developed geometrical deep learning method	• The method handled well the integration of heterogeneous data such as user profile, user activity and social network structure • It is useful in fake news identification as it achieves high accuracy and robust behavior when working with real data on a large scale	[22]
Fake News	Detection experiment with 7 variations of CNN and LSTM	• Acceptable level of accuracy in deep learning methods, between 70 and 90% • Maximum accuracy of 88.78% achieved by the model with CNN and bidirectional LSTM with attention mechanism • Accuracy achieved with machine learning for the same task was below 60%	[18]
Fake News	Experiment with CNN and RNN combination method	• Takes advantage of the ability of CNNs to extract local features, as well as the ability of LSTMs to identify long-term dependencies • The method demonstrated better accuracy, precision and recall than other methods	[23]
Fake News	Experiment with deep learning models and two data mining methods	• Deep learning models perform better with TFIDF instead of Word embedding • CNN algorithm with Word Embedding achieved the best final accuracy of 93.92%	[28]
Fake News	Experiment with Deep learning linguistic model	• The model extracts grammatical, syntax, emotional and readability features of news • Average accuracy of 86% in detection and classification of fake news • Faster performance than other models based on machine learning	[8]

(continued)

Table 1 (continued)

Field	Investigation	Results	Reference
Deepfakes	Experiment with a method based on a convolutional LSTM	• The method performs better than previous methods when using a sequence of consecutive frames as input • Good generalization when working with different databases with deepfakes	[37]
Deepfakes	Experiment with CNN-based method working with mouth features	• The method isolates, analyzes and verifies lip and mouth movement in deepfake videos • The method worked with three databases and obtained better accuracy than other methods • Accuracy of 71.25% with the first database, 98.7% with the second, and 73.1% with the third	[38]
Deepfakes	Research on photo and video detection methods	• Application of Gaussian noise and blur, and hybrid methods for photo detection • Application of CNN and RNN in detection of physiological signals such as blinking or eye movement	[4]
Deepfakes	Experiment with CNN and Vision Transformer (ViT) combination method	• The CNN algorithm extracts local features, which are fed into the ViT for analysis and classification using an attention mechanism • The model achieved an accuracy of 91.5% with videos from the DFDC database	[43]
Deepfakes	CNN and LSTM combination method experiment for deepfake detection	• The method includes the CNN resnext50 algorithm and an LSTM layer • The model achieved an accuracy of 94.21% when working with videos from the CelebDF database	[34]

subsequent studies with a larger number and variety of databases are required in order to validate their functionality. However, the study conducted by Nasir et al. [23] found that although such models tend to perform exceptionally well when working with specific databases, they also tend to present problems when attempting to be generalized to other ones.

8 Discussion

Deep learning has proven to be very useful in the analysis and classification of fake news and Deepfakes. Even in several of the studies and experiments analyzed in Table 1, it was found that Deep learning techniques have a better performance in these tasks than conventional machine learning techniques. The analysis by Islam et al. [14] also acknowledges Deep learning as the main actual trend to detect misleading information in social networks in an efficient, effective way, with good performance and results that resemble human performance.

Apart from the advantages that Deep learning offers in detection tasks, these techniques still require further development in order to improve their performance. Randika [30] also acknowledges Deep learning techniques as the future of fake news detection, although he is also aware that there are still several limitations to be solved through further research. Almars [4] highlights the exceptional performance of Deep learning techniques in detecting Deepfakes. However, he does not ignore the fact that the quality of Deepfakes is constantly improving, and detection techniques shall improve with them.

9 Conclusion

This book chapter explores the evolution of fake news and its exponential growth in recent years as a result of technological progress. This new trend has generated an important need to curb this uncontrolled dissemination of false information on the Internet. Given its harmful effects on society, it is an issue that has gained relevance in recent times, particularly with the COVID-19 pandemic, and whose research must be prioritized in order to effectively curb it.

Considering the extensive spread of fake news in recent years, the search for more efficient methods of automatic detection of fake news has spiked, focusing on artificial intelligence and its derivatives, among which deep learning techniques are currently the most promising. Deep learning has a long way to go to be a reliable tool for fake news detection. However, the potential shown by these techniques make it possible to project that they will play an important role in the future of the war against fake news.

References

1. Adriani, R. (2019). The evolution of fake news and the abuse of emerging technologies. *European Journal of Social Science, 2*(1), 32–38. https://doi.org/10.26417/ejss-2019.v2i1-53
2. Albawi, S., Mohammed, T. A., Al-Zawi, S. (2017). Understanding of a convolutional neural network, in 2017. In *International Conference on Engineering and Technology (ICET).*

Presented at the 2017 International Conference on Engineering and Technology (ICET) (pp. 1–6). https://doi.org/10.1109/ICEngTechnol.2017.8308186.
3. Aldwairi, M., & Alwahedi, A. (2018). Detecting fake news in social media networks. *Procedia Computer Science, 141*, 215–222. https://doi.org/10.1016/j.procs.2018.10.171
4. Almars, A. M. (2021). Deepfakes detection techniques using deep learning: A survey. *Journal Computer Communication, 9*, 20–35. https://doi.org/10.4236/jcc.2021.95003
5. Bondielli, A., & Marcelloni, F. (2019). A survey on fake news and rumour detection techniques. *Information Sciences, 497*, 38–55. https://doi.org/10.1016/j.ins.2019.05.035
6. Cao, W., Yan, Z., He, Z., & He, Z. (2020). A Comprehensive survey on geometric deep learning. *IEEE Access., 8*, 35929–35949. https://doi.org/10.1109/ACCESS.2020.2975067
7. Catalán-Matamoros, D. (2020). La comunicación sobre la pandemia del COVID-19 en la era digital: manipulación informativa, fake news y redes sociales. Review Espanola Comunicaciones EN SALUD 5–8. https://doi.org/10.20318/recs.2020.5531.
8. Choudhary, A., & Arora, A. (2021). Linguistic feature based learning model for fake news detection and classification. *Expert Systems with Applications, 169*, 114171. https://doi.org/10.1016/j.eswa.2020.114171
9. Farhall, K., Carson, A., Wright, S., Gibbons, A., & Lukamto, W. (2019). Political elites' use of fake news discourse across communications platforms. *International Journal of Communication, 13*, 4353–4375.
10. Figueira, Á., & Oliveira, L. (2017). The current state of fake news: Challenges and opportunities. *Procedia Computer Science, 121*, 817–825. https://doi.org/10.1016/j.procs.2017.11.106
11. Gallardo-Camacho, J., & Marta, C. M. (2020). La verificación de hechos (fact checking) y el pensamiento crítico para luchar contra las noticias falsas: Alfabetización digital como reto comunicativo y educativo. *Review Estilos Aprendiz, 13*(26), 4–6.
12. Girgis, S., Amer, E., Gadallah, M. (2018) Deep learning algorithms for detecting fake news in online text. In *2018 13th International Conference on Computer Engineering and Systems (ICCES). Presented at the 2018 13th International Conference on Computer Engineering and Systems (ICCES)* (pp. 93–97). https://doi.org/10.1109/ICCES.2018.8639198.
13. González, M. A. (2019). Fake news: Desinformación en la era de la sociedad de la información. Fake News: disinformation in the information society. *Ámbitos Revista Internacional de Comunicación, 45*, 29–52. https://doi.org/10.12795/Ambitos.2019.i45.03.
14. Islam, M. R., Liu, S., Wang, X., & Xu, G. (2020). Deep learning for misinformation detection on online social networks: A survey and new perspectives. *Social Network Analysis and Mining, 10*, 82. https://doi.org/10.1007/s13278-020-00696-x
15. Kadhim, A. (2018). An evaluation of preprocessing techniques for text classification. *International Journal Computer Science Information Security, 16*, 22–32.
16. Kalsnes, B. (2018). Fake News. *Oxford Research Encyclopedia Communication*. https://doi.org/10.1093/acrefore/9780190228613.013.809
17. Krešňáková, V. M., Sarnovský, M., Butka, P. (2019) Deep learning methods for fake news detection. In *2019 IEEE 19th International Symposium on Computational Intelligence and Informatics and 7th IEEE International Conference on Recent Achievements in Mechatronics, Automation, Computer Sciences and Robotics (CINTI-MACRo)* (pp. 000143–000148). https://doi.org/10.1109/CINTI-MACRo49179.2019.9105317.
18. Kumar, S., Asthana, R., Upadhyay, S., Upreti, N., & Akbar, M. (2020). Fake news detection using deep learning models: A novel approach. *Transactions Emerging Telecommunication Technology, 31*, e3767. https://doi.org/10.1002/ett.3767
19. Lee, D. H., Kim, Y. R., Kim, H. J., Park, S. M., & Yang, Y. J. (2019). Fake news detection using deep learning. *Journal Information Processing System, 15*, 1119–1130. https://doi.org/10.3745/JIPS.04.0142
20. Manaswi, N. K. (2018). RNN and LSTM. In N. K. Manaswi (Ed.), *Deep Learning with Applications Using Python: Chatbots and Face, Object, and Speech Recognition With TensorFlow and Keras* (pp. 115–126). Apress, Berkeley, CA. https://doi.org/10.1007/978-1-4842-3516-4_9.
21. Manzoor, S. I., Singla, J., Nikita (2019). Fake news detection using machine learning approaches: A systematic review. In *2019 3rd International Conference on Trends in Electronics and Informatics (ICOEI). Presented at the 2019 3rd International Conference on Trends*

in Electronics and Informatics (ICOEI) (pp. 230–234). https://doi.org/10.1109/ICOEI.2019.8862770.

22. Monti, F., Frasca, F., Eynard, D., Mannion, D., Bronstein, M. M. (2019). Fake news detection on social media using geometric deep learning. ArXiv190206673 Cs Stat.
23. Nasir, J. A., Khan, O. S., & Varlamis, I. (2021). Fake news detection: A hybrid CNN-RNN based deep learning approach. *International Journal Information Management Data Insights, 1*, 100007. https://doi.org/10.1016/j.jjimei.2020.100007
24. Nguyen, T. T., Nguyen, Q. V. H., Nguyen, C. M., Nguyen, D., Nguyen, D. T., Nahavandi, S. (2021) Deep learning for deepfakes creation and detection: A survey. ArXiv190911573 Cs Eess.
25. Ongsulee, P. (2017). Artificial intelligence, machine learning and deep learning. In *2017 15th International Conference on ICT and Knowledge Engineering (ICT KE). Presented at the 2017 15th International Conference on ICT and Knowledge Engineering (ICT KE)* (pp. 1–6). https://doi.org/10.1109/ICTKE.2017.8259629.
26. Pangrazio, L. (2018). What's new about 'fake news'? Critical digital literacies in an era of fake news, post-truth and clickbait. *Páginas Education, 11*, 6. https://doi.org/10.22235/pe.v11i1.1551
27. Parikh, S. B., Patil, V., Atrey, P. K. (2019) On the origin, proliferation and tone of fake news. In *2019 IEEE Conference on Multimedia Information Processing and Retrieval (MIPR). Presented at the 2019 IEEE Conference on Multimedia Information Processing and Retrieval (MIPR)* (pp. 135–140). https://doi.org/10.1109/MIPR.2019.00031.
28. Pathwar, P., Gill, S. (2021). Tackling COVID-19 infodemic using deep learning. ArXiv210702012 Cs.
29. Qawasmeh, E., Tawalbeh, M., Abdullah, M. (2019). Automatic identification of fake news using deep learning. In *2019 Sixth International Conference on Social Networks Analysis, Management and Security (SNAMS). Presented at the 2019 Sixth International Conference on Social Networks Analysis, Management and Security (SNAMS)* (pp. 383–388). https://doi.org/10.1109/SNAMS.2019.8931873.
30. Randika, B. (2020). *The misinformation era: Review on deep learning approach to fake news detection*. https://doi.org/10.6084/m9.figshare.13299440.v1
31. Rodríguez, C. (2019). No diga fake news, di desinformación: Una revisión sobre el fenómeno de las noticias falsas y sus implicaciones. *Comunicación., 40*, 65–74.
32. Rodríguez-Fernández, L. (2019). Desinformación y comunicación organizacional: Estudio sobre el impacto de las fake news. *Revista Latina de Comunicación Social, 74*(13), 1714–1728.
33. Sahoo, S. R., & Gupta, B. B. (2021). Multiple features based approach for automatic fake news detection on social networks using deep learning. *Applied Soft Computing, 100*, 106983. https://doi.org/10.1016/j.asoc.2020.106983
34. Shende, A., Paliwal, S., & Kumar, T. (2021). Using deep learning to detect deepfake videos. *Turkish J Computer Mathematic Education TURCOMAT, 12*, 5012–5017.
35. Shinde, P. P., Shah, S. (2018). A review of machine learning and deep learning applications. In *2018 Fourth International Conference on Computing Communication Control and Automation (ICCUBEA). Presented at the 2018 Fourth International Conference on Computing Communication Control and Automation (ICCUBEA)* (pp. 1–6). https://doi.org/10.1109/ICCUBEA.2018.8697857.
36. Shrestha, A., & Mahmood, A. (2019). Review of deep learning algorithms and architectures. *IEEE Access, 7*, 53040–53065. https://doi.org/10.1109/ACCESS.2019.2912200
37. Tariq, S., Lee, S., Woo, S. S. (2020). A convolutional LSTM based Residual Network for Deepfake Video Detection. ArXiv200907480 Cs.
38. Tayseer, M., Ababneh, M., Al-Zoube, M., Elhassan, A. (2020). Forensics and analysis of deepfake videos. In *2020 11th International Conference on Information and Communication Systems (ICICS). Presented at the 2020 11th International Conference on Information and Communication Systems (ICICS)* (pp. 053–058). https://doi.org/10.1109/ICICS49469.2020.239493.

39. Thota, A., Tilak, P., Ahluwalia, S., Lohia, N. (2018). Fake news detection: A deep learning approach. *SMU Data Science Review 1*(3), 10. https://scholar.smu.edu/datasciencereview/vol1/iss3/10.
40. Valarezo-Cambizaca, L.-M., Rodríguez-Hidalgo, C. (2019). La innovación en el periodismo como antídoto ante las fake news. *RISTI—Revista Ibérica de Sistemas E Tecnologias de Informação 20*, 24–35.
41. Valero, P. P., Oliveira, L. (2018) Fake news: una revisión sistemática de la literatura. *Observation, 12*. https://doi.org/10.15847/obsOBS12520181374.
42. van der Linden, S., Roozenbeek, J., Compton, J. (2020). Inoculating against fake news about COVID-19. Frontier in Psychology, *11*. https://doi.org/10.3389/fpsyg.2020.566790.
43. Wodajo, D., Atnafu, S. (2021). Deepfake video detection using convolutional vision transformer. ArXiv210211126 Cs.
44. Yang, S., Shu, K., Wang, S., Gu, R., Wu, F., & Liu, H. (2019). Unsupervised fake news detection on social media: A generative approach. *Processing AAAI Conference Artificial Intelligent, 33*, 5644–5651. https://doi.org/10.1609/aaai.v33i01.33015644
45. Zhang, W. J., Yang, G., Lin, Y., Ji, C., Gupta, M. M. (2018). On definition of deep learning. In *2018 World Automation Congress (WAC). Presented at the 2018 World Automation Congress (WAC)* (pp. 1–5). https://doi.org/10.23919/WAC.2018.8430387.
46. Zhou, X., Zafarani, R. (2020). A survey of fake news: Fundamental theories detection methods and opportunities. *ACM Computer Survey 53*, 1–40. https://doi.org/10.1145/3395046.

Machine Learning Techniques and Fake News

Early Detection of Fake News from Social Media Networks Using Computational Intelligence Approaches

Roseline Oluwaseun Ogundokun, Micheal Olaolu Arowolo, Sanjay Misra, and Idowu Dauda Oladipo

Abstract In recent years, misinformation which includes false news (FN) has become a worldwide issue owing to its exponential development, mostly on social media (SM). The broad dissemination of misinformation and FN can create harmful societal repercussions. Despite current improvements, spotting fake news remains difficult owing to its intricacy, multiplicity, multi-modality, and fact-scrutiny or annotation expenses. Therefore, there is a necessity for Computational Intelligence Approaches (CIA) that can identify this fake news automatically. This study projected using dimensionality reduction (DR) approach to decrease the dimensionality of the feature vectors before sending them to the classifiers. The study focuses on a computational intelligence-based fake news detection system and a novel approach of employing three CIA for the detection of FN was proposed. The CIA employed for this study are Genetic Algorithm (GA), K-Nearest Neighbor (KNN), and Bagged Ensembled Learning (BEL). The proposed system performance was evaluated utilizing confusion matrix measures like accuracy, sensitivity, specificity, precision, and f-measure. The system was compared with the existing system and it was deduced that the projected system outperformed that of existing systems with an accuracy of 99.28%, sensitivity of 99.28%, precision of 99.99%, and f-measure of 99.63%. In conclusion, it was discovered that GA + KNN performance in terms of accuracy, sensitivity, specificity, precision, and f-measure sur-passed that of the GA + BEL.

R. O. Ogundokun (✉) · M. O. Arowolo
Landmark University Omu-Aran, Omu-Aran, Nigeria
e-mail: Ogundokun.roseline@lmu.edu.ng

M. O. Arowolo
e-mail: arowolo.olaolu@lmu.edu.ng

S. Misra
Covenant University Ota, Ota, Nigeria
e-mail: Sanjay.misra@covenantuniversity.edu.ng

I. D. Oladipo
University of Ilorin, Ilorin, Nigeria
e-mail: odidowu@unilorin.edu.ng

M. Lahby et al. (eds.), *Combating Fake News with Computational Intelligence Techniques*, Studies in Computational Intelligence 1001,
https://doi.org/10.1007/978-3-030-90087-8_4

Keywords Computational intelligence · Genetic algorithm · Fake news · Social media · K-Nearest Neighbor · Decision tree · Ensemble learning

Abbreviations

FN	Fake news
SM	Social media
CIA	Computational intelligent approaches
GA	Genetic algorithm
KNN	K-Nearest Neighbor
BEL	Bagged ensemble learning
TP	True positives
TN	True negatives
FP	False positives
FN	False negatives

1 Introduction

Fake news (FN) is a sort of propaganda in which untrue information is disseminated to influence the public. Traditional print/visual broadcasting such as social media (SM) platforms can be employed to transmit the word or information. FN is not a recent occurrence, yet has grown more bothersome in recent years as a result of its viral nature on SM. FN has been identified as one of the most serious dangers to democracy by political analysts. FN campaigns have been employed effectively across the globe to sway elections, shape public opinion, change individual's perceptions, and indoctrinate the masses [1].

Every day, a massive quantity of data is created online in the age of technology. Fake news, on the other hand, accounts for an unprecedented quantity of material flooding on the Internet, that are created to draw the attention of the audience, affect people's views and decisions [2–4], boost income earned by clicking [5], and influence significant occasions, for instance, political elections [6]. By purposefully disseminating untrue information, readers are misled. Acquiring and disseminating information via SM platforms has to turn out to be increasingly simple, making it problematic and time-consuming to identify FNs built only on broadcast content. According to some claims, Russia has developed phony accounts and public bots to propagate false information. Conferring to a survey, 64% of Americans believe that FN has generated a "huge lot of uncertainty" regarding the truthfulness of recorded occurrences [7]. Furthermore, large-scale untrue information cascades are having increasingly negative repercussions in the fields of business, advertising, and stock-market trading. For example, in 2013, the stock market lost 130 billion dollars when false rumors were disseminated on Twitter that Barack Obama had been harmed in an eruption [6]. In the 2016 presidential election in the United States, false news was accused of being a major contributor to rising governmental polarization and party strife, including influencing the result [7–10]. It has been discovered that identifying FN is an undeniable issue for the news organizations, journalists, and technologies for detecting FN have been an absolute need. Because fact-checking by hand is a time-consuming and laborious process, the automatic recognition of FN has piqued the interest of the Natural Language Processing (NLP) community as a way to assist and ease the cumbersome and laborious humanoid activity of fact-checking [11, 12]. Even for automated methods, determining the veracity of news remains a difficult process [13]. Comprehending what other news establishments are publishing on a similar issue may be an advantageous initial phase in identifying FN items.

Stance detection has long been a crucial basis for a variety of activities, including evaluating online discussions [14–16], detecting the veracity of bruits on Twitter [17, 18], and comprehending the logical edifice of effective writings [19]. Stance detection is the name for this phase. Pomerleau and Rao [20] established the inaugural Fake News Challenge (FNC-1) [21] to assess what a broadcasting authority is communicating concerning a specific difficulty to stimulate the advancement of computerized fake news recognition systems (FNRS) utilizing artificial intelligence (AI) expertise and machine learning (ML). This competition drew over 50 teams from both business and academics. The FNC-1 challenge is designed to identify a news article's position concerning a particular title. An article can take four different viewpoints. It can approve or disapprove with the title, address the equivalent issue, or be unconnected to the headline. On their official website, you may learn about the FNC-1 mission, its guidelines, the dataset, and the assessment measures [21].

Fake news is now widely considered to be one of the most serious dangers to the republic and media [22]. The spread of FN was superlatively demonstrated throughout the calendar month of the 2016 U.S. constitutional voting drive when the topmost 20 most-deliberated false voting stories produced 8,711,000 allotments, feed-backs, and statement of opinions on Facebook, outnumbering the topmost 20 most-deliberated election stories forwarded by 19 leading news sites. Fake news has been linked to stock market swings and huge trades in our economy as well [23]. In the interim, when confronted with misleading information, people have shown to be incapable of distinguishing between true and untrue information [24]. Several expert-built (e.g., PolitiFact2 and Snopes3) and crowd-sourced (e.g., Fiskkit4 and Text Thresher [25]) labor-intensive fact-scrutiny websites, devices, and platforms have therefore been developed to assist the community. Manual fact-checking, on the other hand, does not scale well with the volume of freshly produced data, particularly on social media [26]. As a result, in recent years, automatic false news identification has been created, with the existing approaches being divided into composition-built and propagation-built approaches. False news identification founded on content tries to identify FN by examining the content of news stories. Researchers frequently detect false news using either dormant (through NN) or non-latent (typically hand-crafted) characteristics of the content [27–30] within a machine learning framework. Fundamental ideas in public and criminological mindsets, however, have not contributed a substantial function in any of these methods. By identifying certain possible false news trends and easing understandable ML techniques for FN identification, such philosophies could greatly enhance FN recognition [15]. According to the Undeutsch hypothesis [31], un-true speech diverges from a sincere one in respect of writing style and quality, such theories can be denoted to either be misinformation [31–34], i.e., information that is deliberately and confirmable untrue, or click-baits [35], captions whose primary motives is to fascinate readers' consideration and inspire them to get on a link to a specific webpage [36]. Using such ideas, as opposed to current features, enables the introduction of explainable features, which may assist the audience to better comprehend false news and investigate the links between FN, misinformation, and click-baits. Misinformation, in theory, is a broader term that encompasses FN stories, false claims, and phony reviews, among other things. As a result, the features associated with deceit/disinformation may or may not be congruent with those associated with fake news, prompting scholars to investigate the connections between FN and other forms of trick. In the interim, clickbait has been linked to the spread of fake news [36, 37]. Click-baits aid false news in attracting further clicks (i.e., discernibility) and gaining community belief, as seen by the concentration prejudice [38], which asserts that the people's faith in a news story increases with increased exposure, which is aided by click-baits. While news items containing click-baits are typically untrustworthy, not all of them are false news, prompting researchers to investigate the links between FN and clickbait. Unlike content-built FN identification, propagation-built FNR looks at how news spreads on SM to identify FN. Novel models demonstrating satisfactory performance have recently been suggested using propagation-based techniques [39–45]. When it comes to detecting false news, however, propagation-based approaches confront a significant difficulty. There are three fundamental steps in the life cycle of each news article: creation, publication on news vent(s), and dissemination on SM (platform) [36]. Predicting false news before it reaches the third phase, i.e., before it has been spread on SM, is challenging using a propagation-built technique that relies on societal context information. To discover fake news early, that is, when it is available on a news vent but has not so far been propagated on social networking sites, to then take initial action for FN interference (i.e., FN early recognition), news material must be extensively mined. Early identification is especially important in the situation of FN because the more people are exposed to it, the more likely they are to believe it [46]. In the interim, it has been shown that it is problematic to alter one's perception once one has been tricked by bogus news (that is., Semmelweis reflex [47, 48], confirmation bias [49], and anchoring bias [50]). In conclusion, the present state of FN identification necessitates the development of algorithms that mine news content in-depth rather than relying on how fake news spreads. For interpretability considerations, such approaches should look into how social and forensic theories may aid in the detection of false news. The authors decided to solve these issues by building a theory-motivated FNR method that focuses on broadcast content, allowing the authors to recognize FN before it spreads on SM. By performing an interdisciplinary investigation, the method signifies news items by a collection of labor-intensive characteristics that apprehends both content structure and style crosswise etymological echelons. The features are then used in a supervised ML technique to identify FN.

This study's motivation is on the enhancement of the existing approaches to discover FN on SM by using a genetic algorithm for dimensionality reduction technique. Hence, this study proposed a novel approach of employing three CIA for the recognition of FN. The CIA employed for this study are Genetic Algorithm (GA), K-Nearest Neighbor (KNN), and Bagged Ensembled Learning (BEL). The proposed system performance was evaluated utilizing confusion matrix measures such as accuracy, sensitivity, specificity, precision, and f-measure.

The remaining part of the article is pre-structured as Sect. 2 presented the related works on fake news identification. Section 3 presented the material and method employed for the implementation of the study. The result and discussion are similarly presented in Sect. 4 and the article was concluded in Sect. 5.

2 Related Works

To detect FN, a variety of approaches have been suggested, including data mining (DM), machine learning, computational intelligence, deep learning, and social network analysis. Aldwairi and Al-wahedi [51] created a method on FNR on SM to identify and re-move FN from a collection of search engine results or social network news feeds. It is a user-downloadable application that may be installed as a supplementary to the user's preferred search engine. When a handler uses a search engine with this tool, which is installed to look for information, the tool will go through all of the search engine's links. The program will categorize the search result as false or real while the words are utilized in the link, the sum of words employed, punctuation marks employed, and bounce rates. The user is given the sites or results that are classified as legitimate, while the results that are categorized as false are banned.

Liao et al. [52] presented a customer comment chat (CCT) innovative erudition system to identify FN. The CCT for FNR was incorporated in their study. This study investigated customer comment information as a key characteristic to develop a method for detecting false news. The characteristic of user comments was extracted, and the context was represented as a network.

A Multi-source system was postulated by Karimi et al. [53]. They employed various aspects of false news for FNR in their multi-class FNR. Their study also provided a concept for categorizing news in addition to the traditional binary classifications of false or authentic news. This article similarly looked at the news, which is a combination of true and false. Bogus news tracker is an instrument for collecting, detecting, and visualizing FN.

Shu et al. [54] proposed the false news tracker, which is a method for detecting bogus news. The technology automatically captures news content and social context, resulting in a massive dataset for false news studies.

Shu et al. [55] investigated FN identification from multiple viewpoints. The title of their article was FNR on SM: A DM viewpoint. Their study pointed researchers in the right path for further research on false news detection.

Vogel and Meghana [56] devised a way for distinguishing people who have posted FN previously and those who haven't publicized any FN before on Twitter and they employed a character N-Grams. Their article investigated several feature extractions approaches as well as learning exercises. Character N-grams are utilized to extract features, and an SVM classifier is employed to categorize the news.

Umer et al. [57] created a hybrid network that incorporated CNN and LSTM, as well as PCA and Chi-square, to detect FN stances by utilizing deep learning (DL) architecture. To obtain the word vector, the feature collection was put over pre-processing.

Xu et al. [58] looked at the domain reputation of Facebook users in identifying FN on electronic SM through Domain Reputations and Content Consideration. The registration character, timing, and characteristics of users that post false news are discovered to be distinct from those of actual users.

Shrivastava et al. [59] utilized differential equations to create a defensive model of bogus news spread through online social networks. It's also calculated for stability and balance.

Another study focused on utilizing an agreement-aware article search to anticipate rumor news. They created a contract-alert quest framework to offer consumers a complete picture of a topic for which the pulverized fact was uncertain. They created a dual-phase method that included a tree-built method with handmade features and an RNN with attention method that focused on solitary a rare important word [60]. TF-IDF was employed to excerpt features to epitomize both headings and body of news items in [61], which is a single, endways ranking-built procedure using MLP. The model achieves an accuracy of 86.66% on FNC-1.

In Borges et al. [62], a deep learning approach was utilized to solve the FNC-1 task's stance identification difficulty. The authors used bi-directional RNNs, as well as max-pooling and neural consideration processes, to create depictions from captions and the form of broadcast items, which are then combined with exterior resemblance characteristics. The utilization of pre-training and a mix of brain representations and external similarity characteristics resulted in an accuracy of 83.8%.

Another study by Bhatt et al. [63] used a deep recurrent model to calculate neural embed-dings, a weighted n-gram bag-of-words method to calculate arithmetical attributes and characteristics engineering methods to excerpt hand-crafted outside features. Lastly, using a deep neural network, all of the characteristics are merged to classify the caption-body news combination as agree, disagree, deliberate, or irrelevant. The accuracy attained was 89.29%.

Zeng et al. [64] proposed a system that employed a neural network (NN) for their study and it was deduced that the NN technique employed for handcrafted characteristics outperformed other ML techniques. The model achieves 86.5% accuracy by integrating bilateral multi-perspective matching models (BiMPM) and enhancing the current Observant Clients with a complete consideration mechanism among captions and body-text terms.

In Pfohl et al. [65], a Restrictive Encrypting LSTM method with consideration achieves a score of 80.8%. A conditioned bidirectional LSTM with universal characteristics was utilized in another paper [66]. The paper shows that using a combination of global and resident word entrenching characteristics to forecast the stance of caption-item pairings is more accurate than using them both alone, with an accuracy of 87.4%. Rather than utilizing a classification-based technique, this study used a ranking-based method to solve the news stance identification problem. The ranking-built technique likens and exploits the difference amid a pair of headlines and article body's true and false stances. This method yielded an accuracy of 86.66% [61].

In [67], a unique stacked Bi-LSTM layers-based method with a method comprising of weighted Bi-LSTM layers was developed, and in [68], an innovative weighted CNN was introduced. The LSTM layer was employed to model sequences. Bi-LSTM incorporates information on both borderlines of the phrase, resulting in considerably improved precision. Many models such as CNN and LSTM were implemented and evaluated on FNC-1 in [69]. The authors similarly suggested a new enlargement of the generic design founded on a matrix of resemblance. Their research reveals that the projected sMemNN with the TF model has a maximum accuracy of 88.57%. CNN + LSTM and LSTM + CNN, on the other hand, exhibited limited outcomes, with 48.54 and 65.36% accuracy, correspondingly. The motive behind this was because data for training were collected at 80% while data for testing was obtained at 20%. Equal examples of all the classes were arbitrarily picked for every area to stabilize the data all through the training phase. Furthermore, the CNN architecture does not refer to a pooling layer, which may have contributed to the small accuracy. A huge-scale verbal method for stance identification that employed transference learning on a Roberta deep bidirectional modifier etymological method was suggested by the researcher [70]. On the FNC-I bench-mark dataset, the method had a 93.71% accuracy.

From the previous literature reviewed, it was deduced that the ac-curacy, detection rate, f1-score, precision, and AUC/ROC of the previous works were low and their FPR were high. To solve and overcome these aforementioned constraints or problems, this study, therefore, proposed to employ the use of a genetic algorithm for dimensionality reduction technique and two ML classification techniques like EL, and KNN.

3 Research Methods

The following is the materials and method that was used to implement the proposed model.

3.1 Data Collection and Variable Definition

This research intends to solve the issue of FN. The "Fake" and "True" datasets from Kaggle were used in this study: https://www.kaggle.com/clmentbisaillon/fake-and-real-news-dataset. This data comprises 17,903 fake news unique values.

3.2 Model Design

Feature selection (FS) is a step that must be completed before DM operations, for instance, classification and clustering can be carried out. The main purpose of FS is to make data ready (preprocess) by minimizing regularity in a dataset's attribute vector. FS reduces the complexity of a dataset by removing noisy symmetrical attributes while keeping the utmost enlightening ones. Features having a strong association with further proportioned features and features with a feeble association with the target class (label of the occurrence) are among the noisy features (irrelevant).

In this investigation, an iteration of the GA was used to produce variation in the existing population by using a series of evolutionary operators (selection, crossover, mutation) to make ready an innovative population in a fashion that mimics normal development. To verify its quality and determine whether it is fit or unsuitable, all chromosome is examined conferring to particular assessment measures. In each iteration, the highest-rated solution (best individual) is kept. In-competent resolutions (worst entities) will be substituted by freshly created progeny. This permits the average suitability value to sky-rocket during the recapitulations.

To test the performance of false news detection classifiers, we combined our suggested Genetic algorithm approach with the following classification learning algorithms: "K-Nearest Neighbor and Ensemble learning". The data splitting method employed for the study was a ratio of 70:30 and the study used 30% of holdouts during the training and testing of the data.

Data gathering, data preprocessing and FS, model creation, estimation, and assessment were all aspects of the projected approach that were used in this study. Figure 1 depicts all of the methodology's proposed processes. The news datasets which are fake and real were collected from the Kaggle database after which they are passed into MATLAB for data preprocessing. Here the data cleansing and FS which is GA was performed on them for selected subset features. These subset features selected from the preprocessing phase are passed into the two classifiers which are KNN and BEL. The datasets are classified into fake or real news. The system performance was evaluated using confusion matrix metrics for instance accuracy, precision, sensitivity, specificity, and F-measure.

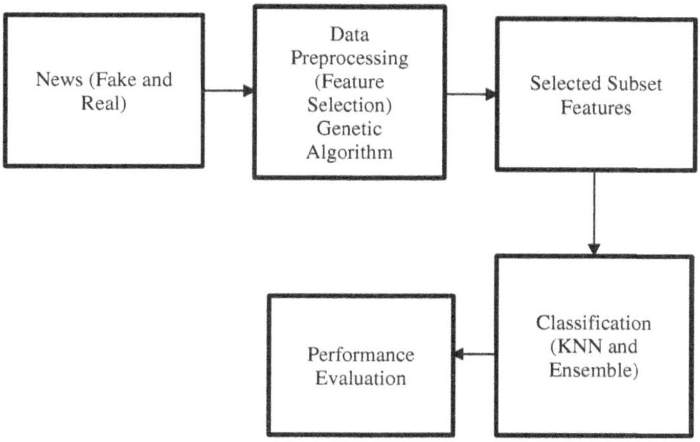

Fig. 1 Methodology processes

3.3 Genetic Algorithm

A Genetic Algorithm (GA) is a metaheuristic method for finding useful solutions to complicated problems that are inspired by genetic rules. The five basic elements that make up genetic algorithms are as follows. The feasible solutions to the optimization issue are represented as chromosomes. The population of plausible solutions at the start; each solution is evaluated using a fitness function. Those who employ hereditary operatives to create an innovative populace from current ones; and those who do not employ hereditary operatives to create an innovative populace from current ones. Control parameters include population size, genetic operator probability, generation number, and so forth [71].

Algorithm 1: Genetic Algorithm [73]

Input:
   Populace Size, n
   Extreme sum of recapitulations, MAX
Output:
   Global superlative result, Y_{bt}
begin
   Generate preliminary populace of n chromosomes Y_i $(i = 1, 2, \ldots, n)$
   Set recapitulation counter $t = 0$
   Calculate the suitability value of all chromosomes
   while $(t < MAX)$
      Select a pair of chromosomes from the preliminary populace built on suitability
      Apply crossover process on carefully chosen pair with crossover likelihood
      Apply mutation on the offspring with a mutation likelihood
      Substitute the longstanding populace with the recently created populace
      Increment the present recapitulation t by 1
   end while
   return the superlative result, Y_{bt}
end

3.4 K-Nearest Neighbor (KNN)

KNN is an unsupervised ML technique that does not necessitate the utilization of a dependent variable to forecast the outcome of a certain group of data. The technique is given sufficient training data and allows it to select which neighborhood a data region fits into. The KNN technique computes the distance among an innovative data region and its contiguous neighbors, and the value of K computes the bulk of its neighbors' polls; if K is 1, the innovative data region is allotted to the class with the shortest distance. KNN identifies new locations based on the majority of noises from the sur-rounding k. According to the function of distance, the position given in the class is extremely mutually exclusive between the closest neighbors K. The following Eqs. (1)–(3) are the mathematical formulas for calculating the distance between two places [72].

$$Euclidean\,distance = \sqrt{\sum_{i=1}^{k}(x_i - y_i)^2} \tag{1}$$

$$Manhattan\,distance = \sum_{i=1}^{k}|x_i - y_i| \tag{2}$$

$$Minkowski\,distance = \left(\sum_{i=1}^{k}|x_i - y_i|^q\right)^{\frac{1}{q}} \tag{3}$$

Algorithm 2: KNN [73]

```
Categorize (X, Y, x);
X: training data, Y: class labels of X, x: unidentified
model
For i = 1 to m
do
Calculate distance d (Xi, x)
end for
Calculate set i comprising indices for k minimum distances
d (Xi, x)
return
bulk label for {Yi where i∈I}
end
```

3.5 Ensemble Learning

Bootstrap aggregating, or bagging classifier (BC), is an initial ensemble technique largely intended to decrease the alteration (overfitting) over a training set. The RF technique is one of the utmost extensively employed as a version of the BC. To reduce overall variance, the bagging model selects the class for a classification problem based on main votes measured by the number of trees, but the data for all the trees are designated employing arbitrary selection with substitutes from the entire dataset. The bagging model, on the other hand, averages numerous estimates for regression issues. For categorization, this experiment employs a bagging method [73].

3.6 Evaluation Measures

The accuracy, sensitivity, specificity, precision, and f-measure were utilized as evaluation metrics in this study. The equations of these evaluation metrics are shown in Eqs. (4)–(8) [72, 74]. The true positives (TPs) signified news that was forecasted as fake and was FN, the true negatives (TNs) signified news that was fore-casted as not fake and was non-fake, the false positives (FPs) signified news that was forecasted as fake but was not fake, and the false negatives (FNs) signified news that was forecast as fake but was not fake [72, 75].

$$\text{Accuracy}: \frac{TP + TN}{TP + TN + FP + FN} \tag{4}$$

$$\text{Sensitivity}: \frac{TP}{TP + FN} \tag{5}$$

$$\text{Specificity}: \frac{TN}{FP + TN} \tag{6}$$

$$\text{Precision}: \frac{TP}{TP + FP} \tag{7}$$

F-measure:

$$\frac{2TP}{2TP + FP + FN} \tag{8}$$

4 Results and Analysis

In this investigation, the fake news dataset was input into the Genetic algorithm, the subset result achieved was then passed into KNN and Ensemble classifiers. All the experiments performed are summarized and display the output of the confusion matrix (CM), and Table 1 demonstrates the classifiers' CM. Figure 2 demonstrates the CM computation for the GA + KNN classifiers while Fig. 3 demonstrates the

Table 1 Values of the classifiers confusion matrix

Confusion matrix	GA + KNN	GA + BEL
TP	32,979	16,492
TN	6422	1
FP	4	0
FN	239	3329

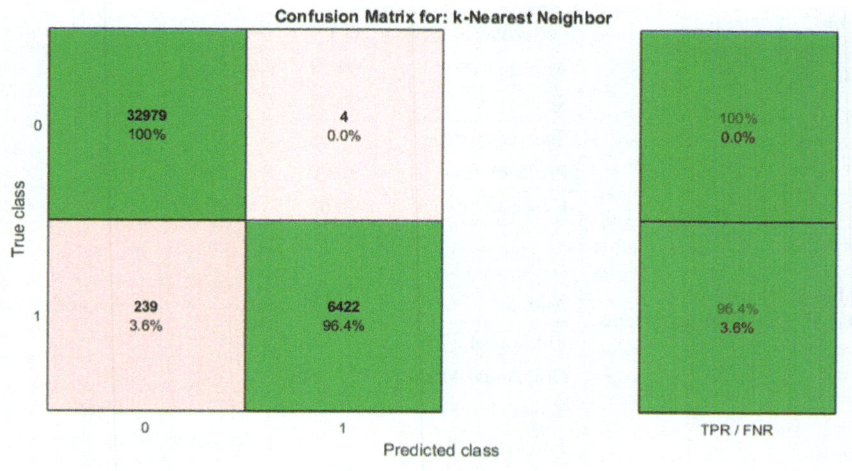

Fig. 2 Confusion matrix showing the dataset with KNN TP = 32,979, TN = 6422, FP = 4, FN = 239

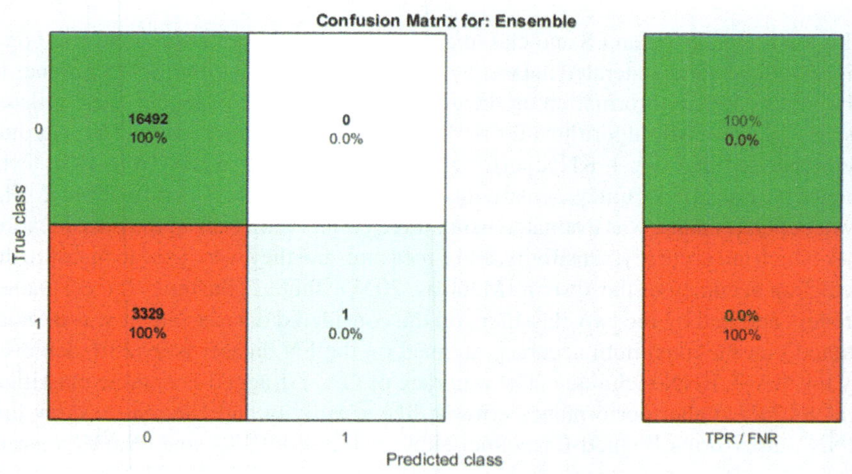

Fig. 3 Confusion matrix showing the dataset with bagged ensemble TP = 16,492, TN = 1, FP = 0, FN = 3329

CM computation for the GA + BEL classifier. Table 3 demonstrate the comparative analysis of the proposed system with existing systems.

Table 2 Predictive performance evaluation summary

Performance metrics	GA + KNN	GA + ensemble learning
Accuracy (%)	99.28	83.21
Sensitivity (%)	99.28	83.20
Specificity (%)	99.38	100
Precision (%)	99.99	100
F-measure (%)	99.63	90.83

Table 3 Comparative analysis with existing systems

Authors	Methods	Accuracy (%)
Rubin et al. [76]	NLP	76
Granik and Mesyura [77]	NB	74
Seo et al. [78]	CNN	86.65
Jain et al. [79]	NB, SVM and NLP	93.50
Proposed CIA	GA + KNN	99.28

4.1 Discussion

This study employed an FS and classification approach and this was conducted on a piece of Fake news generated dataset by utilizing genetic algorithm FS techniques to obtain appropriate information on the given data. KNN and Ensemble were utilized as classification ML algorithms for the classification implementation. The outcome demonstrated that GA + KNN performance surpassed that of the GA + BEL technique, in terms of accuracy, sensitivity, and F-measure as revealed in Table 2. The system performance was evaluated using three common metrics in the classification task which are accuracy, sensitivity, and f-measure, and they were used to measure the detection accuracy of the system (Metaxas, 2021). Table 2 illustrates the evaluation process attained by the two classifiers on the considered datasets and it was demonstrated that the maximum accuracy attained on the FN dataset is 99.28% achieved by the GA + KNN technique. The accuracy of GA + Bagged Ensemble classifiers was 83.21%. Other performance criteria, like sensitivity and f-measure, show that KNN outperforms Bagged Ensemble with a value of 99.28% and 99.63% respectively. The Bagged ensemble method outperformed the GA + KNN technique in terms of precision and specificity, with a score of 100% for both metrics. The model can accurately identify misclassified points and reduce the problem of overfitting.

5 Conclusion and Future Works

Early identification of false news is critical for the public and society, as it allows for the redesign of the 'information ecosystem in the twenty-first century,' which will ultimately lead to the development of a system and culture that values truth [12]. The article shows that using a dimensionality reduction algorithm namely genetic algorithm with the KNN classifier can effectively identify fake news with an accuracy of 99.28%, the sensitivity of 99.28%, specificity of 99.38%, the precision of 99.99%, and f-measure of 0.9963 which surpassed existing ML classifiers and some state-of-the-art the proposed system was compared with.

In the future, more dimensionality reduction could be employed with more ML classifiers and even deep learning algorithms can be used for the classification of fake news datasets to get better accuracy, sensitivity, specificity, precision, and f-measure.

References

1. Deepak, S., & Chitturi, B. (2020). A deep neural approach to Fake-News identification. *Procedia Computer Science, 167*, 2236–2243.
2. Mihaylov, T., Georgiev, G., & Nakov, P. (2015, July). Finding opinion manipulation trolls in news community forums. In *Proceedings of the Nineteenth Conference on Computational Natural Language Learning* (pp. 310–314).
3. Mihaylov, T., Koychev, I., Georgiev, G., & Nakov, P. (2015, September). Exposing paid opinion manipulation trolls. In *Proceedings of the International Conference Recent Advances in Natural Language Processing* (pp. 443–450).
4. Mihaylov, T., & Nakov, P. (2019). Hunting for troll comments in news community forums. arXiv preprint arXiv:1911.08113.
5. Bourgonje, P., Schneider, J. M., & Rehm, G. (2017, September). From clickbait to fake news detection: An approach based on detecting the stance of headlines to articles. In *Proceedings of the 2017 EMNLP Workshop: Natural Language Processing Meets Journalism* (pp. 84–89).
6. Vosoughi, S., Roy, D., & Aral, S. (2018). The spread of true and false news online. *Science, 359*(6380), 1146–1151.
7. Barthel, M., Mitchell, A., & Holcomb, J. (2016). *Many Americans believe fake news is sowing confusion* (Vol. 15, p. 12). Pew Research Center.
8. Chaudhry, A. K., Baker, D., & Thun-Hohenstein, P. (2017). Stance detection for the fake news challenge: Identifying textual relationships with deep neural nets. In *CS224n: Natural Language Processing with Deep Learning.*
9. Chopra, S., Jain, S., & Sholar, J. M. (2017, December). Towards automatic identification of fake news: Headline-article stance detection with LSTM attention models. In *Stanford CS224d Deep Learning for NLP Final Project.*
10. Bhatt, G., Sharma, A., Sharma, S., Nagpal, A., Raman, B., & Mittal, A. (2018, April). Combining neural, statistical, and external features for fake news stance identification. In *Companion Proceedings of the Web Conference 2018* (pp. 1353–1357).
11. Konstantinovskiy, L., Price, O., Babakar, M., & Zubiaga, A. (2021). Toward automated fact-checking: Developing an annotation schema and benchmark for consistent automated claim detection. *Digital Threats: Research and Practice, 2*(2), 1–16.
12. Lazer, D. M., Baum, M. A., Benkler, Y., Berinsky, A. J., Greenhill, K. M., Menczer, F., & Zittrain, J. L. (2018). The science of fake news. *Science, 359*(6380), 1094–1096.
13. Borges, L., Martins, B., & Calado, P. (2019). Combining similarity features and deep representation learning for stance detection in the context of checking fake news. *Journal of Data and Information Quality (JDIQ), 11*(3), 1–26.
14. Walker, M., Anand, P., Abbott, R., & Grant, R. (2012, June). Stance classification using dialogic properties of persuasion. In *Proceedings of the 2012 Conference of the North American Chapter of the Association for Computational Linguistics: Human Language Technologies* (pp. 592–596).

15. Sridhar, D., Foulds, J., Huang, B., Getoor, L., & Walker, M. (2015, July). Joint models of disagreement and stance in online debate. In *Proceedings of the 53rd Annual Meeting of the Association for Computational Linguistics and the 7th International Joint Conference on Natural Language Processing* (Vol. 1: Long Papers, pp. 116–125).
16. Somasundaran, S., & Wiebe, J. (2010, June). Recognizing stances in ideological online debates. In *Proceedings of the NAACL HLT 2010 Workshop on Computational Approaches to Analysis and Generation of Emotion in a Text* (pp. 116–124).
17. Lukasik, M., Srijith, P. K., Vu, D., Bontcheva, K., Zubiaga, A., & Cohn, T. (2016, August). Hawkes processes for continuous-time sequence classification: an application to rumor stance classification in Twitter. In *Proceedings of the 54th Annual Meeting of the Association for Computational Linguistics* (Vol. 2: Short Papers, pp. 393–398).
18. Gorrell, G., Bontcheva, K., Derczynski, L., Kochkina, E., Liakata, M., & Zubiaga, A. (2018). Rumoureval 2019: Determining rumor veracity and support for rumors. arXiv preprint arXiv: 1809.06683.
19. Stab, C., & Gurevych, I. (2017). Parsing argumentation structures in persuasive essays. *Computational Linguistics, 43*(3), 619–659.
20. Pomerleau, D., & Rao, D. (2017). The fake news challenge: Exploring how artificial intelligence technologies could be leveraged to combat fake news. Fake news challenge.
21. Team, F. N. C. (2018). Exploring how artificial intelligence technologies could be leveraged to combat fake news. Available http://www.fakenewschallenge.org/.
22. Zafarani, R., Zhou, X., Shu, K., & Liu, H. (2019, July). Fake news research: Theories, detection strategies, and open problems. In *Proceedings of the 25th ACM SIGKDD International Conference on Knowledge Discovery and Data Mining* (pp. 3207–3208).
23. Rapoza, K. (2017). Can "fake news" impact the stock market? Retrieved from www.forbes.com/sites/kenrapoza/2017/02/26/canfake-news-impact-the-stock-market/ (9. 7. 2018).
24. Rubin, V. L. (2010). On deception and deception detection: Content analysis of computer-mediated stated beliefs. *Proceedings of the American Society for Information Science and Technology, 47*(1), 1–10.
25. Zhang, A. X., Ranganathan, A., Metz, S. E., Appling, S., Sehat, C. M., Gilmore, N., & Mina, A. X. (2018, April). A structured response to misinformation: Defining and annotating credibility indicators in news articles. In *Companion Proceedings of the Web Conference* 2018 (pp. 603–612).
26. Zafarani, R., Abbasi, M. A., & Liu, H. (2014). *Social media mining: An introduction.* Cambridge University Press.
27. Ciampaglia, G. L., Shiralkar, P., Rocha, L. M., Bollen, J., Menczer, F., & Flammini, A. (2015). Computational fact checking from knowledge networks. *PloS One, 10*(6), e0128193.
28. Pérez-Rosas, V., Kleinberg, B., Lefevre, A., & Mihalcea, R. (2017). Automatic detection of fake news. arXiv preprint arXiv:1708.07104.
29. Shi, B., & Weninger, T. (2016). Discriminative predicate path mining for fact-checking in knowledge graphs. *Knowledge-Based Systems, 104*(2016), 123–133.
30. Sitaula, N., Mohan, C. K., Grygiel, J., Zhou, X., & Zafarani, R. (2020). Credibility-based fake news detection. In *Disinformation, Misinformation, and Fake News in Social Media* (pp. 163–182). Springer, Cham.
31. Undeutsch, U. (1967). Beurteilung der glaubhaftigkeit von aussagen. *Handbuch der psychologie, 11*, 26–181.
32. Johnson, M. K., & Raye, C. L. (1981). Reality monitoring. *Psychological Review, 88*(1), 67.
33. McCornack, S. A., Morrison, K., Paik, J. E., Wisner, A. M., & Zhu, X. (2014). Information manipulation theory 2: A propositional theory of deceptive discourse production. *Journal of Language and Social Psychology, 33*(4), 348–377.
34. Zuckerman, M., DePaulo, B. M., & Rosenthal, R. (1981). Verbal and nonverbal communication of deception. In *Advances in Experimental Social Psychology* (Vol. 14, pp. 1–59). Academic Press.
35. Loewenstein, G. (1994). The psychology of curiosity: A review and reinterpretation. *Psychological Bulletin, 116*(1), 75.

36. Zhou, X., & Zafarani, R. (2018). Fake news: A survey of research, detection methods, and opportunities. arXiv preprint arXiv:1812.00315, 2.
37. Chen, Y., Conroy, N. J., & Rubin, V. L. (2015, November). Misleading online content: recognizing clickbait as "false news". In *Proceedings of the 2015 ACM on Workshop on Multimodal Deception Detection* (pp. 15–19).
38. MacLeod, C., Mathews, A., & Tata, P. (1986). Attentional bias in emotional disorders. *Journal of Abnormal Psychology, 95*(1), 15.
39. Castillo, C., Mendoza, M., & Poblete, B. (2011, March). Information credibility on Twitter. In *Proceedings of the 20th International Conference on World Wide Web* (pp. 675–684).
40. Gupta, S., Thirukovalluru, R., Sinha, M., & Mannarswamy, S. (2018, August). CIMTDetect: A community-infused matrix-tensor coupled factorization-based method for fake news detection. In *2018 IEEE/ACM International Conference on Advances in Social Networks Analysis and Mining (ASONAM)* (pp. 278–281). IEEE.
41. Jin, Z., Cao, J., Zhang, Y., & Luo, J. (2016, March). News verification by exploiting conflicting social viewpoints in microblogs. In *Proceedings of the AAAI Conference on Artificial Intelligence* (Vol. 30(1)).
42. Ruchansky, N., Seo, S., & Liu, Y. (2017, November). CSI: A hybrid deep model for fake news detection. In *Proceedings of the 2017 ACM on Conference on Information and Knowledge Management* (pp. 797–806).
43. Shu, K., Wang, S., & Liu, H. (2019, January). Beyond news contents: The role of social context for fake news detection. In *Proceedings of the Twelfth ACM International Conference on Web Search and Data Mining* (pp. 312–320).
44. Zhang, J., Cui, L., Fu, Y., & Gouza, F. B. (2018). Fake news detection with the deep diffusive network model. arXiv preprint arXiv:1805.08751.
45. Zhou, X., & Zafarani, R. (2019). Network-based fake news detection: A pattern-driven approach. *ACM SIGKDD Explorations Newsletter, 21*(2), 48–60.
46. Boehm, L. E. (1994). The validity effect: A search for mediating variables. *Personality and Social Psychology Bulletin, 20*(3), 285–293.
47. Bálint, P., & Bálint, G. (2009). The semmelweis-reflex. *Orvosi hetilap, 150*(30), 1430–1430.
48. Greentree, C. (2021). Semmelweis Reflex. In *Decision Making in Emergency Medicine* (pp. 339–343). Springer.
49. Nickerson, R. S. (1998). Confirmation bias: A ubiquitous phenomenon in many guises. *Review of General Psychology, 2*(2), 175–220.
50. Tversky, A., & Kahneman, D. (1974). Judgment under uncertainty: Heuristics and biases. *Science, 185*(4157), 1124–1131.
51. Aldwairi, M., & Alwahedi, A. (2018). Detecting fake news in social media networks. *Procedia Computer Science, 141*, 215–222.
52. Liao, H., Liu, Q., & Shu, K. (2020). Incorporating user-comment graph for fake news detection. arXiv preprint arXiv:2011.01579.
53. Karimi, H., Roy, P., Saba-Sadiya, S., & Tang, J. (2018, August). Multi-source multi-class fake news detection. In *Proceedings of the 27th International Conference on Computational Linguistics* (pp. 1546–1557).
54. Shu, K., Mahudeswaran, D., & Liu, H. (2019). FakeNewsTracker: A tool for fake news collection, detection, and visualization. *Computational and Mathematical Organization Theory, 25*(1), 60–71.
55. Shu, K., Sliva, A., Wang, S., Tang, J., & Liu, H. (2017). Fake news detection on social media: A data mining perspective. *ACM SIGKDD Explorations Newsletter, 19*(1), 22–36.
56. Vogel, I., & Meghana, M. (2020). Fake news spreader detection on Twitter using character N-grams. In *CLEF*.
57. Umer, M., Imtiaz, Z., Ullah, S., Mehmood, A., Choi, G. S., & On, B. W. (2020). Fake news stance detection using deep learning architecture (CNN-LSTM). *IEEE Access, 8*, 156695–156706.
58. Xu, K., Wang, F., Wang, H., & Yang, B. (2019). Detecting fake news over online social media via domain reputations and content understanding. *Tsinghua Science and Technology, 25*(1), 20–27.

59. Shrivastava, G., Kumar, P., Ojha, R. P., Srivastava, P. K., Mohan, S., & Srivastava, G. (2020). Defensive modeling of fake news through online social networks. *IEEE Transactions on Computational Social Systems, 7*(5), 1159–1167.
60. Shang, J., Shen, J., Sun, T., Liu, X., Gruenheid, A., Korn, F., & Han, J. (2018, October). Investigating rumor news using agreement-aware search. In *Proceedings of the 27th ACM International Conference on Information and Knowledge Management* (pp. 2117–2125).
61. Zhang, Q., Yilmaz, E., & Liang, S. (2018, April). Ranking-based method for news stance detection. In *Companion Proceedings of the Web Conference 2018* (pp. 41–42).
62. Borges, L., Martins, B., & Calado, P. (2019). Combining similarity features and deep representation learning for stance detection in the context of checking fake news. *Journal of Data and Information Quality, 11*(3), 1–26. https://doi.org/10.1145/3287763
63. Bhatt, G., Sharma, A., Sharma, S., Nagpal, A., Raman, B., & Mittal, A. (2018). Combining neural, statistical, and external features for fake news stance identification. In *Proceedings of Companion Web Conference (WWW)*, Geneva, Switzerland (p. 1353). https://doi.org/10.1145/3184558.3191577.
64. Zeng, Q., Zhou, Q., & Xu, S. (2017). Neutral stance detectors for fake news challenge. In *CS224n: Natural Language Processing with Deep Learning.*
65. Pfohl, S. R., Triebe, O., & Legros, F. (2017). Stance detection for the fake news challenge with attention and conditional encoding. In *CS224n: Natural Language Processing with Deep Learning.*
66. Ghanem, B., Rosso, P., & Rangel, F. (2018, November). Stance detection in fake news a combined feature representation. In *Proceedings of the First Workshop on Fact Extraction and Verification (FEVER)* (pp. 66–71).
67. Hanselowski, A., PVS, A., Schiller, B., Caspelherr, F., Chaudhuri, D., Meyer, C. M., & Gurevych, I. (2018). A retrospective analysis of the fake news challenge stance detection task. arXiv preprint arXiv:1806.05180.
68. Umer, M., Sadiq, S., Ahmad, M., Ullah, S., Choi, G. S., & Mehmood, A. (2020). A novel stacked CNN for malarial parasite detection in thin blood smear images. *IEEE Access, 8*, 93782–93792.
69. Mohtarami, M., Baly, R., Glass, J., Nakov, P., Màrquez, L., & Moschitti, A. (2018). Automatic stance detection using end-to-end memory networks. arXiv preprint arXiv:1804.07581.
70. Dulhanty, C., Deglint, J. L., Daya, I. B., & Wong, A. (2019). Taking a stance on fake news: Towards automatic disinformation assessment via deep bidirectional transformer language models for stance detection. arXiv preprint arXiv:1911.11951. [Online]. Available https://arxiv.org/abs/1911.11951.
71. Ghaheri, A., Shoar, S., Naderan, M., & Hoseini, S. S. (2015). The applications of genetic algorithms in medicine. *Oman Medical Journal, 30*(6), 406–416. https://doi.org/10.5001/omj.2015.82
72. Abdulsalam, S. O., Mohammed, A. A., Ajao, J. F., Babatunde, R. S., Ogundokun, R. O., Nnodim, C. T., & Arowolo, M. O. (2020). Performance evaluation of ANOVA and RFE algorithms for classifying microarray dataset using SVM. *Lecture Notes in Business Information Processing, 402*, 480–492.
73. Ahmad, I., Yousaf, M., Yousaf, S., & Ahmad, M. O. (2020). Fake news detection using machine learning ensemble methods. *Complexity, 2020*, 1–11. https://doi.org/10.1155/2020/8885861
74. Adegun, A. A., Viriri, S., & Ogundokun, R. O. (2021). Deep learning approach for medical image analysis. *Computational Intelligence and Neuroscience, 2021*(2021), 6215281.
75. Azeez, N. A., Atiku, O., Misra, S., Adewumi, A., Ahuja, R., & Damasevicius, R. (2020). Detection of malicious URLs on Twitter. In *Advances in Electrical and Computer Technologies* (pp. 309–318). Springer.
76. Rubin, V. L., Chen, Y., & Conroy, N. K. (2015). Deception detection for news: Three types of fakes. *Proceedings of the Association for Information Science and Technology, 52*(1), 1–4.
77. Granik, M., & Mesyura, V. (2017, May). Fake news detection using naive Bayes classifier. In *2017 IEEE First Ukraine Conference on Electrical and Computer Engineering (UKRCON)* (pp. 900–903). IEEE.

78. Seo, Y., Seo, D., & Jeong, C. S. (2018, October). FaNDeR: Fake news detection model using media reliability. In *TENCON 2018–2018 IEEE Region 10 Conference* (pp. 1834–1838). IEEE.
79. Jain, A., Shakya, A., Khatter, H., & Gupta, A. K. (2019, September). A smart system for fake news detection using machine learning. In *2019 International Conference on Issues and Challenges in Intelligent Computing Techniques (ICICT)* (Vol. 1, pp. 1–4). IEEE.
80. Katoch, S., Chauhan, S. S., & Kumar, V. (2021). A review on the genetic algorithm: Past, present, and future. *Multimedia Tools and Applications, 80*(5), 8091–8126. https://doi.org/10.1007/s11042-020-10139-6
81. Khanam, Z., Alwasel, B. N., Sirafi, H., & Rashid, M. (2021). Fake news detection using machine learning approaches. *IOP Conference Series: Materials Science and Engineering, 1099*(1), 012040. https://doi.org/10.1088/1757-899X/1099/1/012040

Fandet Semantic Model: An OWL Ontology for Context-Based Fake News Detection on Social Media

Anoud Bani-Hani, Oluwasegun Adedugbe, Elhadj Benkhelifa, and Munir Majdalawieh

Abstract The detection of fake news on social media has become a very active research area. Several approaches and techniques have been proposed and implemented to address the challenge, across diverse technological domains such as NLP (Natural Language Processing) and machine learning. While substantial progress has been made on these, it remains a daunting task due to complexities in its nature. Therefore, it has become pertinent to significantly explore and integrate other technologies to detect fake news on social media. Hence, this research focuses on further exploring and developing native semantic technology solutions for the discourse space. The initial result is a taxonomy classifying socially contextual features for news articles and then Fandet: an OWL ontology for context-based fake news detection by semantically annotating contextual features of news articles and datasets using the ontology. This provides a basis for patterns recognition, analysis, and identification of news articles on social media as either fake or not.

Keywords Fake news detection · Social media · Social network · Social data analysis · Semantic annotation · OWL ontology · Machine learning · Natural language processing

1 Introduction

Social media has become more of a personal media, democratising news creation and dissemination. However, it has greatly fostered an existing menace, fake news. The fast growth with information technology innovations has resulted in an information overload, which produces difficulties with obtaining relevant, quality information quickly and easily. Based on these, it has become pertinent to ascertain veracity of

A. Bani-Hani (✉) · M. Majdalawieh
College of Technology Innovation, Zayed University, Dubai, UAE
e-mail: Anoud.Bani-Hani@zu.ac.ae

O. Adedugbe · E. Benkhelifa
Cloud Computing and Applications Research Lab, Staffordshire University, Stoke-on-Trent, UK

M. Lahby et al. (eds.), *Combating Fake News with Computational Intelligence Techniques*, Studies in Computational Intelligence 1001,
https://doi.org/10.1007/978-3-030-90087-8_5

news obtained online, especially from social media [1]. With recent and widespread attention on social media and its role in the spread of fake news regarding social and political affairs, it is imperative to understand how social media users interact with information on the media platforms. This has given rise to automated fact-checking systems on the web. However, the need for more such systems has also been stressed [2]. Alongside prominent ones such as PolitiFact, Snopes, and FactChec.org, rumour detection systems such as Tweetcreed, Truthy, Rumorlens also exist [3]. This is in addition to individual efforts by social media platforms. For instance, Twitter aims to curb phishing and spread of fake news by checking URLs on the platform, which has been implemented via an extension on Chrome that detects links hidden on shortened URLs. In addition, tools such as 'Dispute Finder' compares claims and corrects articles in real-time. Some of these systems are fully automated while others still rely on human input for content, thereby incapable of scanning incoming or updated data on their own [4]. Hence, despite these approaches and diverse techniques based on machine learning, deep learning and natural language processing among others, fake news is still very prominent and quite challenging to detect. As the conventional printed media is gradually fading away and replaced with a digital one, each social media account possesses ability to become a news writer or journalist [5].

Semantic technologies possess the potential to enhance analysis of social media data, facilitating exchange, merging, querying and transformation of data extracted from various platforms [6]. Furthermore, they are beneficial in ensuring interoperability and availability of data in a format that can be read by machines, thus enhancing the potentials of big data analysis. They are also vital in the collection, merging and integration of data on social media platforms from heterogeneous sources, in a manner that is easy, robust and interoperable [7], attempting to bridge the gap between humans and computers. According to Bennato [8], media has a vital role in shaping the world's perception, so if fake news keeps spreading, the world's perception could be wrongfully moulded, posing serious risks in civil coexistence, security, and democracy. The scalability, replicability, and persistence of digital content enhances detrimental effects that fake information could have on the society [9], therefore an innovative approach is required which combines semantic technologies with the goal of implementing a digital framework that supports understanding of some social processes underlying the current digital society [10]. Two standards very central in this context provide a model definition for representing and defining associations between resources; RDF (Resource Description framework), and vocabularies used for representing resource semantics; such as RDFS (RDF Schema) and OWL (Web Ontology Language) [6]. In addition, SPARQL query language is utilised for querying semantic vocabularies to define context for data extracted from social media. Driven by the concept of graph databases, it facilitates representation of data descriptions in a graph format with the ability for nodes to be widely connected to each other based on defined properties. These descriptions can be based on RDF; offering a rich typed graph that results into a more powerful representation of data from social media in comparison to conventional models of social media data representation [11]. In the recent past, interactions via web 2.0 social platforms have raised concerns within semantic web communities. Some ontologies are applied for the representation of social platforms, for instance FOAF is used to describe user's online activities, their relationships, and profiles [12]. The RELATIONSHIP ontology has focused on the 'knows' property of FOAF to give a more precise representation of the social media relationships, be it professional, familial or friendship relationships. The SIOC ontology extends FOAF to social media user's online activities such as conversations, forum posts, and blogs [13]. However, these ontologies are generic in nature and scope, as they are designed for other purposes and do not model the social context for news articles on social media towards detecting fake articles. The remainder of this paper is structured as follows: Sect. 2 presents a review of related work and attempts at semantic technologies solutions for fake news detection. Section 3 analyses news articles features both from a content and context perspective. The section concludes with developing a taxonomy for context-based fake news detection. In Sect. 4, an OWL ontology is developed from the taxonomy, modelling context features as ontology classes, sub-classes, object properties, and so on. Section 5 provides a use case and analysis for the Fandet ontology and the paper concludes with a summary in Sect. 6.

2 Related Work

Machine learning techniques are significantly utilised for fake news detection. They are applied in building tools such as classifiers which utilise news content to label it as fake or not. Such tools can review large data volumes and discover trends seemingly impractical for humans to discover [14]. Such capabilities are essential to developing fake news detection systems through pattern identification. For example, a professional social media such as LinkedIn can employ machine learning tools for understanding patterns employers utilise when seeking new applicants. Subsequently, job posts with unusual patterns can be flagged down as falsified information, thus protecting users. Machine learning algorithms are designed to continually improve in efficiency and accuracy by learning from daily tasks. As training datasets keep growing, machine learning tools can make better decisions and more accurate predictions towards fake news detection [15]. It is both a cost-effective, accurate, and proven technique for analysing and scanning large volumes of data [16]. However, a key limitation of machine learning techniques for fake news detection is challenges with linguistics. Fake news authors tend to use irony, crafting messages to imply the opposite of its content, obscuring intended messages within a news article to prevent detection by an algorithm [17]. A study carried out by Georgia Tech Research Institute demonstrated that modelling textual content of news articles can be enough to detect news bias, but not to evaluate its credibility. This is due to the breadth of contextual data that must be analysed and compared against training data sets [18]. In addition, for machine learning to be effective in determining credibility of news articles, it requires a high level of tailoring to specific theories that must be included in training datasets. This is unlike detecting bias, which needs to use keywords to determine if a news article is bias. Machine learning requires high quality, unbiased, and huge datasets to train. Acquiring such data, especially in the news context, can be challenging. Machine learning techniques can be broadly classified as either based on neural or non-neural networks.

Neural networks define a classification for algorithms of computing systems with a high inter-connectedness of diverse nodes, with wide application in deep learning which is a sub-set of machine learning. An example is Long Short-Term Memory (LSTM) which is a type of recurrent neural network that helps in solving challenges of the vanishing gradient, with ability to retrieve longer-term dependencies [19]. Rashkin et al. [20] came up with two types of LSTM model: the first placed simple word embeddings starting with GloVe into LSTM and the second concatenates LSTM output with LIWC (Linguistic Inquiry and Word Count) characteristic vectors prior to going through the activation layer. For both cases, accuracy levels were slightly higher than those of NBC (Naïve Bayes Classifier) and Maximum Entropy (MaxEnt) models. LSTM was also applied by Ruchansky et al. [21] for article representation. Native machine learning techniques are non-neural in nature, with popular classification models such as Support Vector Machine and Naive Bayes Classifier. Other methods include Logistic Regression and Random Forest Classifier [19]. These techniques utilise sets of algorithms for data parsing towards training datasets. The algorithms are designed to continually improve in efficiency and accuracy by learning from daily functions. As the training datasets keep increasing, machine learning tools become more capable of informing better decisions and more accurate predictions towards filtering news articles [15]. Table 1 provides a comparison of machine learning techniques that are native (or non-neural) and techniques neural in nature.

The use of NLP is also highly significant with machine learning techniques towards fake news detection. These involve use or extraction of major linguistic

Table 1 Comparison of neural network-based ML and non-neural network-based ML techniques

Comparison metric	Non-neural ML (Native Machine Learning)	Neural network-based ML (Deep learning)
Feature engineering	Features are identified by an expert and then extracted	Features are learned by the system itself
Hardware dependency	Hardware requirement relatively minimal	Requires high-performance GPUs
Data dependency	Can be minimal	Lots of data required
Problem-solving approach	Solutions provided over several components	End-to-end solutions
Execution time	Shorter training time	Longer training time
Testing time	More testing time	Lesser testing time
Number of available classifiers	Numerous classifiers	Few classifiers
Algorithms	Algorithms parse data and learn from the data to make informed decisions	Structures algorithms in layers to mimic human "neural network" for learning data to make informed decisions

features from news articles. Some of the methods include ngram analysis, punctuation, psycho-linguistic features, readability, and syntax. Ngram Analysis involves extraction of unigrams and bigrams from words in a story and preferably stored as TFIDF (Term Frequency Inverse Document Frequency) values to retrieve information. TFIDF is the numerical statistic for reflection of the importance of a word to the document in which it is utilised [5]. Punctuation aids algorithms for differentiating between deceptive and truthful texts. This feature collects eleven types of punctuations. Psycho-linguistic features foster identification of appropriate word proportions and LIWC lexicon has been recommended for this feature. This helps in determining language tone, statistics of text, and part-of-speech category [5]. Readability helps in extracting content features such as number of characters, complex words, long words, number of syllables, word types, and number of paragraphs [22]. The availability of such content features helps in performing readability metrics, such as Flesch-Kincaid, Flesch Reading Ease, Gunning Fog, and Automatic Readability Index (ARI). Syntax helps in extracting features based on CFG (Context-Free Grammar) [5].

2.1 State-Of-The-Art with Semantic Technologies

The study by Ismailova et al. [23] assumed semantic virus' defeat of a healthy resource model in cases of information forgery, thereby intentionally distorting the original content. The research included formulation of a computational model with channelled effect based on the initial model of interaction between information processes whenever there is generation of an error opinion (bug) for an information resource (page) with its damage (harm). That is, bringing harm through the indication of a false benefit (fake+ , fake−) if the resource user pursues (hit) a specific interest or benefit (gain). When searching for G—gain, the bug virus damages malicious harm content P with the virus. A framework defined as semantic was proposed for each resource, represented by an information process, and formed map for channelled distribution of fake news. The transformation of each semantic map became a map of variable domains, which is a conceptual model made up of vertex processes and process connections in between. Levi et al. [24] on the other hand used pre-trained models of Bidirectional Encoder Representations from Transformers (BERT) and modified it, using fake news datasets and satire articles with Adam optimizer, 3 decay types and 0.01 decay rate. A BERT-based binary classifier was proposed through addition of a new layer in BERT's neural network architecture. The proposed system was also trained to modify BERT and classified fake news and satire articles. Furthermore, Coh-Metrix was utilised with input documents to have 108 indices associated with text statistics such as number of words and sentences, referential cohesion, a variety of readability formulas, different types of connective words and more. A Principal Component Analysis (PCA) was also run on the set of Coh-Metrix indices to account for multicollinearity among other features. In addition, the research by Gomes-Jr and Frizzon [25] utilised Fake. Br Corpus to analyse news classified as fake. The study involved annotation of each news article with DBpedia entities using DBpedia Spot-Light tool, forming a graph for each article and mentioning the entities as vertices. Two separate datasets were built from the graphs of each article, identifying clusters of entities within each. Modularity algorithm was also utilised in defining clusters of entities correlated within the graph. Furthermore, Local Outlier Factor (LOF) was implemented for automatic detection of important events over the period across the news timeline. Brasoveanu and Andonie [26] in their work, deployed semantic features for the 'Liar' dataset in improving fake news recognition accuracy. This was implemented by computing semantic features (sentiment analysis, named entities and relations) and adding them to a set of syntactic features (part-of-speech and noun phrases) as well as to features of the original input dataset. A variety of classifiers were used on the resultant augmented dataset which includes deep learning models; Long-Short Term Memory (LSTM), Convolutional Neural Network (CNN), and Capsule Networks. Wang [27] also applied CNN together with a LSTM layer in detecting fake news via text context and additional metadata. Similar research was conducted by Klyuev [28] proposing a solution based on NLP, text mining for semantic similarity evaluation, machine learning and automatic fact checking.

Bharadwaj and Shao [29] utilised diverse learning techniques as well, presenting comparison between recurrent neural networks, naïve bayes, and random forest algorithms using various linguistic features for fake news detection. The models were evaluated using real or fake datasets from kaggle.com, containing 6256 articles including titles. The defined semantic features involved term frequency (TF), term frequency–inverse document frequency (TFIDF), bigrams, trigram, quadgram, and vectorised word representations. The results of the study had bigrams and random forest algorithm producing higher levels of accuracy. Furthermore, Shu et al. [15] investigated linguistic features such as word count, frequency of words, character count, similarity and clarity score for videos and images towards rumour classification, truth discovery, click-bait detection, and spammer/bot detection. Pan et al. [30] was observed to be one of a few based on native semantic technologies; utilising knowledge graph embeddings for computing semantic similarities towards content-based fake news detection. The research generated background knowledge by producing three knowledge graphs; built, entity and relation, embedding in low-dimensional vector space using B-TransE model for detecting fake news. According to the findings, incomplete and imprecise knowledge graphs can still be beneficial for detecting fake news. Similarly, Sabeeh et al. [31] utilised Wikipedia for addition of semantic features to news articles, proposing a CNIRI-FS (Contextual Negation Handling and Inherent Relation Identification for Enhanced Feature Selection) model for the detection of fake news. The model was aimed at improving the classification process by exploiting Wikipedia infobox data as domain knowledge in covering the complete range of information related to an article. It focused on representing real-world entities, such as organisations and persons, offering real-time and up-to-date information in alleviating the semantic gap. The proposed model employed word2vec model in producing word embeddings for the extraction of event-related features. Rashkin et al. [20] focused on analysis of news datasets, analysing over 70,000 news documents from the English Gigabit corpus. After document classification, 14,000 trusted news articles were retrieved, with 15,000 satire, 12,000 hoax, and 33,000 propaganda. From the diverse related works, it can be observed that while some research efforts have been defined as semantic based, semantic components within such are either from a general semantic perspective or based on semantic features utilised in other domains such as NLP and machine learning. However, three of the research efforts; Gomes Jr and Frizzon [25], Brasoveanu and Andonie [26] and Pan et al. [30] utilise native semantic technologies in their solutions. While these enhance fake news detection from the perspective of semantic solutions, this research provides a novel direction by developing an OWL ontology which uniquely models social context for news articles on social media comprehensively as objects, classes, properties, etc. towards detailed annotation of news articles' contextual data with their corresponding entities within the ontology. Hence, by means of the semantic annotation, detailed analysis can be conducted towards detecting news articles as fake. Furthermore, identification of patterns from large datasets of news articles using the ontology would go a long way in addressing fake news challenges, including diffusion models and speed of spreading. The proposed OWL ontology utilises context-based approach for fake news detection. Table 2 presents a summary of the different related works.

3 Taxonomy for Context-Based Fake News Detection

Pierri and Ceri [32] conducted a data-driven survey regarding approaches towards fake news detection by defining three classifications: content-based, context-based, and a hybrid of both. This is also supported by Shu et al. [15], defining a tri-relationship between social media users, publishers and news articles based on their features. While content-based focuses on features within news articles such as linguistics and style, context-based focuses on features of the social context for news articles, which are predominantly based on social network and user features. Content-based approaches analyse only news articles' textual content such as body and title. Stance detection models are a type of content-based approach introduced during the 2017 Fake News Challenge Stage 12 (FNC-1). The approach classifies stance of an entire news article with respect to its headline; that is, stance detection at document level, utilising neural networks such as Talos by Baird et al. [33], Athene by Hanselowski et al. [34], and UCL Machine Reading by Baird et al. [33]. These systems work by combining lexical features such as bag-of-words, topic modelling and word similarity features. Hanselowski et al. [35] analysed extensively these approaches and experimented with their capacity for generalisation on data. The study by Wang [27] utilised a Liar dataset and applied a multi-label classification task involving labels dwelling on the six degrees of truth from PolitiFact fact-checking system. Content-based approaches also utilise several machine learning and deep learning methods, including logistic regression, convolutional and recurrent neural networks. The study of Horne and Adali [36] was based on deep textual analysis which required examination of body and title of various categories of news

Table 2 Summary of different related works to this research

Research paper	Domain	Defined semantic component	Usage of semantic Technologies	Remarks	Type
Ismailova et al. [23]	Fake News Detection	Constraints Processing	None	Uses variable domains to develop a model to indicate post-truth with fake news channels	Context-Based
Levi et al. [24]	Fake News Detection	BERT (Bidirectional Encoder Representations from Transformers)	None	Utilised for downstream NLP tasks based on pre-training on a dataset	Content-Based
Gomes Jr and Frizzon [25]	Fake News Detection	DbPedia Spotlight for annotating news articles	Semantic Annotation	News articles are real-time events, most likely with entities and relationships not yet defined in a knowledge graph	Content-Based
Klyuev [28]	Fake News Detection	Semantic Filter (suggested)	None	Suggests the use of NLP, text mining, machine learning and automatic fact-checking for fake news detection	Content-Based
Sabeeh et al. [31]	Fake News Detection	Wikipedia Infobox and web pages for feature enrichment	None	Utilises Wikipedia Infobox and other external sources to enrich corpus features for analysing news articles	Content-Based

(continued)

Table 2 (continued)

Research paper	Domain	Defined semantic component	Usage of semantic Technologies	Remarks	Type
Brasoveanu and Andonie [26]	Fake News Detection	Metadata collection, relation extraction and embeddings	Relation extraction and embeddings	DBpedia knowledge graph for metadata embedding	Content-Based
Pan et al. [30]	Fake News Detection	Knowledge Graph using B-TransE model	Knowledge Graph based on triples semantic format	Use of knowledge graph for content-based fake news detection	Content-Based
Bharadwaj and Shao [29]	Fake News Detection	Text Pre-processing	None	Text Analytics prior to implementing machine learning techniques	Content-Based
FANDET Semantic Model	Fake News Detection	OWL Ontology for Discourse Space Classification and Description	OWL Ontology based on turtle format	Development and utilisation of an ontology for fake news detection	Context-Based

articles (true, false and satire), extracting complexity, psychological and stylistic characteristics.

Unlike content-based which considers content of news articles, context-based approaches focus on contextual data for news articles. These include data retrieved from social interactions between users, such as likes, comments and re-tweet towards detecting fake news [32]. The research by Tacchini et al. [16] focused on identifying fake news based on user likes on Facebook, proposing a technique which collects a set of posts and users from both conspiracy theories and scientific pages, as well as building a dataset where every characteristic vector stands for the set of users who liked a page. Likewise, Volkova et al. [37] addressed challenges of prediction for four sub-types of suspicious news; satire, hoaxes, click-bait and propaganda. This involved manually constructing lists of trusted and suspicious Twitter news accounts and collecting tweets during the Brussels bombing in 2016. The study also involved incorporation of tweet texts, linguistic cues and user interactions in a fused neural network model for comparing ad-hoc baselines trained on similar characteristics. The linguistic cues included bias, subjectivity, and moral foundations. In addition, Wang et al. [38] proposed a multi-modal neural network model for extracting both textual and visual characteristics from Twitter and Weibo conversations towards detection of fake news articles. The proposed model was based on adversarial settings and utilised an event discriminator. The research findings suggested the model could eliminate event-specific characteristics and generalise to unseen scenarios, in which the number of events is detailed as a parameter. Furthermore, research efforts have modelled the spread of messages characterised by malicious content on social networks. Liu and Wu [39] proposed building of custom datasets that reflect both true and fake news. These can leverage APIs such as Twitter API for Twitter and fact-checking websites, such as Snopes. The initial step involved inferring embeddings for users from the social graph and utilisation of a neural network model for classification of news articles. The research also produced a sequence classifier through LSTM networks for analysing propagation pathways of messages. The results reveal that the proposed model outperforms other embedding approaches. For both content and context-based approaches, comparisons drawn are presented in Table 3.

Content and context-based approaches consider both news content and associated social context interactions for fake news detection [32]. Ruchansky et al. [21] introduced a neural network model that integrates text of news articles, social network users' responses and source users promoting them. The testing of the model was on Twitter and Weibo while it was evaluated in comparison with other approaches. The study by Shu et al. [15] suggested that detection of fake news was by a tri-relationship among publishers, news articles and users. The context-based component focused on embedding interactions between users and news articles, as well as relations between publishers and news articles with non-negative matrix factorisation and user's credibility scores. The content-based component on the other hand focused on building classifiers based on emergent characteristics and performances from the context-based component, followed by evaluation on FakeNewsNet dataset in comparison with other advanced information credibility algorithms. According to the results, it was discovered that effective exploitation of social context could enhance the process of detecting fake news. From Table 3, it can be observed that both content and context-based have strengths and challenges. The comparison does not suggest or imply adoption of either, rather it reveals a role for each to play within the discourse space. Considering huge potentials with semantic technologies yet to be fully exploited, this research focuses on context-based fake news detection. Within the scope of this paper, the following steps are defined towards this:

- Identification and classification of diverse entities within social media platforms that have a relationship with news propagation on the platforms. The classification is presented as a taxonomy.
- Development of a semantic model: an OWL ontology from the taxonomy. The ontology models context for news objects on social media in a notation that

Table 3 Comparison of content-based and context-based fake news detection approaches

	Content-based approaches	Context-based approaches
Focus	News content	Social context for news
Extracted data	Lexical and syntactic features	User-based, network-based and impact-based features

machines can understand and process. News objects and datasets are annotated with the ontology to achieve this.
- Development of a use case for the ontology, providing description and analysis for its usage as well as evaluating it.

The taxonomy defines a news object with four major components: Data, Metadata, Advertiser and User. Data and Metadata are classified under 'Content', representing content for news objects while Advertiser and User are classified under 'Publisher', in terms of the entity where Data and Metadata originated from. So, in this context, Publisher can be an advertiser on social media or a personal social media user. In line with contextual nature that the taxonomy represents for fake news detection, both Data and Metadata are from a contextual perspective, rather than actual content of news objects. Hence, Data is further classified in terms of its format such as Text, Image, Audio, Video, Animation and Multimedia. For Audio and Video, data streaming is classified as either Live or Recorded. This is also applicable to Multimedia, which contains two or more other formats under 'Data'. The graphical format, or Image is classified based on standard image formats on social media and the web at large, such as PNG (Portable Network Graphics), JPEG (Joint Photographic Experts Group), GIF (Graphics Interchange Format) or WebP. Metadata on the other hand defines three major classifications: Date, Title and URI (Uniform Resource Identifier). Date stands for publication date of news objects. This can be very vital for monitoring other data and statistics such as speed of propagation, in relation to news objects' diffusion models. It can also identify if a news object is recent or recycled to appear as recent. Date is further classified as either String or Numeric. While String is not further classified due to the diverse representational possibilities, Numeric is further classified, with delimiter type utilised for classification. Delimiters for numeric date formats are either hyphen (-) or a forward slash (/) in most cases. However, the taxonomy provides a third classification, for instances where delimiter is neither of both. Title under Metadata is classified as either alphanumeric or non-alphanumeric. The third classification under Metadata: URI identifies address of news objects. Oftentimes, this will be a URL (Uniform Resource Locator), which is a type of URI. Different parameters are further captured within the URI, such as Protocol, Type and TLD (Top Level Domain). Common protocols for URIs and web content include HTTP (HyperText Transfer Protocol), HTTPS (HyperText Transfer Protocol Secure), FTP (File Transfer Protocol) and SFTP (Secure File Transfer Protocol). URI type for news objects can be vital in terms of the object's nature and its intended purpose when social media users engage with it. However, some news objects, such as comments on Facebook and Twitter may not belong to any of the standard URIs. URIs are further classified as either Default or Shortened, where the latter refers to utilisation of URL shortening technique [40] to define shorter version of a URL, which still redirects to the required page or content. The third classification under URI is TLD. This is of importance in the overall analysis of news objects, as certain TLDs such as '.gov' and '.ac.uk' are less likely to be utilised for fake news propagation, in comparison to generic TLDS such as '.com' could be used.

For Advertiser, it is classified as a type of Publisher that publishes Adverts on social media. Such Adverts are classified based on Type, Model and Platform. Type defines Adverts as either of commercial intent or not. Platform defines Adverts published on a business platform, such as LinkedIn or on a personal one, such as Facebook. Model on the other hand, defines business model for Adverts. These can be PPC (Pay Per Click), Paid Influencers and Paid Content Promo among others [44]. User category defines four classifications. Firstly, Privacy: in terms of a user's privacy status which can be open or closed. It also defines Status, which can be verified or non-verified. This is a common feature on platforms such as Facebook and Twitter to confirm authenticity of certain accounts based on verification of its owner. It also defines Circus, in terms of a smaller unit for specific users on a platform. This can be a group, friends, followers, or a page on a social media platform. Lastly, it defines Engagement in terms of type of interaction between a user and a news object on social media. This is of keen interest as news objects can target specific interaction types. Figure 1 further illustrates this.

Popular interaction types include via a chat, post, repost (or retweet), likes, notifications, subscriptions, search, mentions, tagging, share and comment. The capture of contextual data for news objects on social media based on these classifications would provide a basis for analysis and identification of patterns for different types of news objects, as well as detecting fake ones. The availability of news datasets from sources such as PolitiFact, BuzzFeed, GossipCop and others would facilitate analysis based on this classification. Figure 2 presents a graphical representation of the taxonomy.

While the taxonomy illustrated in Fig. 2 describes context for news objects on social media, it requires development into a software model that can constitute an interface for annotating and analysing instances of social media news objects. With

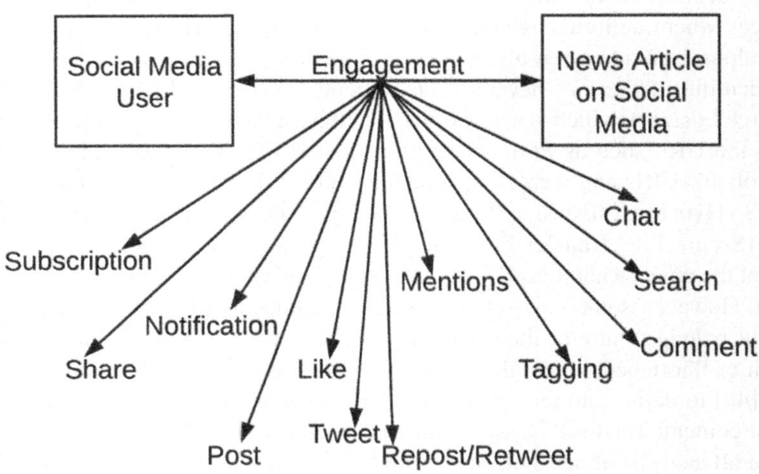

Fig. 1 Engagement types between users and news articles on social media

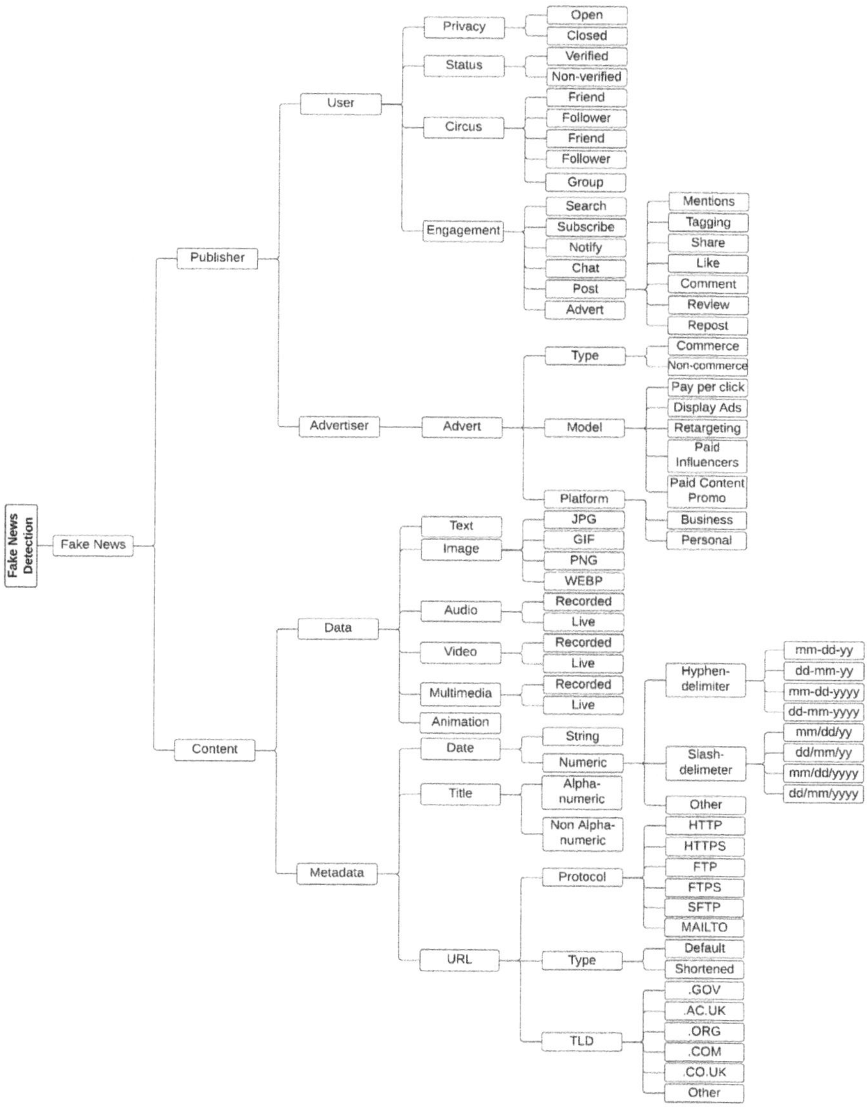

Fig. 2 Taxonomy for context-based fake news detection on social media

semantic technologies capabilities to provide context, presenting information in notations processable by machines and based on gap analysis from Sect. 2 (summarised in Table 3), the next section focuses on development of the taxonomy into a semantic model. This will be an ontology developed using OWL (Web Ontology Language).

OWL is a computational logic-based language for modelling and representing knowledge for machine-level processing, facilitating consistency within the represented knowledge, and making implicit knowledge explicit, among several other capabilities. An Ontology is defined as an "explicit, formal specification of a shared conceptualisation". The term emanated from philosophy where it refers to a logical account of existence [41]. It defines a representation for a knowledge domain, providing a formal description of concepts and their relationships resulting in a shared understanding [42]. With the current evolvement of a semantic web, the need for standards to facilitate it is very vital. Ontologies provide this by means of defining data model schemas, which are utilised by annotation data in the semantic annotation process [43]. With ontologies being developed using scientific programming languages, it also implies that machines can easily understand the annotation they provide to web documents and further assist humans with information usage, extraction, and retrieval on the web.

4 Fandet Ontology

OWL can be used to explicitly represent meaning of terms in vocabularies and relationships between the terms. This representation of terms and their interrelationships is called ontology. Web ontology language (OWL) has more facilities for expressing meaning and semantics than XML, RDF, and RDFS, and thus OWL goes beyond these languages in its ability to represent machine interpretable content on the Web. OWL is a revision of the DAML + OIL web ontology language, incorporating lessons learned from the design and application of its predecessor. Social media are widely used for news consumption and diffusion. In contrast to traditional news outlets, they support low cost easy access to information as well as the capacity for rapid dissemination. However, since social media lacks norm and processes for ensuring the accuracy and credibility of information, the rise of fake news has become a global concern. Automated fake news detection on social media is aimed at preventing exposure of individuals to fake news. Fake news detection is recognised to be a challenging task that requires analysis not only of the published content but also of its contextual information, such as user social engagements and spread type. Fandet ontology is designed to represent the latter, by providing a method to structure the otherwise complex and often incomplete data and facilitate its analysis. Fandet ontology provides the main concepts and properties required to describe information that contextualises the spread of fake news in social media, and hence it enables semantic structuring for data analysis and automated reasoning. In contrast to other well-known ontologies for modelling social network environments, Fandet ontology is particularly tailored to representing news transmission events, hence introducing a process-oriented approach. The scope of the ontology is the characterisation of context and characteristics in which fake news instances are found in social media, as well as focus on their diffusion models. Therefore, Fandet ontology provides a lighter specification of the content and structure of social media networks and, instead, concentrates on the description of such events, on the modelling of the relevant actors (not only social media users but also advertisers) and on sequential structures that can be extracted from the news diffusion models. The ontology consists of 47 concepts and 25 roles, structured around the main components of a News_Transmission_Event, to which they are related via roles that extend the standard part-hood relation. Components do not only model participants of the event, such as the social media object and the social media agents involved, but also features of the process, such as engagement types and user relationships that facilitated a fake news diffusion. Roles are generally provided with their inverses to support easy querying, yet for those extending the part-hood relation, the general isPartOf inverse role is employed to keep a manageable set of roles.

Fandet ontology is organised under the overarching concept NewsConcept, which is equipped with the attributes hasDescription and UID, so that every instantiated object can be given a unique ID and description. Immediately subsuming NewsConcept are some of the main concepts of the ontology, namely News_Transmission_Event, Social_Media_Agent, Social_Media_Object, Social_Media_Relation, Social_Metadata, Engagement and Platform. News_Transmission_Events are the fundamental concepts that are modelled in this ontology, where news is any kind of content that is being transmitted in social media. However, they are not only characterised by some content but also by the context of occurrence. The latter includes the user that encountered the news, their origin, the possible engagement of the user with it and the possible advertisers that promote the content among others. Social_Media_Agent is a high-level concept that denotes different types of agents in social media. In the context of a News_Transmission_Event, a social media agent encounters or promotes news content. Such a promotion can be made in the form of a traditional advertisement or as a result of the interaction of the user with the news content, for instance via search, like or reposting content. Social_Media_Agent is defined as the disjunction of its subclasses, which are not mutually exclusive. That is, a Social_Media_Agent can be simultaneously a Social_Media_User and an Advertiser. This would be the case, for instance, with a business having a user profile on social media. Figure 3 presents central roles of the ontology with NewsConcept as the overarching ontology concept.

Social_Media_Object is also a high-level concept representing information that is published on a social media platform. The concept is further specified by its subclasses. Social_Media_Object is the disjoint union of the possible data types, namely Animation, Image, Audio, Multimedia, Text, and Video. Other non-disjoint sub-concepts further specify nature of the object, which can be an Advert or a Streaming_Media_Object. Furthermore, Social_Media_Object are associated to at most one title and one URL, and they can be associated to further elements of the class Social_Metadata, such as publication and creation dates; hasCreatedDate and hasPublishedDate respectively. Moreover, they can be part of a News_Transmission_Event. Finally, they can be linked to the file or text to which they refer to via the property socialMediaContent. Social_Media_Relation denotes a relationship between a social media agent and another social media entity, be that another

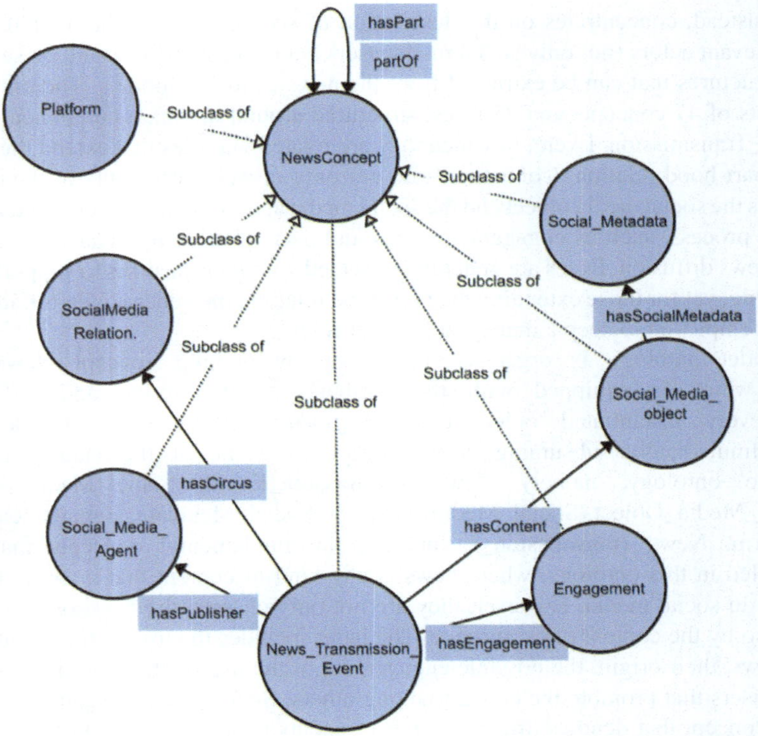

Fig. 3 Central roles within Fandet ontology

agent, a group, or a channel. When linked to a News_Transmission_Event it determines the circus or origin of the news object. Social_Metadata is a concept containing one item of metadata that is relevant to exactly one Social_Media_Object, to which it is linked via the property hasSocialMetadata. Social_Metadata is further refined in type by its disjoint sub-concepts. Engagement is another high-level concept that is typically used in the context of a News_Transmission_Event and characterises how a user engages with a news object. For instance, it may be via liking a post containing the news object. Subclasses of Engagement can be used to further specify the type of interaction. Platform concept is used to characterise the platform in which an advert is published. The tailored attributes hasPlatformDescription and hasPlatformTitle are used to provide the relevant information, and the two sub-concepts are used to refine the type of platform. A UID can also be given, like for any NewsConcept. The class is defined as the disjunction of its sub-classes, which are mutually exclusive. Table 4 summarises classes with a central role, alongside their sub-classes, domain and range.

Whenever a news transmission contains fake news, the subsumed concept Fake_News_Transmission_Event is utilised. To model this,

Table 4 Central role classes and their sub-classes, domain and range within the ontology

Class name	Super/sub class(es)	Domain	Range
Engagement	**Super Class** NewsConcept **Sub-Classes** ChatComment Notify PostEngagement SearchResult SubscriptionItem		hasEngagement
News transmission event	**Super Class** NewsConcept **Sub-Classes** FakeNewsTransmissionEvent	hasCircus hasContent hasEngagement hasPublisher	
News concept	**Sub-Classes** Engagement NewsTransmissionEvent Platform SocialMediaAgent SocialMediaObject SocialMediaRelation SocialMetadata	hasDescription hasPart isRelatedToUser partOf uID	hasPart partOf
Platform	**Super Class** NewsConcept **Sub-Classes** BusinessPlatfom PersonalPlatform	hasPlatformDescription hasPlatformTitle	publishedOn
Social media agent	**Super Class** NewsConcept **Sub-Classes** Advertiser SocialMediaUser	advertises hasEmail hasName	hasChannelAdmin hasGroupMember hasPublisher hasRelationOriginUser hasRelationTargetUser isRelatedToInfluencer
Social media object	**Super Class** NewsConcept **Sub-Classes** Advert Animation Audio Image Multimedia StreamingMediaObject Text Video	hasCreatedDate hasPublishedDate hasSocialMetadata hasUrl socialMediaContent	hasContent
Social media relation	**Super Class** NewsConcept **Sub-Classes** ChannelRelation ConnectionRelation GroupMembershipRelation	hasRelationOriginUser hasRelationTargetUser	hasCircus

(continued)

Table 4 (continued)

Class name	Super/sub class(es)	Domain	Range
Social metadata	**Super Class** NewsConcept **Sub-Classes** Date Title Url		hasSocialMetadata

News_Transmission_Event is linked to different key concepts of Fandet ontology which subsequently models the different components. Overall, it is organised so that each concept serves a role that responds to one of the five questions: who, what, when, where and how as follows:

- Who is responsible for the news transmission event? News_Transmission_Event has exactly one publisher that is the main participant of the event and belongs to class Social_Media_User, which is linked through the hasPublisher role.
- What is the content that constitutes news? News_Transmission_Event is linked to exactly one Social_Media_Object through the hasContent role.
- When and where did the news transmission occur? The context of a News_Transmission_Event can be retrieved through different components. On the one hand, Dates and URLs are queried via the Social_Media_Object. Moreover, beyond origin of the actual content, the news object is characterised by the property hasCircus, which maps to at most one Social_Media_Relation, linking to the social connection through which the news object was accessed.
- How did the transmission occur? The Engagement determines the way in which a user transmits the news, with a News_Transmission_Event having at most one Engagement linked via the role hasEngagement. Figure 4 presents a screenshot of the overall ontology.

4.1 Fandet Ontology Classes

Classes within an OWL ontology are the standard means of grouping meaningful resources together. The Fandet ontology comprises of 47 classes, with a NewsConcept class at its base, alongside a few other classes with central roles as presented in Fig. 3. Classes Social_Media_User and Advertiser belong to the central role: Social_Media_Agent. The Social_Media_User is a Social_Media_Agent with user profile on a platform, interacting with some content. News_Transmission_Events are associated to exactly one Social_Media_User which is responsible for such transmission or interaction. Social_Media_User can be further characterised by some personal information properties such as hasAge, and by some features of its social media account, particularly hasPrivacy and hasStatus. Advertiser on the other hand, is a Social_Media_Agent that may advertise content. This may be

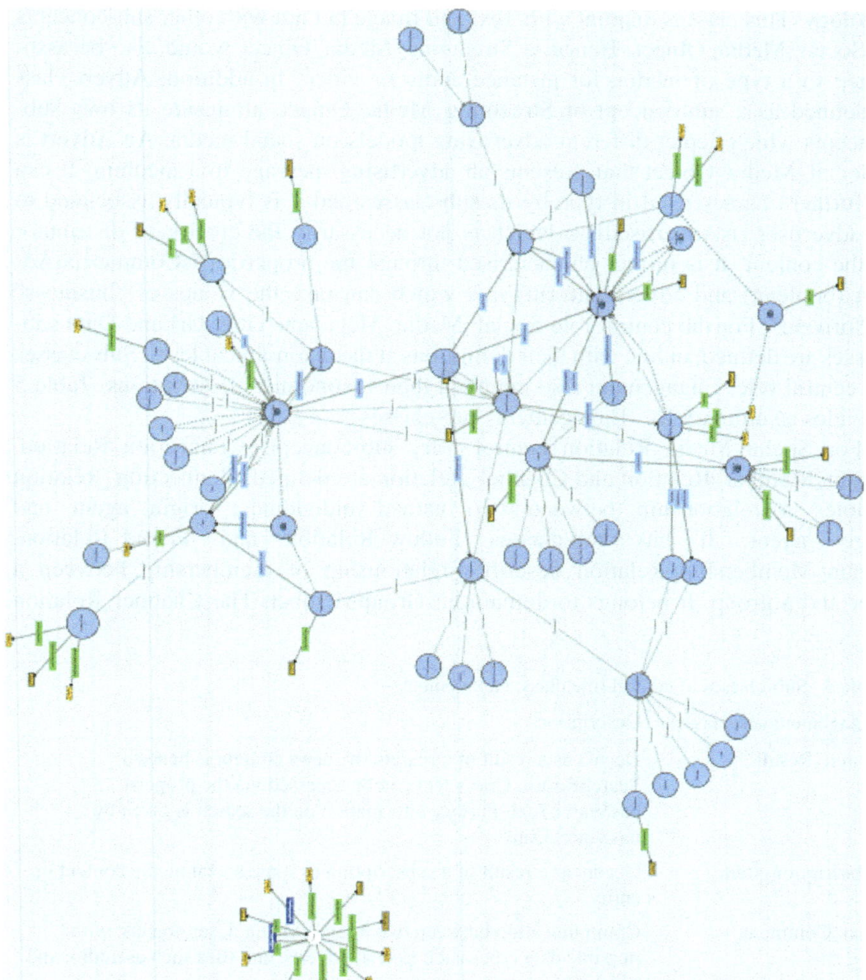

Fig. 4 Screenshot of the Fandet ontology

publicised by the advertiser or via any other Social_Media_User. Axiomatically, any Social_Media_Agent that advertises content is an Advertiser. Sub-properties for central role Social_Media_Object define diverse sub-classes for different types of content such as text, audio, video, images, animation and multimedia, all of which are direct representations of their descriptions at the taxonomical level in Sect. 3. In contrast however, a Streaming_Media_Object class is defined for the ontology. This class is disjoint with Text and Image but not with other sub-concepts of Social_Media_Object. Hence, a Streaming_Media_Object would also be associated to a type of media, for instance audio or video. In addition, Advert class is defined as a sub-concept of Streaming_Media_Object, alongside its own sub-concepts which depict different advertising models on social media. An Advert is a Social_Media_Object that presents an advertising message in a medium. It can be further characterised in type by its sub-classes and it is typically associated to an advertiser (isAdvertisedBy) which is not necessarily the creator or distributor of the content. It is further characterised through the properties isCommertialAdvert (boolean) and advertPlatformType, which can take the values of "business" or "private". For the central role Social_Media_Metadata; Title, Url and Date sub-classes are defined, in line with their definitions at the taxonomical level. Sub-classes for central role: Engagement also maintain their taxonomical descriptions. Table 5 provides a summary for 'Engagement' sub-classes.

For Social_Media_Relation central role, sub-concepts Connection_Relation, Group_Member_Relation and Channel_Relation are defined. Connection_Relation denotes a relationship between two named individuals, origin agent and target agent. It has sub-classes Follow_Relation and Friend_Relation. Group_Membership_Relation describes relationship of membership between a user and a group. It belongs to domain hasGroupMember. The Channel_Relation

Table 5 Sub-classes of central role class: engagement

Engagement sub-classes	Description
Search_Result	Occurs as a result of a search; the news content is hence a SearchResult. Query text can be accessed via the property hasSearchText. Further information on the search is given by hasSearchDate
Subscription_Item	Occurs as a result of a subscription to some social media content or entity
Chat_Comment	Communication between two Social_Media_User objects, with possible data types such as text, adverts, and files such as audio, and video
Notify	Engagement with News_Transmission_Event object via notification. Such as notifications on Facebook and for new videos on YouTube
Post_Engagement	High-level concept that includes all engagement instances occurring in the context of a post. Subclasses of Post_Engagement should be used to further specify nature of the engagement, as well as to link additional information such as linked users. It has sub-classes Comment, Like, Mentions, Repost, Review, Share and Tag

describes relationship of subscription between a user and a channel. A channel is an internal environment within a social media platform, providing content for users who are part of the channel, by means such as subscription. An example is YouTube channels. This class belongs to domain hasChannelAdmin, hasChannelDescription and hasChannelTitle. Lastly, the News_Transmission_Event central role defines a sub-class of Fake_News_Transimission_Event, which can be utilised for named individuals already detected as fake news. Figure 5 presents the hierarchical class structure for Fandet ontology.

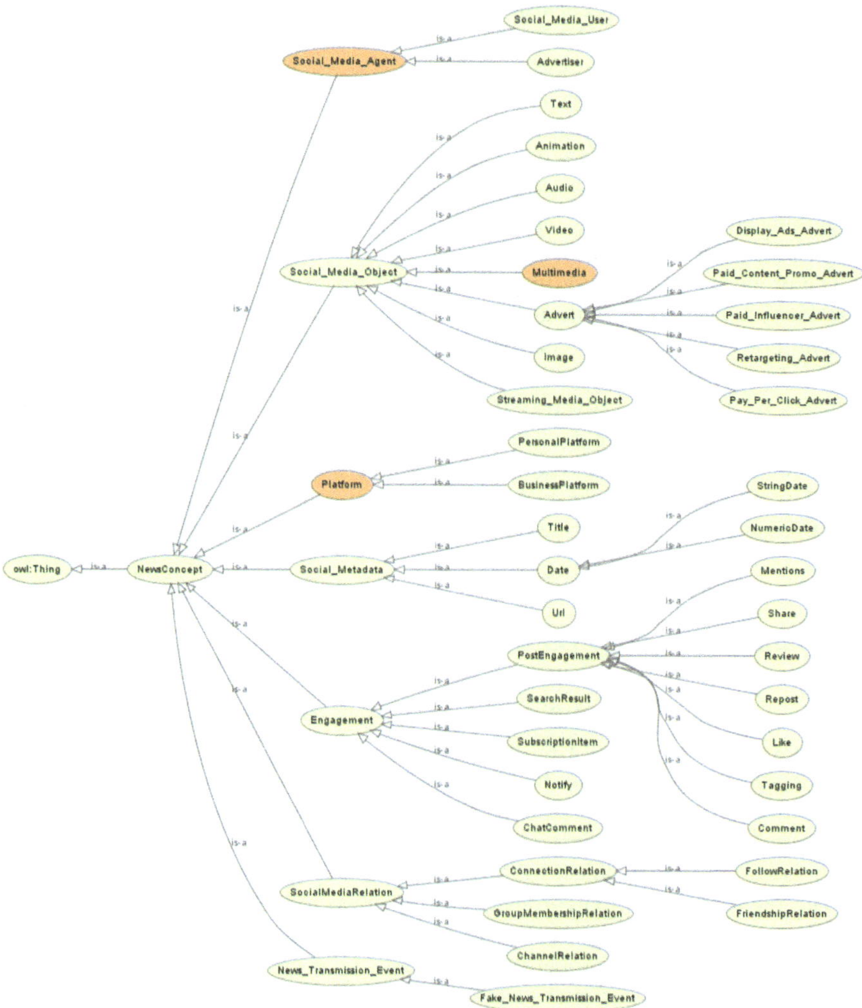

Fig. 5 Hierarchical class structure for Fandet ontology

4.2 Fandet Ontology Object Properties

Object properties within an OWL ontology refer to instances of the built-in OWL class. They relate objects to other objects. Object properties link individuals to individuals. The Fandet ontology comprises of 25 object properties, with 'topObjectProperty' at its base. A 'publishedOn' property belonging to domain 'Advert' links an advert to some platform where it is published. 'hasMention' property refers to social media mention for Social_Media_User object. For instance, if USER01 is 'part of' a News_Transmission_Event NTE01, with engagement of type 'Mention' MEN01 which consists of the user mentioning another user USER02, then USER02 'hasMention' MEN01. The 'hasMention' property is an inverse of 'mentionsUser', which refers to Social_Media_User object that a 'Mentions' engagement is directed towards. Both properties belong to domain 'Mentions' and has range 'Social_Media_User'. The 'hasPart' property is interpreted in the standard mere topological sense. Its sub-properties are as detailed in Table 6.

The 'isRelatedToUser' property is a high-level one that subsumes diverse properties linking any sort of NewsConcept to a Social_Media_User object. Table 7 defines its sub-properties.

The property 'isAdvertisedBy' refers to the advertiser of an advert. The advertiser does not need to be same as the user that created, published, or interacted with content. For instance, a user can interact with some content ADVERT01 posted by an influencer, that contains advertising of product of company ADVERTISER01. In this case, ADVERT01 isAdvertisedBy ADVERTISER01. It is the inverse of 'advertises'. Hence, a user can interact with some content ADVERT01 posted by an influencer, containing product advertising for company ADVERTISER01. In this case, ADVERTISER01 advertises ADVERT01. Figure 6 further presents a hierarchical object property structure for the ontology.

4.3 Fandet Ontology Data Properties

Data properties within an OWL ontology relate objects to data type values and link individuals to data values. The Fandet ontology comprises of 35 data properties, with 'topDataProperty' at its hierarchical base. A 'hasUrlProperty' is defined as a high-level property that subsumes asset of properties relating to the 'Url' class. These are 'hasTld' determining the top-level domain of a URL, 'urlAddress' as a string value for a news object full URL. Others are 'urlFormat', 'urlProtocol' and 'urlTld' representing format, protocol and type respectively as defined under 'URL' within the taxonomy in Fig. 2. The 'hasAgentProperty' is also a high-level property that subsumes the set of properties belonging to a social media agent. It has sub-properties; 'hasAge', 'hasEmail' and 'hasName' for age, email address and name of social media agent. It also has 'hasPrivacy' determining privacy of the social media user, with possible values "open" and "closed". Its fifth sub-property is 'hasStatus'

Table 6 The sub-properties for 'hasPart' property

Sub-property	Description	Domain
hasCircus	Links News_Transmission_Event to a Social_Media_Relation, that determines the sort of relationship between Social_Media_User and context where associated Social_Media_Object appears Because it is a sub-property of 'hasPart', the inverse relation can be accessed via 'partOf'. That is, if: NTE01 hasCircus CIR01 then: CIR01 partOf NTE01	News_Transmission_Event
hasContent	Links News_Transmission_Event to exactly one Social_Media_Object. Because it is a sub-property of 'hasPart', the inverse relation can be accessed via 'partOf'. That is, if: NTE01 hasContent CON01 then: CON01 partOf NTE01	News_Transmission_Event
hasDateMetadata	Links a SearchResult or a Social_Media_Object to a date metadata object. It has range: date with hasCreatedDate, hasPublishedDate and hasSearchDate sub-properties	Search_Result Social_Media_Object
hasEngagement	Links News_Transmission_Event to an engagement object, that determines the sort of engagement that Social_Media_User had with its associated Social_Media_Object. Because it is a sub-property of 'hasPart', the inverse relation can be accessed via 'partOf'. That is, if: NTE01 hasEngagement EN01 then: EN01 partOf NTE01	News_Transmission_Event
hasPublisher	Links News_Transmission_Event to one Social_Media_Agent, to determine agent for news transmission. Because it is a sub-property of 'hasPart', the inverse relation can be accessed via 'partOf'. That is, if: NTE01 hasPublisher PUB01 then: PUB01 partOf NTE01. It has range: Social_Media_Agent	News_Transmission_Event
hasSocialMetadata	Links any Social_Media_Object to some Social_Metadata. Because it is a sub-property of 'hasPart', the inverse relation can be accessed via 'partOf'. That is, if: SMO01 hasMetadata MET01 then: MET01 partOf SMO01. It has sub-property: 'hasUrl'	Social_Media_Object

Table 7 The sub-properties of high-level property: isRelatedToUser

Sub-property	Description	Domain
hasRelationOriginUser	For Social_Media_User that originated Social_Media_Relation. For instance, if USER01 is 'partOf' a News_Transmission_Event NTE01, with a circus of any type of Social_Media_Relation SMR01 (for instance friendship), then SMR01 hasRelationOriginUser USER01	Social_Media_Relation
hasRelationTargetUser	Social_Media_User to which Social_Media_Relation applies. For instance, if USER01 is 'partOf' News_Transmission_Event NTE01, with a circus of any type of Social_Media_Relation SMR01 (for instance friendship), then SMR01 hasRelationTargetUser USER02 if USER02 is the friend of USER01 for the circus of the news. It has sub-properties: hasChannelAdmin and hasGroupMember	Social_Media_Relation
isRelatedToInfluencer	Links adverts of class Paid_Influencer_Advert to the Social_Media_Agent that is influencer of such an advert	Paid_Influencer_Advert
mentionsUser	Social_Media_User object that a 'Mentions' engagement is directed towards, belonging to domain 'Mentions' and has range 'Social_Media_User'	Mentions
tagsUser	Social_Media_User that a tagging engagement is directed towards. For instance, if TAG01 involves the action of tagging users USER02 and USER03, then: TAG01 tagsUser USER02 and TAG01 tagsUser USER03	Tagging

determining status of social media user, with possible values "non-verified" and "verified". The 'hasDateProperty' is another high-level property that subsumes the set of properties relating to dates. Its sub-properties are described in Table 8.

A 'uId' data property is also defined as a unique ID that can be assigned to any instance of the type NewsConcept. 'hasTitle' refers to title of an object. It is noteworthy to mention that this property does not refer to an object of the class 'Title', subclass of Social_Metadata, but rather contains a string value for a less expensive characterisation. It belongs to the domain 'Channel_Relation' or 'Platform' and has sub-properties; 'hasChannelTitle' and 'hasPlatformTitle'. Furthermore, the 'hasDescription' data property is description for any instance of type NewsConcept. This can be description for a channel (hasChannelDescription) or platform (hasPlatformDescription). The 'hasSocialMediaObjectProperty' is also high-level that subsumes set

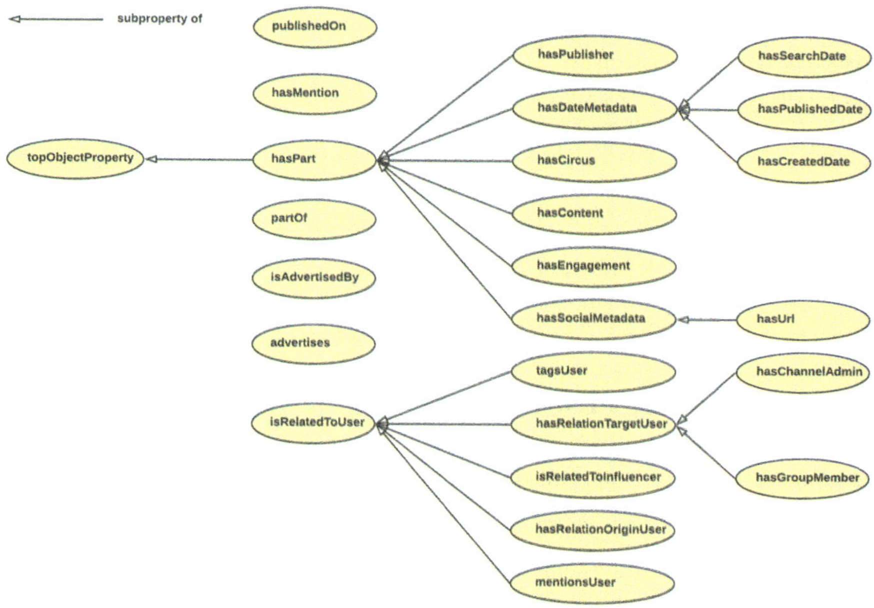

Fig. 6 Hierarchical object property structure for Fandet ontology

Table 8 The sub-properties for high-level property: hasDateProperty

Sub-property	Description	Domain
hasDate	Date associated to an object, with sub-properties; 'hasFormattedDate' (date in datetime format) and 'hasOriginalDate' (date in original format in which it was collected)	Date
hasDateOrderFormat	Format regarding order of elements day, month, and year. A common format is "dmy" and some other valid formats are {"dmy", "mdy", "ydm", "ymd"}	Numeric_Date
hasDateSeparator	Character used as a separator in a numerical date format. Two common values are "/" and "-", for example "01/09/2018" and "01–09–2018". It has a range of 'string'	Numeric_Date
hasYearFormat	Format of year in a numerical date, which can either be two or four digits	Numeric_Date

of properties belonging to social media objects. Its sub-properties are described in Table 9.

Other data properties are 'hasComment' for textual value of a comment engagement, 'hasSearchText' for searched string of a 'SearchResult' engagement and 'isAlphaNumeric' belonging to domain 'Title' for boolean property determining whether title is only composed of alphanumeric characters or not. Figure 7 presents a hierarchical structure for the ontology data properties.

Table 9 The sub-properties for high-level property: hasSocialMediaObjectProperty

Sub-property	Description	Domain
hasImageFormat	Attribute determining the format of an image, with possible values as: {"GIF", "JPEG", "PNG", "WEBP"}	Image
isCommercialAdvert	Boolean property determining if an advert is commercial or not	Advert
socialMediaContent	Content of social media object. It can either be file or text associated to the object. It has sub-property 'socialMediaTextContent' for social media content of text type	Social_Media_Object

5 Fandet Ontology Use Case

The Fandet ontology is specifically developed for semantically annotating social media news articles and news datasets, based on the social context and schema modelled within the ontology. A use case scenario, though minimal is described in this section to demonstrate some of the ontology's capability as well as evaluate it. For this purpose, a dummy news article on social media is conceptualised, based on excerpt from an online story telling service; 'dumburl.co.uk'. For the news article, users; USER001, USER002, USER003 and USER004 are defined as social media users. USER002 is further defined as an influencer, in terms of circulating the news article among other users. Content of the news article is defined as CONTENT001, with other formats such as IMAGE001 for image content, VIDEO001 for video content and TEXT001 for textual content. Furthermore, the social media internal environment, such as a group on FaceBook or channel on YouTube is defined as REL001 while tagging; a form of engagement is defined as ENGTAG001. The news article itself is defined as NEWS001. Considering an example of instantiation for a news event, NEWS001 where a social media user USER001 (publisher) promoted news content CONTENT001 by tagging other users ENGTAG001 (tagging engagement). The news content is itself found in a social media channel REL001 (channel circus). They are all subsumed by NewsConcept, so a uID and description can be given, as done for NEWS001. The semantic annotation is depicted in Fig. 8 based on turtle syntax.

The graphical visualisation for Fig. 8 is presented in Fig. 9.

The concept Social_Media_Object captures the actual content that make up a news article. Objects belonging to this class are of a type, which is encoded with a set of mutually exclusive sub-concepts: Image, Audio, Video, Animation and Multimedia. The latter is defined to have as parts at least two other types of Social_Media_Object; Live and Recorded. In addition, the class Streaming_Media_Object can be used in conjunction with certain other types. This is achieved by stating its disjointedness with non-compatible types. Another important subclass of Social_Media_Object is Advert. Adverts can be any type of content, and they advertise products of an agent. It is worth noting that such an agent is an Advertiser, in the case that this

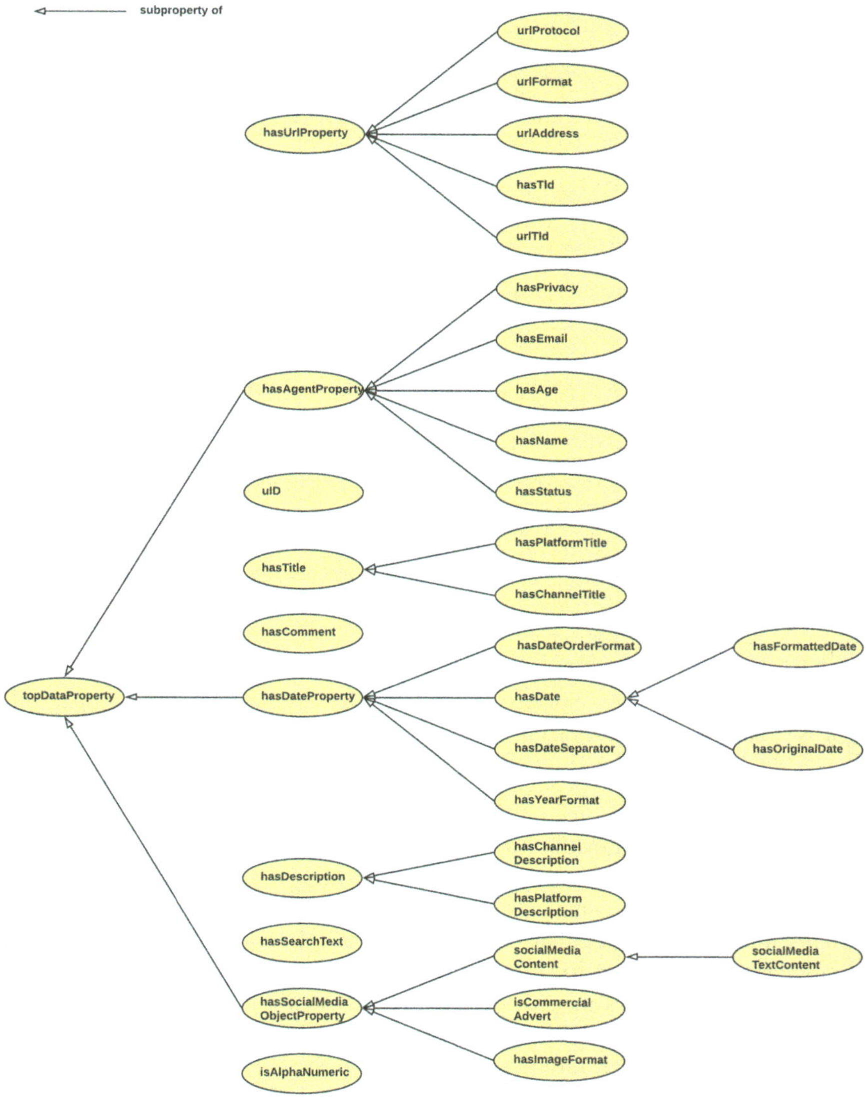

Fig. 7 Hierarchical data properties structure for Fandet ontology

was not previously known. Finally, social media objects have Social_Metadata, that can be linked with the appropriate properties. There are three types of metadata, namely Title, Date and Url. For additional non-explicit types, the general class must be used. Expanding on the use case, CONTENT001 has three parts, IMAGE001, TEXT001 and VIDEO002. TEXT001 contains a description of the news article in text format and VIDEO002 is a live video, implying that the influencer; USER002INF

```
NEWS001 rdf:type owl:NamedIndividual,
                 :News_Transmission_Event;
        :hasCircus :REL001;
        :hasContent :CONTENT001;
        :hasEngagement :ENGTAG001;
        :hasPublisher :USER001;
        :UID :000001;
        :hasDescription "Latest news about Galactic Silver Razor Wars".

USER001   rdf:type owl:NamedIndividual,
                   :Social_Media_User.

CONTENT001 rdf:type owl:NamedIndividual,
                    :Multimedia,
                    :Paid_Influencer_Advert;
           :advertises :USER001;
           :hasPart :IMAGE001,
                    :TEXT001,
                    :VIDEO002;
           :isRelatedToInfluencer :USER002INF;
           :hasTitle "Galactic Silver Razor Wars".

REL001    rdf:type owl:NamedIndividual,
                   :ChannelRelation;
          :hasChannelAdmin :USER002INF;
          :hasRelationOriginUser :USER001;
          :hasChannelTitle "Talk about the latest news with Influencer002".

ENGTAG001 rdf:type owl:NamedIndividual,
                   :Tagging;
          :tagsUser :USER003,
                    :USER004.
```

Fig. 8 Semantic annotation excerpt for use case (part 1)

is streaming the video content as it happens real-time. It can be inferred that the news article in this case is multimedia content, in addition to being an advertisement, as an instance of Paid_Influencer_Advert. The semantic annotation is presented in Fig. 10.

Social_Media_Agent is the class that represents any agent involved in the news transmission event. It can either be a Social_Media_User, an Advertiser, or both. Agents can be associated to different concepts withinin the ontology, fulfilling different roles. Some roles are general for the class Social_Media_Agent while others are specific to either Social_Media_User or Advertiser. From the existence of the role instantiation, one can infer the relevant subclass that an instance belongs to. Within the overall ontology, Fig. 11 presents graphical visualisation for classes and properties related to high-level class: Social_Media_Agent.

Based on these, and in continuation of the use case, USER001 is the social user that engages in the news transmission. Moreover, because the content is an Advert of the subclass Paid_Influencer_Advert, we can associate it to the influencer USER002INF that participates in it, via the isRelatedToInfluencer property. Furthermore, adverts can be linked to their Advertiser with the property ‘advertises’, which in this case turns out to be USER001 itself. Finally, the Tagging engagement can be linked to

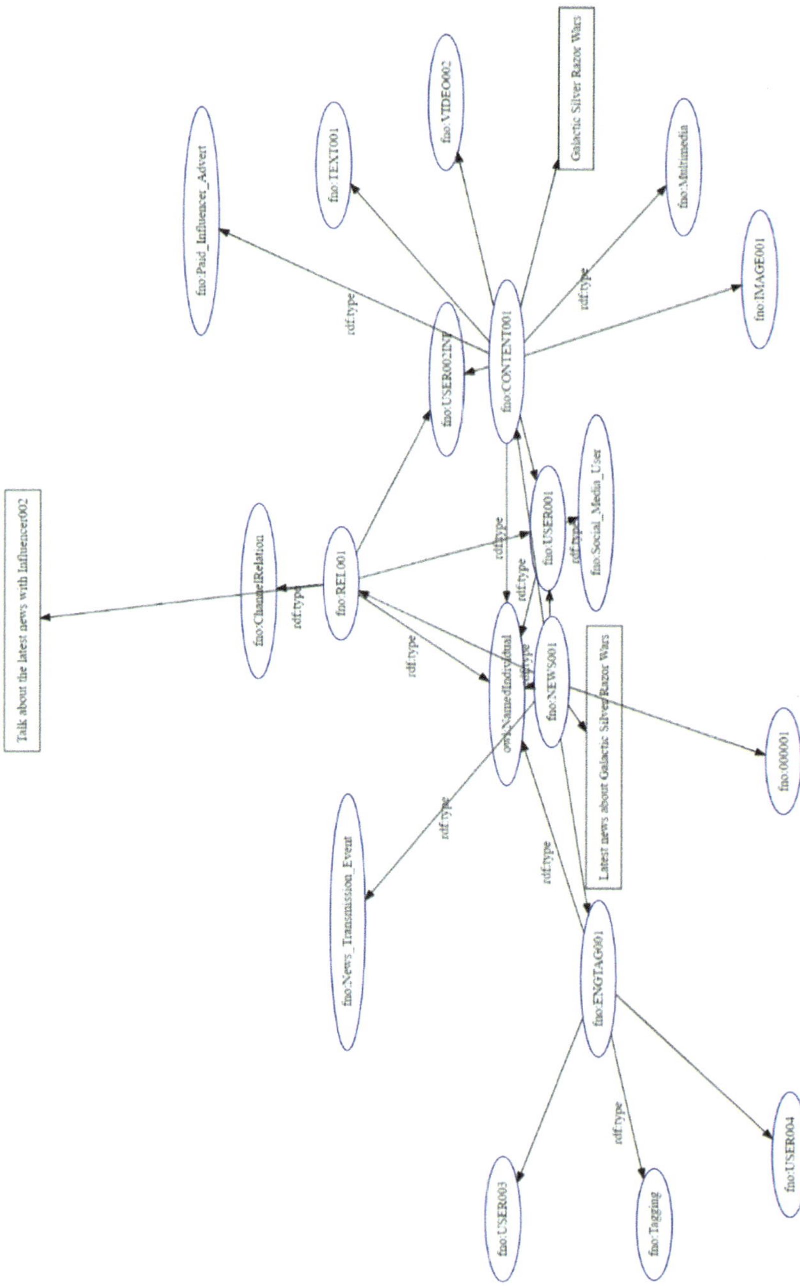

Fig. 9 Graphical visualisation for semantic annotation use case (part 1)

```
:CONTENT001 rdf:type owl:NamedIndividual,
                     :Multimedia,
                     :Paid_Influencer_Advert;
            :advertises :USER001;
            :hasPart :IMAGE001,
                     :TEXT001,
                     :VIDEO002;
            :isRelatedToInfluencer :USER002INF;
            :hasTitle "Galactic Silver Razor Wars".

:MET001     rdf:type owl:NamedIndividual,
                     :NumericDate;
            :partOf :CONTENT001;
            :hasDate "23-05-2018";
            :hasDateSeparator "-";
            :hasOriginalDate "23-05-2018"^^xsd:date.

:MET002     rdf:type owl:NamedIndividual,
                     :Title;
            :partOf :CONTENT001;
            :hasTitle "Galactic Silver Razor Wars";
            :isAlphanumeric "true"^^xsd:boolean.

:MET003     rdf:type owl:NamedIndividual,
                     :Url;
            :partOf :CONTENT001;
            :hasTld "co.uk";
            :urlAddress "https://www.dumburl.co.uk";
            :urlProtocol "https".

:IMAGE001   rdf:type owl:NamedIndividual.

:VIDEO002   rdf:type owl:NamedIndividual,
                     :Streaming_Media_Object,
                     :Video;
            :partOf :CONTENT001.

:TEXT001    rdf:type owl:NamedIndividual,
                     :Text;
            :hasDescription "A long, long time ago in a silver, silver galaxy.
            After leaving the idyllic planet Mooyani, a group of pixies fly
            toward a distant speck. The speck gradually resolves into a cosy,
            space mill. They encounter a tribe of elves".
```

Fig. 10 Semantic annotation excerpt for use case (part 2)

tagged users USER003 and USER004 with tagsUser property, as presented through the semantic annotation excerpt in Fig. 12.

Hence, overall the named individual depicted with this use case describes an agent (USER001) that has a social media profile and is also an advertiser. This agent advertises news article (NEWS001) through an influencer (USER002INF), which published the content on a channel (REL001). Subsequently, the agent has interacted with advert via social user profile and tagged more users, USER003 and USER004. Several other data captured in this instance include news article date, date format, URL (and data for its sub-classes) and news title, among others. From this use case scenario, the importance of context-based fake news detection approaches can be observed. This is based on the data modelling of social media context such as presented by the ontology and demonstrated by semantically annotating entities of

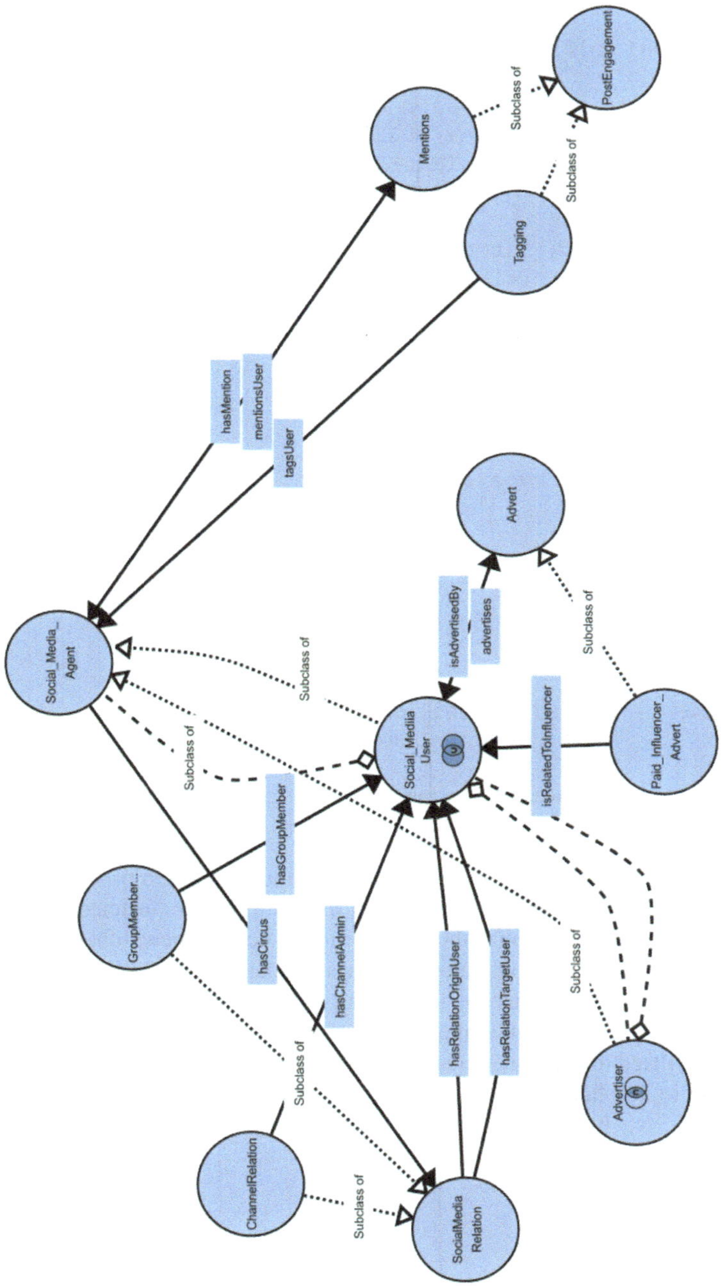

Fig. 11 Main classes and properties related to agents in Fandet ontology

```
:USER001     rdf:type owl:NamedIndividual,
                   :Social_Media_User.

:CONTENT001 rdf:type owl:NamedIndividual,
                      :Multimedia,
                      :Paid_Influencer_Advert;
            :advertises :USER001;
            :hasPart :IMAGE001,
                      :TEXT001,
                      :VIDEO002;
            :isRelatedToInfluencer :USER002INF;
            :hasTitle "Galactic Silver Razor Wars".

:ENGTAG001  rdf:type owl:NamedIndividual,
                      :Tagging;
            :tagsUser :USER003,
                       :USER004.

:USER002INF rdf:type owl:NamedIndividual,
                      :Social_Media_User.


:USER003     rdf:type owl:NamedIndividual,
                   :Social_Media_User.


:USER004     rdf:type owl:NamedIndividual,
                   :Social_Media_User.
```

Fig. 12 Semantic annotation excerpt for use case (part 3)

a sample news articles with the ontology. The semantic annotation provides a basis for identifying different related entities within the sample news articles as well as making them context aware. With the availability of several news datasets—both fake and real, the use case demonstrates that news articles from these datasets can be annotated with the ontology, thereby facilitating analysis of the datasets. By virtue of the analysis, features for both real and fake news articles on social media can be identified and utilised to categorise news articles when they appear on social media, with the ability to flag fake news articles and curb their spread.

6 Conclusion

The challenge of fake news detection on social media and limited exploitation of semantic technology solutions for context-based approaches to addressing the issue has been the focus for this research. The semantic technology suite is believed to possess huge capabilities for filling this gap and has been leveraged by this research. From investigation into news articles social context to identifying relevant entities with respect to the articles, a classification for the entities is developed as a taxonomy. Furthermore, classes are extracted from the taxonomy to create the Fandet OWL ontology, alongside their relations and axioms, with a use case study to wrap up the discussion. However, further research is suggested into analytical methodologies for news datasets; both fake and real that will be annotated by the ontology towards detection of individual fake news articles on social media. In addition, a hybrid approach to fake news detection, which integrates deep learning solutions is expected to produce significantly more accurate results in the classification of social media news articles as either fake or not.

References

1. Barbera, P., Tucker, J. A., Guess, A., Vaccari, C., Siegel, A., Sanovich, S., Stukal, D., & Nyhan, B. (2018). Social media, political polarization, and political disinformation: A review of the scientific literature.
2. Chen, T., Li, X., Yin, H., & Zhang, J. (2018, June). Call attention to rumors: Deep attention based recurrent neural networks for early rumor detection. In *Pacific-Asia Conference on Knowledge Discovery and Data Mining* (pp. 40–52). Springer.
3. Bode, L., & Vraga, E. K. (2015). In related news, that was wrong: The correction of misinformation through related stories functionality in social media. *Journal of Communication, 65*(4), 619–638.
4. Singhal, S., Shah, R. R., Chakraborty, T., Kumaraguru, P., & Satoh, S. I. (2019, September). SpotFake: A multi-modal framework for fake news detection. In *2019 IEEE Fifth International Conference on Multimedia Big Data (BigMM)* (pp. 39–47). IEEE.
5. Parikh, S. B., & Atrey, P. K. (2018, April). Media-rich fake news detection: A survey. In *2018 IEEE Conference on Multimedia Information Processing and Retrieval (MIPR)* (pp. 436–441). IEEE.
6. Bereta, K., Koubarakis, M., Pantazi, D. A., Stamoulis, G., Caumont, H., Daniels, U., Dirk, D., Ubels, S., Venus, V., & Wahyudi, F. (2019, January). Providing satellite data to mobile developers using semantic technologies and linked data. In *2019 IEEE 13th International Conference on Semantic Computing (ICSC)* (pp. 348–351). IEEE.
7. Lanza, J., Sánchez, L., Gómez, D., Santana, J. R., & Sotres, P. (2019). A semantic-enabled platform for realizing an interoperable web of things. *Sensors, 19*(4), 869.
8. Bennato, D. (2017). The shift from public science communication to public relations. The Vaxxed case. *Journal of Science Communication, 16*(2), C02.
9. Shao, C., Ciampaglia, G. L., Flammini, A., & Menczer, F. (2016, April). Hoaxy: A platform for tracking online misinformation. In *Proceedings of the 25th International Conference Companion on World Wide Web* (pp. 745–750).
10. Gordon, R. (2020). Better fact-checking for fake news. [online] MIT News. Available at http://news.mit.edu/2019/better-fact-checking-fake-news-1017. Accessed 16 January 2020.
11. Bansal, S. K., & Kagemann, S. (2015). Integrating big data: A semantic extract-transform-load framework. *Computer, 48*(3), 42–50.
12. Cambria, E., Howard, N., Xia, Y., & Chua, T. S. (2016). Computational intelligence for big social data analysis [guest editorial]. *IEEE Computational Intelligence Magazine, 11*(3), 8–9.
13. Horrocks, I., Giese, M., Kharlamov, E., & Waaler, A. (2016). Using semantic technology to tame the data variety challenge. *IEEE Internet Computing, 20*(6), 62–66.

14. Stahl, K. (2018). *Fake news detection in social media*. California State University Stanislaus.
15. Shu, K., Sliva, A., Wang, S., Tang, J., & Liu, H. (2017). Fake news detection on social media: A data mining perspective. *ACM SIGKDD Explorations Newsletter, 19*(1), 22–36.
16. Tacchini, E., Ballarin, G., Della Vedova, M. L., Moret, S., & de Alfaro, L. (2017). Some like it hoax: Automated fake news detection in social networks. arXiv preprint arXiv:1704.07506.
17. Ahmed, H., Traore, I., & Saad, S. (2017, October). Detection of online fake news using N-gram analysis and machine learning techniques. In *International Conference on Intelligent, Secure, and Dependable Systems in Distributed and Cloud Environments* (pp. 127–138). Springer.
18. Sherburne. (2018). "News", "News", "Fake News": Can machine learning help identify fake news on Facebook? Technology and Operations Management (2019). Available at https://digital.hbs.edu/platform-rctom/submission/news-news-fake-news-can-machine-learning-help-identify-fake-news-on-facebook/. Accessed: 29 December 2019.
19. Oshikawa, R., Qian, J., & Wang, W. Y. (2018). A survey on natural language processing for fake news detection. arXiv preprint arXiv:1811.00770.
20. Rashkin, H., Choi, E., Jang, J. Y., Volkova, S., & Choi, Y. (2017, September). Truth of varying shades: Analyzing language in fake news and political fact-checking. In *Proceedings of the 2017 Conference on Empirical Methods in Natural Language Processing* (pp. 2931–2937).
21. Ruchansky, N., Seo, S., & Liu, Y. (2017, November). CSI: A hybrid deep model for fake news detection. In *Proceedings of the 2017 ACM on Conference on Information and Knowledge Management* (pp. 797–806).
22. Perez-Rosas, V., Kleinberg, B., Lefevre, A., & Mihalcea, R. (2018). Automatic detection of fake news. In *Proceedings of the 27th International Conference on Computational Linguistics* (pp. 3391–3401). Association for Computational Linguistics.
23. Ismailova, L., Wolfengagen, V., Kosikov, S., Maslov, M., & Dohrn, J. (2020). Semantic models to indicate post-truth with fake news channels. *Procedia Computer Science, 169*, 297–303.
24. Levi, O., Hosseini, P., Diab, M., & Broniatowski, D. A. (2019). Identifying nuances in fake news vs. satire: Using semantic and linguistic cues. arXiv preprint arXiv:1910.01160.
25. Gomes Jr, L., & Frizzon, G. (2019, November). Fake bews and Brazilian politics–temporal investigation based on semantic annotations and graph analysis. In *Anais do XXXIV Simpósio Brasileiro de Banco de Dados* (pp. 169–174). SBC.
26. Braşoveanu, A. M., & Andonie, R. (2019, June). Semantic fake news detection: A machine learning perspective. In *International Work-Conference on Artificial Neural Networks* (pp. 656–667). Springer.
27. Wang, W. Y. (2017). "Liar, liar pants on fire": A new benchmark dataset for fake news detection. arXiv preprint arXiv:1705.00648.
28. Klyuev, V. (2018, August). Fake news filtering: Semantic approaches. In *2018 7th International Conference on Reliability, Infocom Technologies and Optimization (Trends and Future Directions) (ICRITO)* (pp. 9–15). IEEE.
29. Bharadwaj, P., & Shao, Z. (2019). Fake news detection with semantic features and text mining. *International Journal on Natural Language Computing (IJNLC), 8*.
30. Pan, J. Z., Pavlova, S., Li, C., Li, N., Li, Y., & Liu, J. (2018, October). Content based fake news detection using knowledge graphs. In *International Semantic Web Conference* (pp. 669–683). Springer.
31. Sabeeh, V., Zohdy, M., & Al Bashaireh, R. (2019, December). Enhancing the fake news detection by applying effective feature selection based on semantic sources. In *2019 International Conference on Computational Science and Computational Intelligence (CSCI)* (pp. 1365–1370). IEEE.
32. Pierri, F., & Ceri, S. (2019). False news on social media: A data-driven survey. *ACM SIGMOD Record, 48*(2), 18–27.
33. Baird, S., Sibley, D., & Pan, Y. (2017). Talos targets disinformation with fake news challenge victory. Fake News Challenge.
34. Hanselowski, A., Avinesh, P. V. S., Schiller, B., & Caspelherr, F. (2017). Description of the system developed by team athene in the FNC-1. Fake News Challenge.

35. Hanselowski, A., PVS, A., Schiller, B., Caspelherr, F., Chaudhuri, D., Meyer, C. M., & Gurevych, I. (2018). A retrospective analysis of the fake news challenge stance detection task. arXiv preprint arXiv:1806.05180.
36. Horne, B. D., & Adali, S. (2017, May). This just in: fake news packs a lot in title, uses simpler, repetitive content in text body, more similar to satire than real news. In *Eleventh International AAAI Conference on Web and Social Media.*
37. Volkova, S., Shaffer, K., Jang, J. Y., & Hodas, N. (2017, July). Separating facts from fiction: Linguistic models to classify suspicious and trusted news posts on twitter. In *Proceedings of the 55th Annual Meeting of the Association for Computational Linguistics* (Vol. 2: Short Papers, pp. 647–653).
38. Wang, Y., Ma, F., Jin, Z., Yuan, Y., Xun, G., Jha, K., Su, L., & Gao, J. (2018, July). Eann: Event adversarial neural networks for multi-modal fake news detection. In *Proceedings of the 24th ACM SIGKDD International Conference on Knowledge Discovery and Data Mining* (pp. 849–857).
39. Liu, Y., & Wu, Y. F. B. (2018, April). Early detection of fake news on social media through propagation path classification with recurrent and convolutional networks. In *Thirty-Second AAAI Conference on Artificial Intelligence.*
40. Padmanabhan, S., Maramreddy, P., & Cyriac, M. (2020). Spam detection in link shortening web services through social network data analysis. In *Data engineering and communication technology* (pp. 103–118). Springer.
41. Gruber, T. (2007). Ontology of folksonomy: A mash-up of apples and oranges. *International Journal on Semantic Web and Information Systems (IJSWIS), 3*(1), 1–11.
42. Munir, K., & Anjum, M. S. (2018). The use of ontologies for effective knowledge modelling and information retrieval. *Applied Computing and Informatics, 14*(2), 116–126.
43. Horsch, M. T., Chiacchiera, S., Seaton, M. A., Todorov, I. T., Šindelka, K., Lísal, M., Andreon, B., Kaiser, E. B., Mogni, G., Goldbeck, G., & Kunze, R. (2020). Ontologies for the virtual materials marketplace. *KI-Künstliche Intelligenz,* 1–6.
44. Kayalvizhi, R., Khattar, K., & Mishra, P. (2018). A survey on online click fraud execution and analysis. *International Journal of Applied Engineering Research, 13*(18), 13812–13816.

Fake News Detection Using Machine Learning and Natural Language Processing

Mansi Patel, Jeel Padiya, and Mangal Singh

Abstract In the present world, online news platform greatly influences our society and culture both in positive and negative ways. Being dependent on social media there is widespread fake news with misleading information leading to the chances where the reputation of the company is threatened. The influence of media has led to heights of depression and mental health issues as they don't find the real cause of it. This makes it an important issue that needs to be explored, analyzed and resolved to maintain peace and harmony in the world. Herein, this kind of pandemic as well where everything is unpredictable there are many cases where false news are been circulated and due to which the fear and panic have increased in the people. To resolve the issue, the chapter elaborates on developing a system using Machine Learning and Natural Language processing that uses RNN and its techniques like LSTM and Bi-LSTM for the detection of misleading information. The implementation is done for general fake news and purely Covid-19 fake news.

Keywords Fake news · Covid-19 · Machine learning · Natural language processing · LSTM · Bi-LSTM

1 Objective

Before we can assess the effect of false news, we need to be able to measure it. As, the misleading information is easy to spread due to the presence of various platforms like- social media, websites, newspapers. Moreover, because of various kinds of structure and presentation of the content, it's difficult to identify fake news. So, opting for machine learning using a sufficient amount of dataset can lead to a system that can easily detect false news and serve mankind in a way.

M. Singh (✉)
Department of Electronics and Telecommunication Engineering, Symbiosis Institute of Technology, Symbiosis International (Deemed University), Pune, Maharashtra, India

M. Patel · J. Padiya
Department of Electronics and Communication Engineering, Nirma University, Ahmedabad, Gujarat, India

M. Lahby et al. (eds.), *Combating Fake News with Computational Intelligence Techniques*, Studies in Computational Intelligence 1001,
https://doi.org/10.1007/978-3-030-90087-8_6

In the present technology crises, the idea is to contribute to society by developing a system that can aid in the detection of fake information. The objective is to use a stylistic-based approach using natural language processing for the text analysis to differentiate between real and fake news.

2 Introduction

The spreading of rumors and fake news in this social media addict society has opened up as a universal crisis with unpredicted socio-economic-mental challenges disturbing our lives. As news is our daily source of inspiration, its integrity should be maintained. As the propaganda of spreading fake news is been for centuries, but the only change today is the widespread of fake news is circulated via an internet communication channel. Justifying the need of having a system for detecting fake news and real news by statistical data like, by the year 2020 the number of fake news websites has grown up to 750 across 116 countries. Also, during this pandemic where peace is a vital part of every livelihood, the virus of fake news is spreading a lot and humiliating humans in various ways. Once during the pandemic in 2020 UNESCO said "Barely an area left untouched by disinformation".

While forwarding news through technology there are chances of people misleading the news and make viral which affects the integrity of news, as well as people, who may have problems due to this [1]. Also due to inefficiency of humans to differentiate between true and false news or facts poses fake news as harm to logical truth, which diminishes various areas of journalism, democracy and also the credibility of government institutions. This gambling of spreading fake news in recent times is done for attracting more readers, influencing opinions and also for generating revenue out from the click on websites. Also due to the spreading of fake news, there may be confusion or have glitches in decision making.

Considering the prevailing situation of widespread of a novel coronavirus from last 1.5 years, people are compelled to stay home. In this condition, the spread of misinformation has caused major damage to the society in form of fear making them mentally weak. So, to maintain integrity and peace in society a computer-aided detection has been a research area attracting interest in the last decade. Machine Learning algorithms are utilized greatly to provide a systematic approach for detection.

The chapter discusses the stylistic-based approach of linguistics for creating a particular type of text pattern, and let that be learnt by neural networks and based on training, we can detect the news whether it is fake or real. Here, the preprocessing of text and natural language processing algorithms like word embedding and RNN techniques like LSTM and Bi-LSTM are elaborated.

3 Previous Approaches

De Oliveira et al. [2] has explained the basic goal was to detect fake news ensemble techniques like LSTM and GRU also for easy user interface an android app was also developed for easy use and according to their survey nowadays people rely on social media for getting news so for better efficiency they have developed a hybrid model of LSTM and GRU and named it F-NAD. They have obtained data using web scraping and obtained a dataset of around 45,000 fake and real news and results are obtained in form of graphs.

Lahlou et al. [3] tried to develop an automated detection model using discourse segment structure analysis and here they have used a style-based approach, here discourse means there is a connection developed between surface features and document-level properties. And at last, by this model, the accuracy of around 74.6% accuracy is obtained having 0.76 F1 scores.

Barua et al. [4] used computational stylistic analysis based on NLP and they have used news from social media: Twitter. And the paper describes what all approaches have been done like automatic proof of facts known, the identity of the user, analysis of news spreading, etc. And the methodology used is reduction methodology with training, Matrix transformation methodology and Radial limit methodology. And the results obtained are of 86% accuracy and about 94% precision.

Sero et al. [5] implemented various NLP algorithms to get the most efficient way for detecting Fake news. Here the dataset has been taken from around 11,000 articles taken from signal media and opensource co. and here the NLP algorithm applied is TF-IDF and PCFG. And for classification purposes, SVM, gradient descent, bounded decision tree, random forest, etc. were used. And at last, a comparison was made among all using various results and graphs.

Hiramath et al. [6] introduced a quick and effective false news recognition model which can sort out whether the given suggestion is valid or not from an article by utilizing grammatical transformation in light of deep learning. The model is consisting of four layers and also using LSTM as a neural network to train the model and dataset from the parallel corpus and DeepMind is taken for evaluation. They evaluated the model by calculating the perplexity of the generated sentences.

4 Design Tools and Technologies

4.1 Machine Learning

Machine learning is a technology where a computer takes actions or thinks similar to humans. Here, for automation and fast operation as well as for smart solution machine learning is used in the world in various areas. So, ML is a type of algorithm which uses statistics to find out some particular pattern available in a large amount of data.

First of all, the data available is converted to digital form then some specific algorithms are used for detection, classification and various other applications. Machine learning is a subset of Artificial Intelligence [7]. Here, to imitate humans, machine learning uses neural networks to train the system and perform a particular function. There are various types of neural networks available and are being developed, which can be classified depending upon their structure, neurons used, data flow and their density and also on the used activation function. The basic main neural networks are—Artificial Neural Network (ANN), Convolutional Neural Network (CNN), and Recurrent Neural Network (RNN).

4.2 Deep Learning

Deep learning is a branch of machine learning that enables computers to replicate human behavior. Its architecture works by generating the right combination of neural networks and a huge collection of data for classification. The network comprises various deep that is to say hidden layers which contribute to achieving the required output. Deep learning often uses low, medium, or high-level highlights and classifiers in a multilayer approach. Each level of deep learning learns to turn the data it receives into a little more abstract and composite representation. Unsupervised learning problems can benefit from deep learning methods. This is a significant advantage because unlabeled data is available more than labeled data [8].

4.3 Recurrent Neural Network

A recurrent neural network, part of a deep learning network that uses time series of data or sequential data. The mechanism of RNN is quite different from CNN or feedforward network where RNN uses memory, where this is used to store the prior output of the system. They are influenced by the prior inputs for performing a particular function. A recurrent neural network shares the same weight parameter within each layer of the network. In traditional feedforward and CNN, there is mapping of one input to output, whereas in RNN it can vary in length of inputs and outputs. There are different types of RNN like one-to-one, one-to-many, many-to-one and many-to-many. Also, we can view RNN as a sequence of neural networks that allows us to train one after another with backpropagation [9]. Also, RNN needs to tackle two obstacles that of exploding gradients and vanishing gradients. RNN is a looping constraint on the hidden layers of ANN. While using a non-linear function, it can learn weights that map any input to the output. Discussing upon the applications where RNN is used are-

Time series data. This type of data has dependencies between observations in the form of time. Here the readings are noted at various particular instances of time. This data is used in applications like the stock market, financial analysis, etc.

Text Data. This type of data is in form of bytes to megabytes. In machine learning, this data is used by some preprocessing of data and used in prediction. In this project, text data in the form of a CSV file is used to extract the desired output.

Audio Data. This type of data has uncompressed monaural pulse code modulation data. This data is nowadays used in applications like music classification, voice recognition, generating and tagging data, etc.

The RNN algorithms used for detection purposes in the discussed chapter are

LSTM. One of the popular RNN architecture is LSTM, which is the long short-term memory that was introduced in 1997 by Schmidhuber and Hochreiter. This network is designed to avoid the problem of long-term dependency. This is because remembering a lot and that too from a long period was becoming quite tedious and prone to errors. As the recurrent neural network is a looping network and it has various repeating layers in it, LSTM looks like a chaining structure but all the modules have a different architecture in it. In this, there are four layers for interacting between them and produce the desired result. Here, Fig. 1 shows vector passing from input to output node. In LSTM the major part is of cell state as it is the straight line in the figure along which some linear functions are performed [10]. By using gates in LSTM, we can add or remove information from the cell. These gates are made up of a sigmoid function and a pointwise multiplication. This sigmoid function fluctuates between 0 and 1 to whether to take data or not. Firstly, that part is considered which needs to be thrown off from the cell state. This is done by the sigmoid layer and it is known as the "forget gate layer". The function implemented is shown in Eq. 1.

$$f_t = \sigma\left(W_f \cdot \left[h_{t-1}, x_t\right] + b_f\right) \tag{1}$$

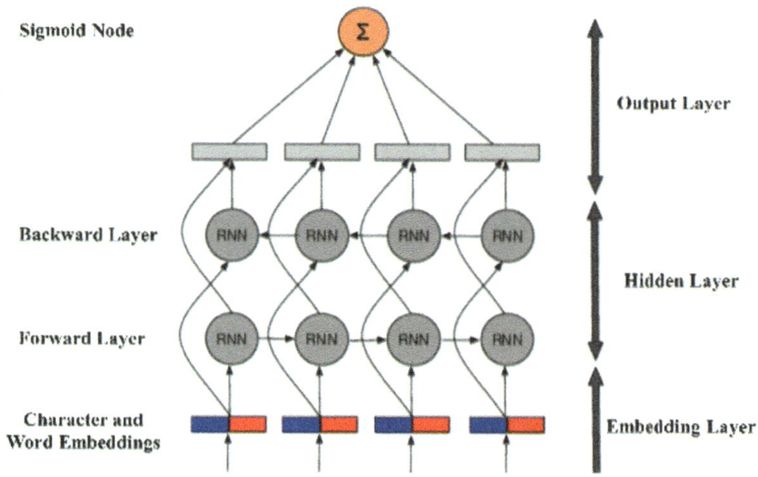

Fig. 1 Recurrent neural network

Further, the decision is made on whether new information needs to be added or not. Here, the sigmoid layer is known as the "input gate layer" and decides which data needs to be updated. Then tanh layer creates a vector for the cell state as shown in Eq. 2.

$$i_t = \sigma\left(W_i.\left[h_{t-1}, x_t\right] + b_i\right) \tag{2}$$

$$\widetilde{C}_t = \tanh\left(W_C \cdot \left[h_{t-1}, x_t\right] + b_C\right) \tag{3}$$

The cell state is upgraded to Ct after forgetting the old state and multiplying it with f_t, the Eq. 4 is obtained as-

$$C_t = f_t * C_{t-1} + i_t * \widetilde{C}_t \tag{4}$$

At last, the decision is to be made as output as depicted in Eqs. 5 and 6.

$$o_t = \sigma\left(W_o\left[h_{t-1}, x_t\right] + b_o\right) \tag{5}$$

$$h_t = o_t * \tanh(C_t) \tag{6}$$

There are other variations available in LSTM like- peephole connections, forgotten and input filters, and GRU. Mainly the application area of LSTM is Question–Answer, classification, Speech Recognition, Part of Speech tagging, etc (Fig. 2).

Bi-LSTM. One of the improved RNN architectures that are introduced after LSTM is Bi-LSTM, which is Bidirectional long short-term memory that was introduced in 1997 by Schuster and Paliwal. The objective for introducing it with bi-direction was to increase the amount of input data for the training of the system. To overcome the limitation of LSTM that is it works only for the forward direction that is it focuses on previous content while training, and does not consider the future or later for training [11]. So Bi-LSTM is a combination of two LSTM cells whose working is similar to LSTM but it works in both directions giving similar output with opposite directions (Fig. 3).

Here when the information processes from left to right then the hidden state of the cell shows the working as shown in Eq. 7.

$$\overrightarrow{h_t} = LSTM\left(x_t, \overrightarrow{h_{t-1}}\right) \tag{7}$$

whereas the information that is processed from right to left is considered as backward propagation its hidden state cell as shown in Eq. 8.

$$\overleftarrow{h_t} = LSTM\left(x_t, \overleftarrow{h_{t+1}}\right) \tag{8}$$

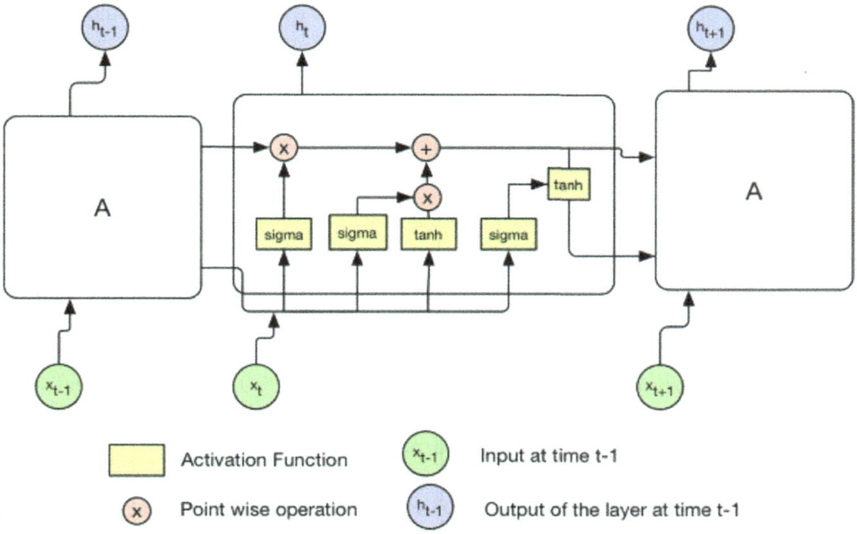

Fig. 2 LSTM

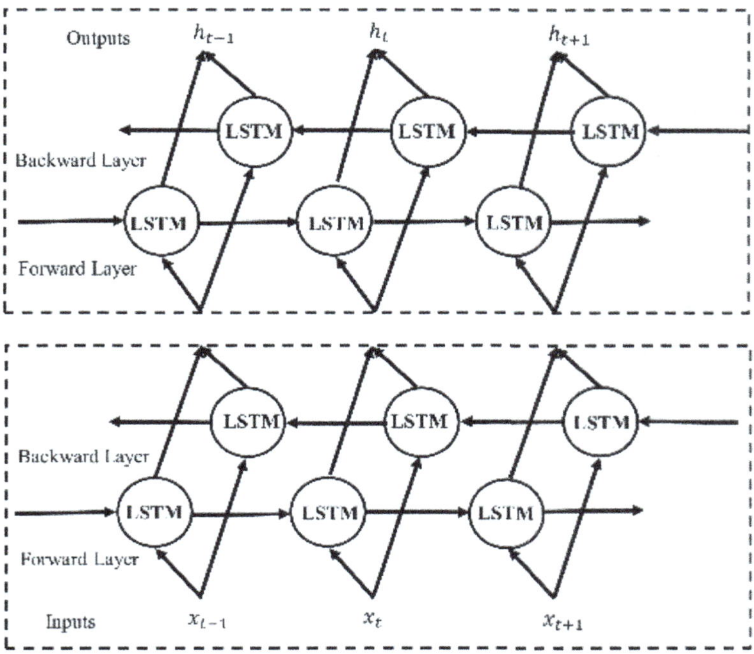

Fig. 3 Bi-LSTM algorithm

Finally, the output of Bi-LSTM can be summarized by concatenating and obtained as shown in Eq. 9.

$$h_t = \overrightarrow{h_t} \oplus \overleftarrow{h_t} \tag{9}$$

4.4 Natural Language Processing

Natural language processing is a field that combines the characteristics of both computer and linguistic to create a smart system. This system will be able to understand, analyze and extract meaning from text. Here the computer learns the human language by understanding and analyzing the meaning of language, its structure, semantics, syntax, morphology and pragmatics [12]. There are several tasks performed by the NLP algorithm like grammar and spell checking, text generation, topic modeling, named entity reorganization, text classification, machine translations, summarization, etc. Also, there are various benefits of using NLP algorithms like it is used where we need to perform large-scale analysis, automate processes in real-time, for industrial purposes, etc (Fig. 4).

The flow of NLP algorithm process as

Entering the text data. Here the dataset available in the form of CSV is being fed into the system.

Tokenization and Segmentation. It is used to break up a string of words into semantically useful units called tokens. And segmenting the words into components.

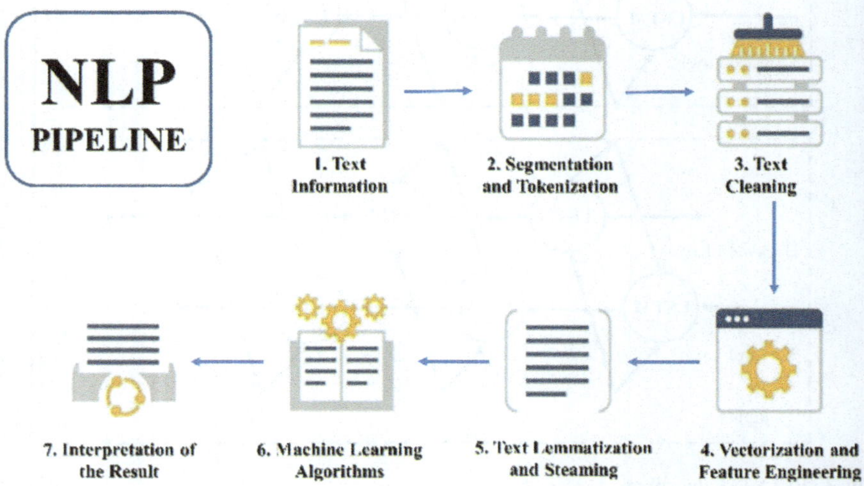

Fig. 4 Natural language processing

Cleaning of text. Here the terms important to the model are kept, others are discarded like full-stop, punctuation marks, prepositions, articles, etc.

Vectorization and feature engineering. Here the words are converted to numerical vectors and form matrices and perform the algorithm.

Lemmatization and stemming. Here lemmatization is the word dictionary-based and picks the suitable lemma dependent on setting, stemming works on single words without thinking about the unique situation.

ML algorithm. Here particular algorithm of a neural network like LSTM, Bi-LSTM and GRU has applied and also the classifier.

Results. Here the classified output is obtained and also the performance of the model is calculated (Fig. 5).

Also, as we are using text as our data for performing the NLP algorithm the model can be trained based on two criteria that are syntactic analysis also called style-based approach and another one is semantic analysis also called knowledge-based approach. Below is the explanation of the two processing task that is to break human language into machine-understandable chunks.

Style Based Approach. This approach is also called parsing, which finds the syntactic structure of a text and also depending on the words. In our project where we need to look upon the news, this method can assess news intention. This fake intention can be recognized by machines is by training like the false news prefers to be exaggerated and written in a special style to attract people to read. The characteristic taken into consideration for the detection of fake news are semantic, lexicon, discourse and syntax. In this, the work of lexicon is that it identifies the number of

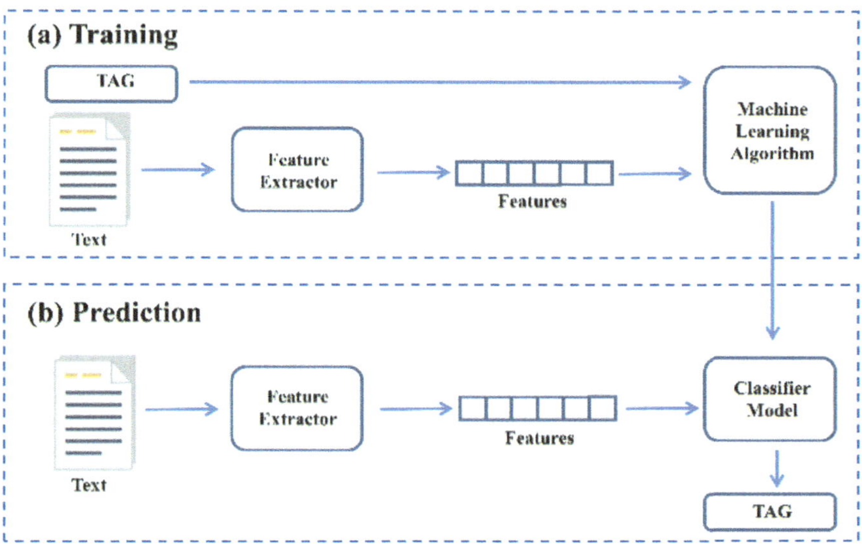

Fig. 5 Training and prediction of NLP

words used in that data or technical terms performs a task like Bag of Words [13]. Also, for examining the syntax the tasks like Parts of Speech and for deep syntax reference probabilistic context-free Grammar (PCFG) parse tree is used. The latest feature is extracted using tasks like word2vec and doc2vec.

Knowledge-Based Approach. This approach is based on the meaning of text available in the dataset. It analyzes the word interactions, structure of sentences and other related tasks are done. Also, this approach is more challenging as well as an automated model for performing knowledge-based concepts [13]. Also, it focuses on assessing the news by its authenticity by comparing the text with verified text content. But mainly we don't use this approach for detecting purposes as it is more complex and harder to train the model according to the knowledge or fact of the news or any other textual data.

Various NLP algorithms are

Edit Distance. In the text data available we need to look into how similar or different the strings are there. For comparison, metrics are taken for different words. This technique is used where it estimates the similarity of words value. It is done through how many operations are required for the conversion of one value to another. This algorithm is used in applications where we need to insert characters, detect or replace a character, or substitute characters in a string.

Cosine similarity. To find text-similarity in a document we can use the cosine similarity technique. The similarity measured here is represented in form of cosine or angle values. The formula for cosine similarity for the vector is shown in Eq. 10.

$$\text{Cos}\theta = \frac{A.B}{|A|.|B|} \tag{10}$$

Vectorization. It is the process of converting words into digits for extracting text features and further apply machine learning algorithms. Here words are converted to numerical vectors.

There are two approaches for vectorization that are

Bag of Words. Here each word is assigned with a particular index for developing a dictionary of words that has the same index values. And further, it calculates the no. of times it appears and stores that number with the appropriate index value (Fig. 6).

TF-IDF. This stands for term frequency and inverse document frequency. It defines the importance of words in the document. In TF means it depicts the frequency of the word in the text compared to the total words available in the text. Whereas IDF is where it signifies the importance of a particular word in the text and then logarithm of no. of texts upon no. of text having this term or word. This technique evaluates the stop words and important terms in the text data.

$$W_{x,y} = tf_{x,y} \times \log\frac{N}{df_x} \tag{11}$$

Part-of-Speech. Part-of-speech labeling (PoS) includes adding a part of speech classification to every token inside a content. Some regular PoS labels are action word,

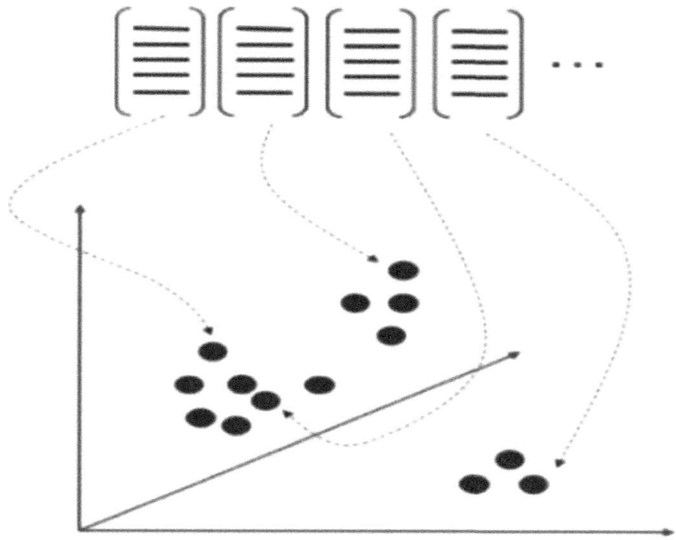

Fig. 6 Bag of words (BOW)

descriptor, thing, pronoun, combination, relational word, crossing point, among others. Generally, POS is used for understanding the meaning of terms in the text.

Name Entity Recognition. Named entity recognition is quite possibly the most well-known undertakings in semantic examination and includes separating elements from inside a container. Elements can be names, places, associations, email locations, etc. Relationship extraction, another sub-part of NLP, goes above and beyond and discovers connections between two things.

Text Normalization. This approach is usually used for pre-processing of text in the document to give a better result of the model used for machine learning-based prediction. In that the following things can be done in it like:

Context-independent normalization. Removal of non-alpha-numeric words is performed.

Canonicalization. Conversion of data in canonical and normal form.

Stemming. In preprocessing root word is extracted.

Lemmatization. It transforms word to its lemma, which is in dictionary form.

Naive Bayesian Analysis. The NBA is a grouping calculation that depends on the Bayesian Theorem, with the speculation on the component's autonomy. All in all, the NBA accepts the presence of any component in the class that doesn't associate with some other element. That is the reason such a methodology is classified as "Naive". The benefit of this classifier is the little information volume for model preparation, boundaries assessment, and classification. Generally, this algorithm is used for text classification where it uses the concept of maximum likelihood method to extract the parameter.

Word Embedding. Here in word embedding the similar words in the dataset are segmented using one identity for comparison purposes. For testing the NLP of machine learning for dealing with the addressed words. Word embeddings are indeed a class of methods where individual words are addressed as real esteemed vectors in a predefined vector space. Each word is planned to one vector and the vector values are learned in a manner that takes after a neural organization, and henceforth the procedure is regularly lumped into the field of machine learning. The key to the methodology is utilizing a dense disseminated portrayal for each word. Each word is addressed by a real vector, regularly tens or many measurements [14]. This is differentiated to the large numbers or a great many measurements needed for scanty word portrayals, like a one-hot encoding. There are various word embedding techniques available like:

Embedding layer. An embedding layer is a word inserting that is adapted for language demonstrating or archive grouping. It is necessary that recorded text be clean and arranged with the end goal that each word is one-hot encoded. The size of the vector space is indicated as a component of the model, like 50, 200, or 300 measurements. The vectors are instated with little irregular numbers [14]. It is fit in a supervised way utilizing the Backpropagation technique or algorithm.

Word2Vec. To calculate word embedding neural network is used based on words' context. Afterward, for the efficient performance of the model, the two approaches were introduced wherein if model does prediction based on context for the current word it is called Continous Bag of Words (CBOW), and if the prediction is based on surrounding words for the particular word it is called Continuous Skip-Gram Model (Fig. 7).

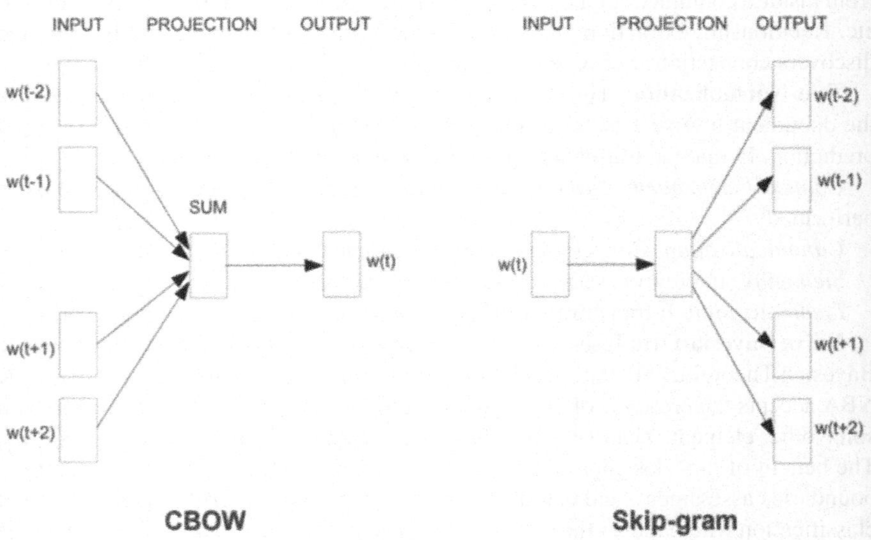

Fig. 7 CBOW and skip-gram model

GloVe. This version is the better version of word2vec. Using Latent Semantic Analysis for representation of words in vector space model and also it is not as efficient as word2vec. GloVe is a way to deal with match both the worldwide statistics of matrix factorization methods like LSA with the neighborhood setting-based learning in word2vec. Maybe than utilizing a window to characterize neighborhood setting, GloVe builds an express word-setting or for utilizing measurements across the entire content corpus. The outcome is a learning model that may bring about commonly better word embeddings.

5 Methodology

5.1 Web Scraping and Dataset Development

Dataset refers to the large collection of data in form of images, text, audio, etc. In this project for developing a model for fake news detection, the model is implemented for general news based on the articles collected from the US and it is accessed from Kaggle. While the detection of false news based upon COVID-19, has been developed using the technology of web scraping and extension-based scraping is done. The open web scraper used in this project is for scraping purposes and the website used is Poynter. Where we can access false or misleading news and a dataset is developed and stored as a CSV file and the same is used for experimental purposes and further the results are obtained.

Web Scrapping. It is perhaps the most established strategy for separating Web contents, is as yet in a position to offer substantial and important assistance to a wide scope of bioinformatics applications, going from straightforward extraction robots to online meta-workers. Also, it is a method used to acquire information from websites whereby the information is extracted and recorded in a PC or to a data set in a table or spreadsheet design. Web scraping is basically like it reduces the manual work for replication, then by using this software or program we can easily do in a small amount of time. The program here does the work like loading numerous amount of pages and scrape them and gives the information [15]. It is either exceptionally worked for a particular website or is one that can be designed to work with any website. With the snap of a catch, we can undoubtedly save the information accessible on the website to a document on a PC. Also for doing scraping, there are various ways like using various software available like visual web Ripper, WebHarvy, etc., also it can be done by providing an extension to the web browser and also by performing programming we can do so (Fig. 8).

Dataset. The model uses the dataset collected from the repository Kaggle along with some custom-generated data. The data used for the training of the US-based general fake news is specifically collected from Kaggle while the COVID-based fake news data is gathered using the available news. This data is obtained from the Poynter website and using the web scraping technique, it is converted into a database [16].

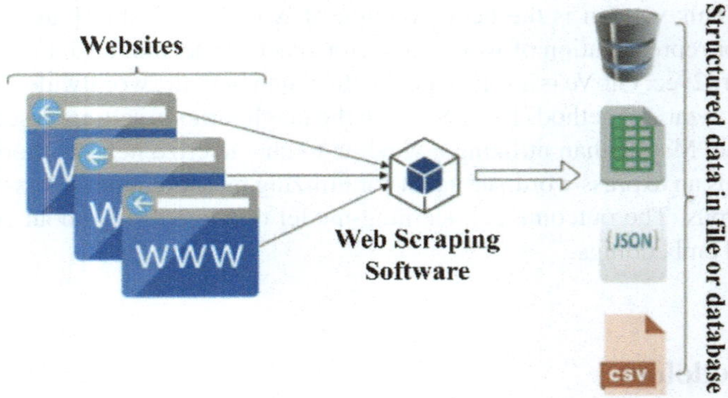

Fig. 8 Web scrapping

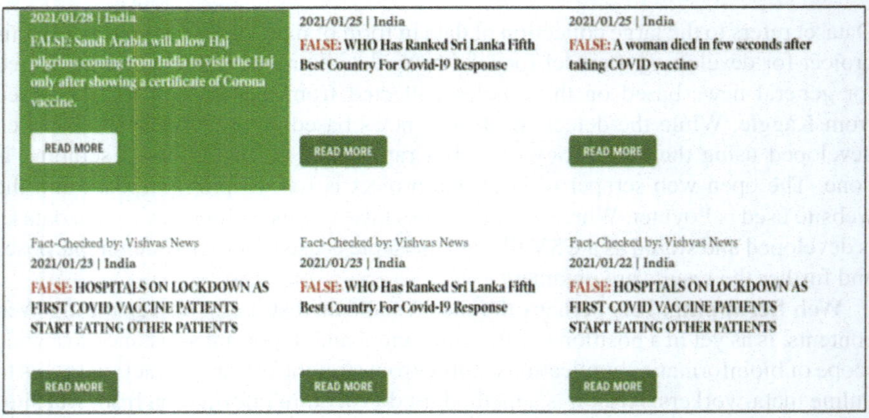

Fig. 9 Poynter—dataset generation

The model is trained on a set containing 31,461 various articles of US news. While the COVID-based dataset consists of 2834 different news from various countries (Fig. 9).

5.2 *Proposed Work*

Initially, by using MATLAB's Text Analytics Toolbox the loading of the dataset obtained from Kaggle and others from the Poynter website using web scraping is done. The dataset has 2 classes labeled as zero and one for fake and real news respectively. Herein, the data path is provided in the program to load the CSV file for analysis along with its labels. Then for analysis purposes, a plot for class distribution is plotted. Further, the dataset is distributed into two parts for carrying out training and validation. Herein, for this distribution in two parts, the hold out is kept as 20%.

Now extraction of text data is done and further labeling is carried from the portioned table of train and validate table. For training to visualize data, the word cloud is plotted. Now, the pre-processing of text data is done where we first need to create a function for pre-processing that tokenizes that data. In preprocess function, the things performed are tokenization, giving them a particular index, converting the text data to lower case and also erase the punctuation marks. As NLP is working on sequence data so text needs to be converted to sequence data which is done using word encoding function in MATLAB.

Further, there arises a need to truncate the documents in equal lengths. And training options give us access to put them in sequence automatically. According to the padding provided the graph of several documents vs length is plotted. For converting sequence to numeric indices, the function used is doc2sequence. Further converting the same for the validation documents. Now the deep neural network needs to be created that is either LSTM or Bi-LSTM network. Several layers are embedded in the developed network and the hyper-parameters are also defined for data processing. Based on the defined hyper-parameters, the training is done and respective accuracy is obtained in the graph of accuracy vs iterations and loss vs iterations which are shown in the result section. Now, for predicting the fake news an array is formed for which it has to undergo pre-processing that converts it into sequential order. By using the pre-defined classify function we can classify real (1) or fake (0) using trained LSTM or Bi-LSTM network.

6 Results and Discussion

The described algorithm has been used to create the system for creating a Fake News detection model based on the concepts of deep neural networks and natural language processing. The generated results were studied to understand the performance of the module. The results obtained were in the form of various plots and graphs. The graph of percentage accuracy versus iterations and that of percentage loss versus iterations was plotted. The plot describes the rate of change in the accuracy of the model concerning the number of iterations taking place in the system. Here, the number of iterations was decided based on the values of the maximum epoch and the batch size provided at the time of training. The batch size here means the parameter that decides the number of samples to work on, before the updating of the ongoing internal parameters of the model. At the time of completion of a single batch, the done predictions are compared with the output variables which are expected and finally the error calculations are done. While Epoch decides the number of times the learning algorithm that is dumped on the system will work through the complete training dataset. The plot below depicts the output at the end of the twofold validation done on the fed dataset. The plot is of the final fold which provides the improved accuracy of classification. Here the dataset is having fake news labeled as ‘0’ and real news labeled as '1'. Through the implemented algorithm in MATLAB the graph of class distribution is plotted for the US-based fake news dataset and COVID-19 based fake news dataset as shown in Figs. 10 and 11 .

In Figs. 12 and 13, the graph plotted which depicts the length of the document for both the dataset respectively is shown. The word cloud which provides the information about the important words extracted from the dataset required for the training purpose is as shown in Figs. 14 and 15 respectively.

The graph plotted in Figs. 16 and 17, shows the Accuracy versus Iteration and Loss versus Iteration plot generated through the training of the US-based general

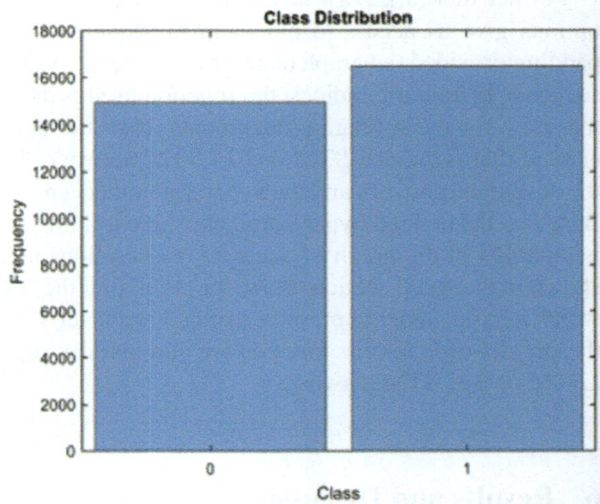

Fig. 10 Class distribution of US-based fake news

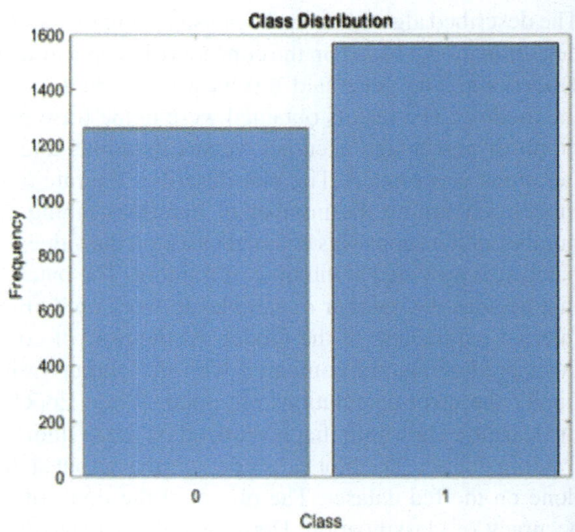

Fig. 11 Class distribution of COVID-19 based fake news

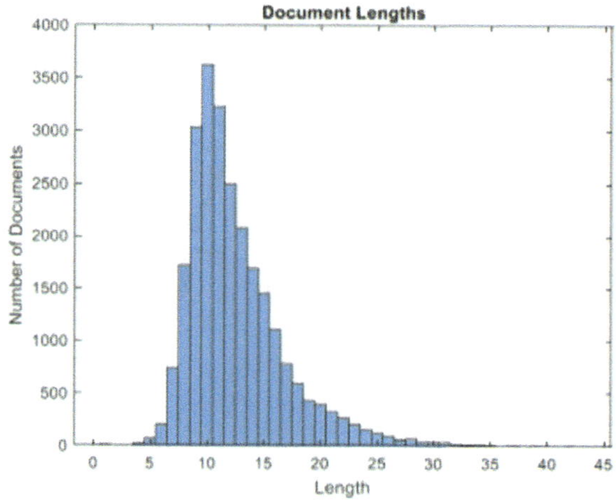

Fig. 12 Document length of US-based dataset

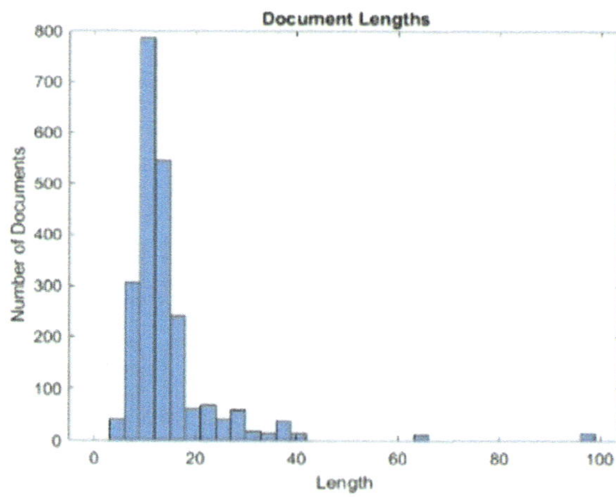

Fig. 13 Document length of COVID-19 based dataset

fake news and the application-specific scenario of COVID-19 respectively (Figs. 18 and 19).

Finally, for testing, 2 sets of news were feed to the model which were correctly detected. This was carried out for both the dataset US-based fake news and COVID-19 based fake news respectively as shown in Figs. 20 and 21.

For the analysis of RNN techniques, the achieved accuracy through the implemented algorithm on both the datasets is as shown in Table 1.

Fig. 14 Word cloud for US-based dataset

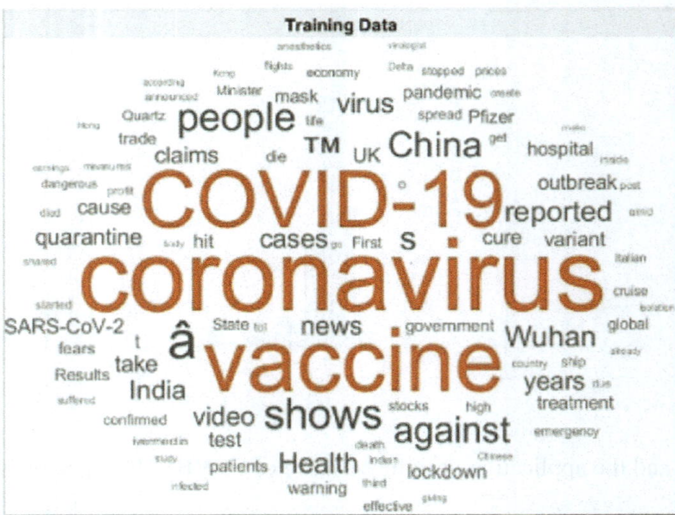

Fig. 15 Word cloud for COVID-19 dataset

7 Conclusion

It is evitable that the detection of fake news plays a vital role for society to reduce the panic and fear spreading through fake news. With the presence of flourishing technology, Artificial intelligence and deep learning have reached great limits. Using

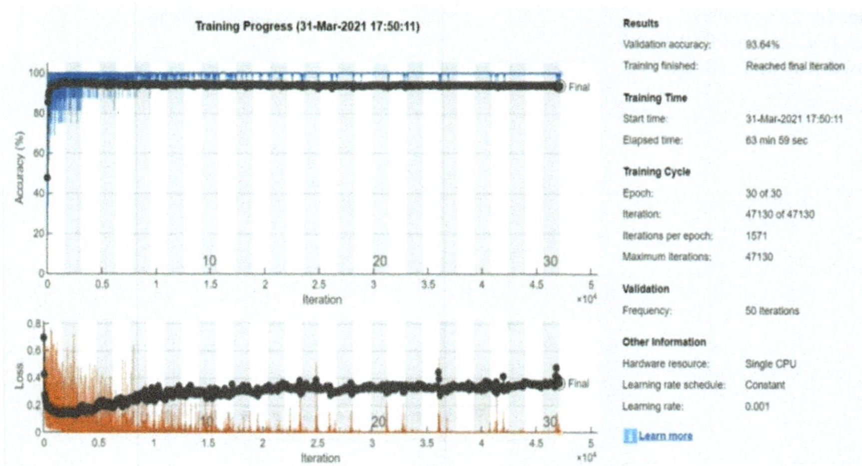

Fig. 16 Accuracy plot for US-based fake news (LSTM)

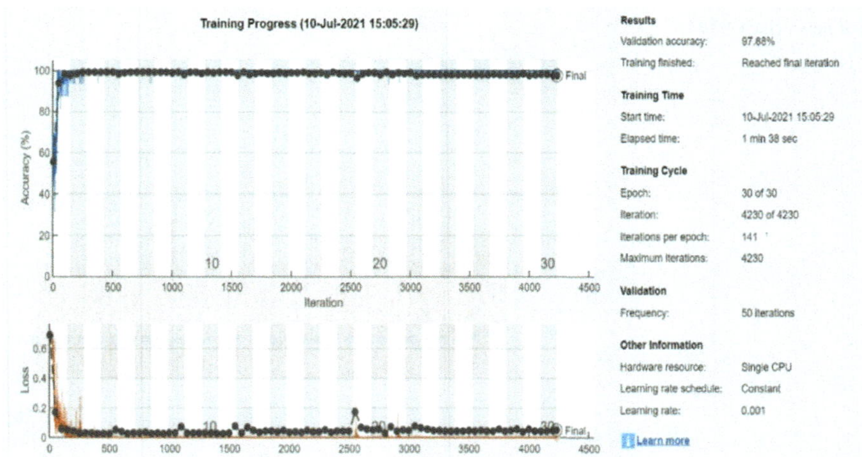

Fig. 17 Accuracy plot for COVID-19 based fake news (LSTM)

the Neural networks, the task of recognition of the text has been simplified to some extent. In this project, we have developed a system for fake news using the Deep neural networks supported by the Text analytics toolbox. The implementation has been carried out in MATLAB software. The concepts of LSTM and Bi-LSTM have been implemented for the development of the algorithm by using the pre-trained model made available. The dataset contains the images of two different classes Fake (0) and Real (1) which have been differentiated by training the model on these texts or news.

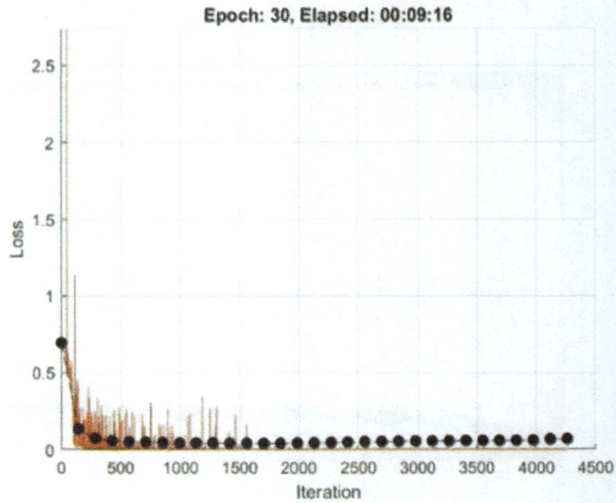

Fig. 18 Loss versus Iteration curve of COVID-19 based fake news (Bi-LSTM)

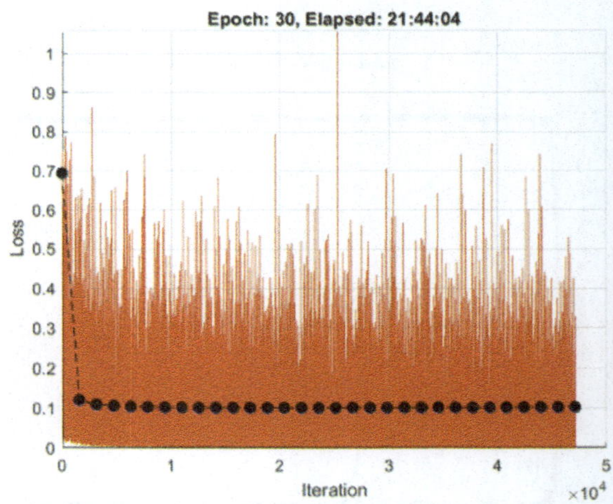

Fig. 19 Loss versus Iteration curve of US-based fake news (Bi-LSTM)

The chapter portrays a deep network model designed by stacking up the layers of network architecture. It is capable enough to train the data based on the user-defined parameters. Various techniques of data augmentation along with pre-processing methods have been implemented for achieving better results. The classification technique is also introduced at the end to classify and predict the class of random news provided to the system. Finally, by training and validating the accuracy of 93.64% for the US-based dataset through LSTM and for Covid-19 97.88% accuracy is obtained through LSTM. While by using Bi-LSTM the US-based fake news was detected with nearly 95.72% accuracy and COVID-19 based fake news gets detected with nearly 98.41% accuracy. Hence, it proves that Bi-LSTM is more efficient as it works in both

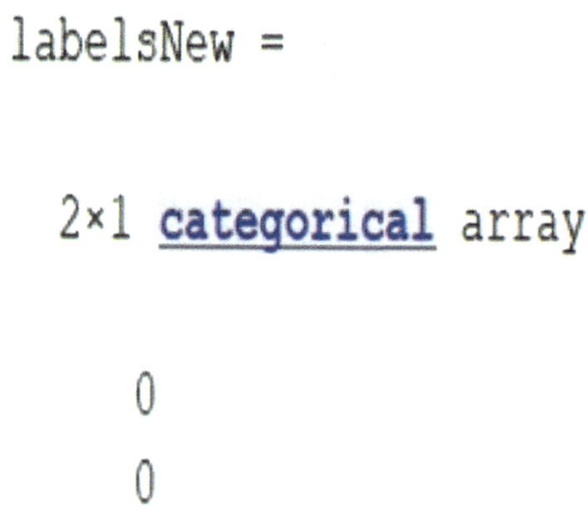

Fig. 20 Classified output of US-based dataset

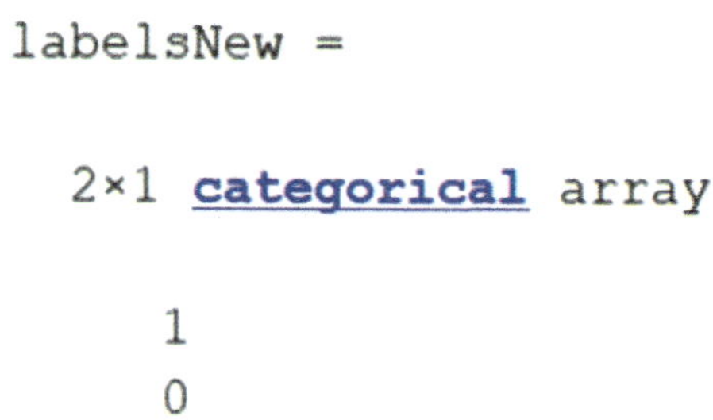

Fig. 21 Classified output for COVID-19 based dataset

Table 1 Accuracy table

RNN technique	Dataset	Accuracy (%)
LSTM	US based general Fake news	93.64
LSTM	COVID-19 based fake news	97.88
Bi-LSTM	US-based general fake news	95.72
Bi-LSTM	COVID-19 based fake news	98.41

forward and backward direction, unlike LSTM which function only in the forward direction.

The system developed has a great scope of improvement in the future using the various techniques for enhancing the obtained results. Also, integration of a larger dataset into the system can be fulfilled and accordingly great results can be obtained. Moreover, the idea of developing an Android-based application can be considered using the concepts of web development. This can provide services to the users on an on-demand basis.

References

1. Quandt, T., Frischlich, L., Boberg, S., & Schatto-Eckrodt, T. (2019). Fake news. *The International Encyclopedia of Journalism Studies*, 1–6.
2. De Oliveira, N. R., Medeiros, D. S., & Mattos, D. M. (2020). A sensitive stylistic approach to identify fake news on social networking. *IEEE Signal Processing Letters, 27*, 1250–1254.
3. Lahlou, Y., El Fkihi, S., & Faizi, R. (2019, October). Automatic detection of fake news on online platforms: A survey. In *2019 1st International Conference on Smart Systems and Data Science* (ICSSD) (pp. 1–4). IEEE.
4. Barua, R., Maity, R., Minj, D., Barua, T., & Layek, A. K. (2019, July). F-NAD: An application for fake news article detection using machine learning techniques. In *2019 IEEE Bombay Section Signature Conference* (IBSSC) (pp. 1–6). IEEE.
5. Seo, Y., & Jeong, C. S. (2018, November). FaGoN: fake news detection model using grammatic transformation on neural network. In *2018 Thirteenth International Conference on Knowledge, Information and Creativity Support Systems* (KICSS) (pp. 1–5). IEEE.
6. Hiramath, C. K., & Deshpande, G. C. (2019, July). Fake news detection using deep learning techniques. In *2019 1st International Conference on Advances in Information Technology* (ICAIT) (pp. 411–415). IEEE.
7. Jordan, M. I., & Mitchell, T. M. (2015). Machine learning: Trends, perspectives, and prospects. *Science, 349*(6245), 255–260.
8. He, K., Zhang, X., Ren, S., & Sun, J. (2016). Deep residual learning for image recognition. In *2016 IEEE Conference on Computer Vision and Pattern Recognition* (CVPR), Las Vegas, NV (pp. 770–778). https://doi.org/10.1109/CVPR.2016.90
9. Li, S., Li, W., Cook, C., Zhu, C., & Gao, Y. (2018). Independently recurrent neural network (indrnn): Building a longer and deeper rnn. In *Proceedings of the IEEE Conference on Computer Cision and Pattern Recognition* (pp. 5457–5466).
10. Understanding LSTM Networks (2021) Github.io. [Online]. Available https://colah.github.io/posts/2015-08-Understanding-LSTMs/. Accessed 22 June 2021
11. Sun, Q., Jankovic, M. V., Bally, L., & Mougiakakou, S. G. (2018, November). Predicting blood glucose with an lstm and bi-lstm based deep neural network. In *2018 14th Symposium on Neural Networks and Applications* (NEUREL) (pp. 1–5). IEEE.
12. Hassan, A., & Mahmood, A. (2018). Convolutional recurrent deep learning model for sentence classification. *IEEE Access, 6*, 13949–13957.
13. Zhou, X., & Zafarani, R. (2020). A survey of fake news: Fundamental theories, detection methods, and opportunities. *ACM Computing Surveys (CSUR), 53*(5), 1–40.
14. Ge, L., & Moh, T. S. (2017, December). Improving text classification with word embedding. In *2017 IEEE International Conference on Big Data* (Big Data) (pp. 1796–1805). IEEE.
15. Glez-Peña, D., Lourenço, A., López-Fernández, H., Reboiro-Jato, M., & Fdez-Riverola, F. (2014). Web scraping technologies in an API world. *Briefings in bioinformatics, 15*(5), 788–797.
16. CoronaVirusFacts alliance Poynter. (2020) Poynter.org, 18-Mar-2020. [Online]. Available https://www.poynter.org/coronavirusfactsalliance/. Accessed 08 July 2021

Fake News Detection Using Ensemble Learning and Machine Learning Algorithms

Sanaa Elyassami, Safa Alseiari, Maryam ALZaabi, Anwar Hashem, and Nouf Aljahoori

Abstract Digital news becomes widely accessible to a large community of users with the advancement of several channels of communication and the progression of technology and thus, contributes to the increase of spreading of fake news. The current study experiments and investigates machine learning models that classify news as either fake or real. Five classifiers were implemented using Random Forest, Support Vector Machine, Gradient Boosting, Logistic Regression, and Naïve Bayes algorithms. Models were trained using merged open-source datasets extracted from online sources covering different domains. Text lemmatization, vectorization, and tokenization were applied to extract useful information from news text and to improve the generalization capabilities and the performance of fake news classification models. The impact of the voting strategy on the performance of ensemble learning models were explored. The performance of the five classifiers was evaluated using the accuracy, the F1-Score, the recall, and the precision. The attained results are promising. The ensemble classifier trained using random forest algorithm and gradient boosting algorithm outperform the other classifiers and thus it might be used effectively against fake news spreading.

Keywords Ensemble learning · Feature extraction · Fake news · Text mining · Machine learning · Natural language processing

1 Introduction

With the large rate of the production of digital news, effectively detecting fake news has become a challenging task. The challenge is further compounded by the presence of malicious accounts for spreading propaganda, as well as the echo chamber effect. Accordingly, it is imperative to develop and deploy machine learning (ML) based models with the capability of detecting fake news and identifying the features that will be used in delineating fake news from legitimate ones. There have been several

S. Elyassami (✉) · S. Alseiari · M. ALZaabi · A. Hashem · N. Aljahoori
Abu Dhabi Polytechnic, Abu Dhabi, UAE
e-mail: sanaa.elyassami@adpoly.ac.ae

M. Lahby et al. (eds.), *Combating Fake News with Computational Intelligence Techniques*, Studies in Computational Intelligence 1001,
https://doi.org/10.1007/978-3-030-90087-8_7

studies in the detection of fake news using artificial intelligence techniques. However, many improvements should be incorporated to get accurate results.

The current study investigates different ML techniques and thus to identify the most suitable model that accurately classify the news into fake or real. ML techniques have been used to solve a wide range of real-world problems [1–4].

Vergeer [5] He studied the effect of the perceived credibility of online information on actual verification behavior by applying regression analysis on the Dutch Association of Journalists database. The analysis results demonstrate that journalists' verification behavior is not affected by journalism education.

According to Nagi [6], new online communication channels led to the increase of fake news spreading and the analysis of fake news related issues is widely linked to the availability of data collected from reliable organizations. Tips for analyzing news sources and tackle fake news menace were discussed and different surveys results were presented explaining the impact of fake news on the society.

Rodríguez [7] applied deep learning techniques to analyze and identify online fake news by studying the news text and implementing three different neural network architectures. The dataset used is composed of 20,015 labeled news gathered from three sources: "Getting Real About Fake" dataset, The New York Times, and The Washington Post. LSTM based model produced an accuracy of 0.91, CNN based model produced an accuracy of 0.93, while BERT based model produced an accuracy of 0.98.

Federico et al. [8] proposed a geometric deep learning-based model to learn fake news propagation patterns using convolutional neural networks and the scaled exponential linear unit as a non-linear activation function. The model analyzes 446, 284 tweets from 1, 129 URLs and attempted to predict the true/fake label and then associate that label to all the retweets by using a graph structure. the model was evaluated using the aggregated measure of accuracy (area under the ROC curve) and the model achieved a mean ROC AUC of 92.70%.

Lyu [9] reported that the difficulties come from the non-automated identification of fake news and the semantics of natural languages. Several machine learning tools have been used such as Support Vector Machine, doc2vec, FakeNew-sTracker, and decision trees to detect fake news. The results indicate that decision tree-based model and SVM-based model are more accurate than the others reaching an accuracy of 95%.

Ruchansky [10] proposed a hybrid deep model to detect fake news. This model is combining the behavior of users reading articles, articles, and people propagating fake news. Two real-world datasets were used; Weibo and Twitter and a tool called CSI was implemented to capture user's activity pattern in dealing with articles using a Recurrent Neural Network. Experiments have shown encouraging results and an acceptable accuracy of CSI in classifying fake news articles.

Abiodun [11] stated that with the coronavirus pandemic, most social media platforms were committed to promoting fact-based information on the pandemic. Google and Facebook implemented advanced algorithms to effectively enable users to surface information that they want to get. While Instagram introduced a new feature to detect fake news, apply a label and inform the user to prevent the spread of misinformation and automatically apply the label to articles with the same content. The author also gave insight on the tools that might be used to combat fake news like Spike, Hoaxy, etc. and reported that AI is widely used to identify word patterns and help in giving clues to fake news detection. AI it is used to get automatically the meaning of online articles using natural language processing techniques.

Nikhil [12] carried out binary classification models of online news articles using artificial intelligence techniques, natural language processing algorithms, and Machine Learning concepts. The goals of the study were to enable the user to identify if the news is fake or real and to evaluate the authenticity of the website that published that news. The used dataset consists of 12,836 human-labelled statements and was collected from the fact-checking website. Three algorithms were used to train the model. They are Logistic Regression, Random Forest, and Naïve Bayes. Performance was measured using precision, recall, F1-score, and accuracy. The best model used to classify articles that achieved the highest accuracy is the Logistic Regression classifier and it produced an accuracy of 65%.

In this work we will be investigating several machine learning and ensemble learning models to classify news in order to contrast results and examine the efficiency of each classifier. NLP techniques were applied to extract information from the news text. This chapter is organized as follows: In Sect. 2, we present an overview of the research materials and methods. Section 3 focuses on the implementation, results, and discussions. In Sect. 4, we provide the conclusion and threats to validity.

2 Materials and Methods

This section defines the datasets used to perform our empirical studies, followed by the description of the proposed process to build our intelligent fake news detection models. It also presents the set of implemented algorithms and the evaluation criteria adopted to measure the predictive power of the proposed machine learning-based models.

2.1 Datasets

The current study is based on three open-source datasets accessible online from Kaggle website [13]. The first dataset contains 44,919 news, where 23,502 news are fake and 21,417 are real [13]. Dataset 1 news are characterized by a title, a text, a subject and a date when the news was posted as shown in Table 1. The politics news constitutes 53% of the overall of Dataset 1 and 47% are world news.

The second dataset includes 20,791 news articles from several domains from Internet [14]. Dataset 2 is constituted of 10,378 reliable news and 10,413 unreliable news. Each news article is described by a unique id, a title, an author, and a text of the article as shown in Table 2.

Table 1 Extract from dataset 1

Title	Text	Subject	Date
As U.S. budget fight looms, Republicans flip their fiscal script	WASHINGTON (Reuters)—The head of a conservative Republican faction in the U.S. Congress, who voted…	Politics News	December 31, 2017
Nigeria says U.S. agrees delayed $593 million fighter plane sale	ABUJA (Reuters)—The United States has formally agreed to sell 12 Super Tucano A-29 planes and weap…	World news	December 27, 2017

Table 2 Extract from dataset 2

Id	Title	Author	Text
1	FLYNN: Hillary clinton, big woman on campus—Breitbart	Daniel J. Flynn	Ever get the feeling your life circles the roundabout rather than heads in a straight line toward th…
5	Jackie Mason: hollywood would love trump if he bombed North Korea over Lack of Trans Bathrooms (Excl…	Daniel Nussbaum	In these trying times, Jackie Mason is the Voice of Reason. [In this week's exclusive clip for Breit…

The third dataset comprises of 4,009 news articles extracted from trusted and untrusted online sources covering different domains such as sports, politics, entertainments and others [15]. The true articles were collected from New York Times, BBC, CNN, Reuters, while fake articles were collected from untrusted websites. Dataset 3 is constituted of 2137 of real news and 1872 of fake news (Table 3).

Table 3 Extract from dataset 3

URL	Headline	Body
http://www.bbc.com/news/world-us-canada-41419190	Four ways bob corker skewered donald trump	Image copyright getty images on sunday morning, donald trump went off on a Twitter tirade against a …
https://www.reuters.com/article/us-filmfestival-london-lastflagflying/linklaters-war-veteran-comedy-…	Linklater's war veteran comedy speaks to modern America, says star	LONDON (Reuters)—"Last Flag Flying", a comedy-drama about Vietnam war veterans, will resonate with…

2.2 Models Design

The training of ML classifiers to be able to detect fake news and accurately classify news into either fake or real is a crucial step in the current study. We have established and followed the process illustrated in Fig. 1 to build, train and evaluate our classifiers. The three datasets used in the current study were pre-processed to avoid causing any discrepancies to the classification process.

The process of cleaning data is required to remove the noises within the news text [16]. Discarding all non-English or unmeaningful words and letters in the news is an essential step to keep only useful text within the news. Replacing the slang words with their standard forms. And for the uniformity, texts were converted to lower case.

NLP techniques can be used for the analysis of news texts and the extract high-quality information in the text. Several NLP techniques were applied prior to train our models. The used techniques are lemmatization, tokenization, vectorization.

Text Lemmatization

Stemming and lemmatization are two NLP techniques that can be used to extract the root form of the infected words (words derived from other words). Lemmatization is different than stemming in the approach used to generate the root forms of words. The difference between these two approaches is that lemmatization produces actual language words whereas stemming might not produce actual words. Thus, lemmatization approach was used in the current study, as it reduces the inflected words properly ensuring that the lemma which is the root word belongs to the language. An example of Word lemmatization—is illustrated in Fig. 2.

Stop-words Removal

Stop words should be removed from the text before training machine learning-based classifiers since stop words have a very little meaning and are abundant in texts. To remove stop words from the news text, each sentence in the text is tokenized and divided into words/tokens. The algorithm iterates through all the tokens and check if the word exists in the list of stop words, so the word is removed and if not, the word is kept. Thus, all words such as ‘to’, ‘he’, ‘is’, ‘an’, and ‘the’, are removed from the text. An example of stop-words removal is illustrated in Fig. 3.

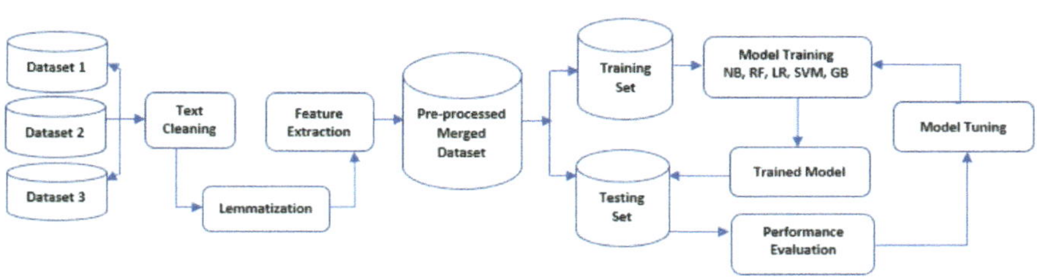

Fig. 1 Fake news detection proposed process

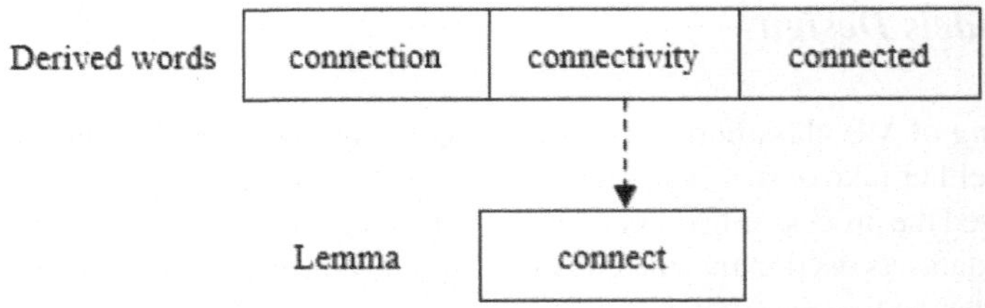

Fig. 2 Lemmatization of derived words

Text with stop words	He is applying the technique of bagging and boosting								
Tokens	He	is	applying	the	technique	of	bagging	and	boosting
Text without stop words	applying technique bagging boosting								

Fig. 3 Stop words removal

Feature Extraction

Feature extraction called also vectorization is the process of encoding words as numbers as textual data cannot be used to feed machine learning algorithms [17]. In order to build our classifiers, textual data should be turned into numerical feature vectors, thus all news text were converted to term/token counts following a strategy of three phase; tokenization, counting and normalization which is called the Bag of n-grams and used in the current study to extract numerical features from news text. Where tokenization is the process of identifying the words within a text and assigning an integer id to each word/token. Then counting the occurrences of tokens/words in each news text. Normalization and weighting by decreasing the importance of tokens that appear frequently within news text and thus each news is described by word occurrences as illustrated in Fig. 4 and tokenizes into a vector of term counts by finding the frequency of a word in an instance and creating a dictionary of term: count pairs that are then fed to our classifiers.

After cleaning, lemmatizing and vectorizing of the news text, we also performed data normalization since we are using different non-scale-invariant algorithms. Therefore, the data were scaled to be within the ranges of [0, 1]. The classifiers

Sentence = ['Ever', 'get', 'the', 'feeling', 'your', 'life', 'circles', 'the', 'roundabout']

	Ever	get	the	feeling	your	life	circles	roundabout
Sentence	1	1	2	1	1	1	1	1

Fig. 4 Text vectorizing

were trained using different algorithms and 70% of the dataset samples were used for training and the remaining samples were used for testing. Our machine learning-based models were trained several times and tuned by investigating different values for the hyperparameters to obtain optimized models and achieve ideal results. The developed classifiers were evaluated using different metrics to measure their performance.

2.3 Classification Algorithms

Different algorithms were investigated for fake news detection. The extracted features were fed into five different classifiers. We have used Random Forests, Support Vector Machine, Naïve Bayes, Logistic Regression, and Gradient Boosting algorithms to build our classifiers and investigate the use of ensemble learning and machine learning algorithms.

Random Forests (RF)

The random forest is a particularly effective algorithm in terms of predictions in the field of machine learning, deep learning and artificial intelligence. A random forest works on the principle of bagging where the first step is to split a dataset into subsets to produce a set of decision trees, after training the trees, the produced results are combined to obtain the most robust forecast. The bagging method is using several samples of data rather than just one sample and the outputs produced by the decision trees are ranked, and the highest value is selected as the final output [18]. There are two methods to determine the end result. A regression random forest consists in calculating the average of the forecasts obtained by taking into account all the predictions coming from the decision trees. A classification random forest is also based on the bagging technique, but the final estimate is made by choosing the most frequent response category rather than using all the obtained results.

The decision trees created by the random forest classifier were trained using the Gini impurity used to split branches and select the nodes that reduce the uncertainty in the decision trees, thus the best split is selected by minimizing the Gini impurity when splitting each node. The Gini impurity of a node represents the probability that a randomly selected sample in a node is incorrectly labeled according to the distribution of samples in the node. The Gini Impurity of a node n is defined as the given formula.

$$I_G(n) = 1 - \sum (p_i)^2 \tag{1}$$

where p_i is the probability of samples belonging to class i at a given node.

A low value of Gini Impurity means that nodes are pure and there is no chance that a sample randomly selected from that node would be misclassified.

Support Vector Machine (SVM)

Support Vector Machines [19] are a set of supervised learning techniques designed to solve classification and regression problems and are a generalization of linear classifiers. SVM were developed in the 1990s and they were quickly adopted for their ability to work with large data, the low number of hyperparameters, their theoretical guarantees, and their good results in practice. Unlike the other learning algorithms, SVM algorithm try to learn the most similar examples between classes to construct a set of support vectors and based on that, the SVM algorithm investigates the optimal hyperplane that splits the classes by calculating the best margin of the hyperplane.

SVM can be used to solve classification problems by deciding which class a sample belongs to, or regression problems by predicting the numerical value of a variable. The resolution of these two types of problems involves the construction of a function f which has an input vector X and matches an output Y.

$$Y = f(X) \tag{2}$$

Kernel functions are used by SVM algorithms. In our study, we have used linear kernel which is commonly recommended for text classification problems [20]. Linear kernel function is using fewer parameters and is faster than most of the other kernel functions like polynomial and radial functions. The decision boundary that the SVM returns is defined by the linear kernel function given in the below formula.

$$f(X) = w^T x + b \tag{3}$$

where w is the weight vector to minimize, x is the data to classify, and b is the estimated linear coefficient. The two parameters w and b are used to define the hyperplane.

Naïve Bayes (NB)

Naive Bayes is a type of simple probabilistic Bayesian classification based on Bayes' theorem given in the below formula. It implements a Naive Bayes classifier that belongs to the family of linear classifiers with a strong independence of assumptions. A NB classifier assumes that the existence of a characteristic for a class is independent of the existence of other characteristics [21]. Even if these characteristics are related.

$$P(A|B) = (P(B|A)P(A))/P(B) \tag{4}$$

The advantage of the NB classifier is that it requires relatively small training data to estimate the necessary parameters for the classification. The algorithm computes the Term-Document Matrix for each class (fake, real). This matrix includes a list of word frequencies existing in a set of documents. The entry (m, n) of the Term-Document Matrix consists of the frequency of the word "m" in the document "n". The frequency is calculated as the number of times each term/word exist within all documents.

In our study, we have used Multinomial Naïve Bayes which is commonly used for text classification problem and the data are represented as word vector counts [22].

Logistic Regression (LR)

Logistic regression [23] is a predictive analysis algorithm based on the concept of probability and used mainly for classification problems to assign samples to a discrete set of classes. Logistic regression is a linear algorithm with a non-linear transform on output, thus it assumes a linear relationship between the input variables and the target variable. The output is transformed using the logistic sigmoid as a cost function that return a probability value that ranges between 0 and 1. The hypothesis of logistic regression leans towards to minimize the cost function. Therefore, using linear functions as cost function is not suitable as it can produce a value greater than 1 which is not acceptable as per the hypothesis of logistic regression. The sigmoid function has been used to map each predicted value that might be any real value represented by x to its probability that should be between 0 and 1. The sigmoid function is defined as below.

$$f(x) = 1/\left(1 + e^{-x}\right) \quad (5)$$

To reduce the error in the probabilities predicted by the model and produce accurate predictions, the optimization of the cost function was performed using Gradient Descent [24]. The Beta coefficients for the logistic regression equation were estimated from the training data using maximum-likelihood estimation so that the model will be able to predict a value very close to 1 for the "fake news" class and a value very close to 0 for the "real news" class.

Gradient Boosting (GB)

Gradient Boosting [25] uses boosting technique which is the hypothesis that a weak learner can be improved to become better. It is an additive model that adds each time weak learners to optimize the loss function. The logarithmic loss function was used as it is most suitable for classification problems. For the weak learners, decision trees are used in gradient boosting where their outputs were added to correct the residuals in the predictions.

Gradient boosting operates by adding one sub-model at a time, while the existing trees in the model are not altered. The model is trained by incrementally enhancing every single tree. Samples are reweighted after each iteration based on their previous prediction. The weights are updated in a way to minimize the error. Higher weights are assigned to challenging instances and lower weights are assigned to samples properly handled and classified, thus sample that is difficult to classify receive increasing larger weights until finding a model that classifies all samples accurately and minimize loss when adding the weak learners. The output for each tree is incorporated to the output of the existing trees to improve the final model accuracy. The final model is a weighted sum of all generated sub-models.

2.4 Evaluation Metrics

In the current study, we have used balanced datasets where the number of samples in negative class is close to the number of samples in the positive class to avoid any overoptimistic or exaggerated results on the majority class. When experimenting our binary classifiers (fake news, real news), each record falls in one of the four following possibilities. True-Negative "TN" where the model correctly predicts the negative class and thus, real news are correctly identified as real. True-Positive "TP" where the model correctly predicts the positive class and thus, fake news are correctly identified as fake. False-Positive "FP" where the model incorrectly predicts the negative class and thus, real news are incorrectly identified as fake. False-Negative "FN" where the model incorrectly predicts the positive class and thus, fake news are incorrectly identified as real. The correct predictions include True-Positives and True-Negatives, whereas the False-Positives and False-Negatives are the incorrect predictions made by the classifiers.

To gauge the performance of our models, various evaluation metrics were used such as the accuracy [26], the F1-Score [27], the sensitivity, the precision, and the specificity [28].

The accuracy is defined as the ratio between the number of correctly classified samples and the overall number of samples.

$$Accuracy = (TP + TN)/(TP + FP + TN + FN) \tag{6}$$

The recall is called also sensitivity and it measures the proportion of actual positives that are correctly classified as positives.

$$Recall = TP/(TP + FN) \tag{7}$$

The precision called also positive predictive value is defined as the results classified as positive by the model.

$$Precision = TP/(TP + FP) \tag{8}$$

The F1-Score was also used to evaluate the model precision and recall rates collectively in order to provide better understanding of the mis-classified records.

$$F1 - Score = 2 \times (Precision \times Recall)/(Precision + Recall) \tag{9}$$

A high F-Measure score will reflect that the models have low False Negative and False Positive cases which in turn means the performance is good. The F1-Score is the balance between the precision and the recall and it highest value is 1 which indicate a perfect precision and recall.

3 Results and Discussion

The current study aims to investigate the use of different machine learning and ensemble learning algorithms in the prevention of fake news spreading specially in social media channels. Five classifiers were implemented using RapidMiner Studio Platform [29] to predict the class of the news and classify them as fake or real. The used classifiers are Random Forest, Gradient Boosting, Support Vector Machine, Logistic Regression, and Naïve Bayes. The classification results of the different classifiers evaluated using the accuracy, the F1-Score, the recall, and the precision were reported in Table 4. We have found that the Random Forest model produce the highest accuracy of 98.3% and the highest precision of 97.8% in classifying news. While for the F1-score and recall, Gradient Boosting achieved the highest scores of 97.7 and 98.7% respectively. Gradient Boosting has a very good balance between precision and recall while Random Forest has a very good classification power.

Ensemble learning is a set of machine learning techniques based on the paradigm that training multiple models and combining weak learners enable to obtain accurate and stable models. Random Forest and Gradient Boosting are both ensemble methods. Gradient boosting uses boosting technique and thus it builds trees one at a time, in a way that each tree benefit from the previously trained tree and correct its errors without any need to use of voting. However, Random Forest uses bagging strategy and thus it uses random sample of data and trains each tree separately and requires voting for model's aggregation.

When building the random forest classifier, each child decision tree produces a class and then their classification results are combined to predict the final classification and bootstraps the votes to obtain the better accuracy from the Random Forest classifier. There are mainly two voting techniques that might be used: confidence vote and the majority vote. The confidence vote selects the class that has the highest accumulated confidence. While the majority vote selects the class that was predicted by the majority of tree models. We investigated the impact of the selection of the voting strategy on the Random Forest classifier performance and we reported the results when applying confidence vote and majority vote in Table 5.

The most accurate and intelligent learning model is the one developed and trained with ensemble learning algorithm using the random forest with confidence vote as voting strategy. Incorporating text mining techniques allowed this model to improve

Table 4 Classification results

Classifier	Accuracy (%)	Precision (%)	F1-Score (%)	Recall (%)
Logistic regression	96.63	97.48%	97.34	96.43
Gradient boosting	97.65	96.42	**97.71**	**98.73**
Random forest	**98.34**	**97.85**	97.03	97.59
Support vector machine	95.23	95.32	93.57	93.44
Naïve Bayes	89.36	89.27	91.45	91.35

Table 5 Random Forest classification results based on the voting strategy

Voting Strategy	Accuracy (%)	Precision (%)	F1-Score (%)	Recall (%)
Confidence vote	98.38	97.49	97.47	98.89
Majority vote	97.62	96.74	97.43	97.52

the generalization capabilities and the performance of the fake news classifier. Therefore, this model has the potential to generate a knowledge-rich environment that can meaningfully help to minimize the spreading of fake news on social media platforms by accurately classify news. When comparing our model to Neural Networks-based models using deep learning, we found out that the result reported in [30], showed that the recurrent neural networks architecture has achieved an accuracy of 97.21% which is slightly less than our ensemble learning classifiers: Random Forest and Gradient Boosting that achieved an accuracy of 98.83 and 97.65% respectively.

4 Conclusion

Classifying news manually is a costly task that requires high expertise and in-depth knowledge of the domain to be able to catch anomalies within the news text. Incorporating intelligent models in the process of evading fake news spreading is crucial as intelligent models based on ML algorithms are so promising. In the current study, we investigated the use of machine learning and ensemble learning algorithms for fake news detection problem. The study objective is to identify patterns in the news text and differentiate between fake news and real ones.

Five classifiers were implemented using Support Vector Machine, Naïve Bayes, Logistic Regression, Random Forest, and Gradient Boosting. Naïve Bayes is based on the theory of naïve Bayes and works with probabilities to determine the classification. Logistic regression is based on the concept of probability and uses the sigmoid function to map each predicted value to its probability in the range 0 and 1. Random Forest on the other hand creates multiple decision tree classifiers and feeds them random subsets of the features and then combines their classifications to produce a final prediction. Gradient Boosting uses boosting technique and trains the model by incrementally enhancing every single tree and reweighting samples after each iteration based on previous predictions in a way to minimize the error and optimize the loss function. Support Vector Machine aims to spot hyperplanes into a high-dimensional space using the kernel function to find the optimal hyperplane for the sample classification.

To train our classifiers, we have merged three datasets into one larger dataset which includes news from different domains and covers large range of news. We applied several NLP techniques such as lemmatization, tokenization, and vectorization to preprocess the used datasets and extract high-quality information from the news text. The dataset was split into training set and testing set. The training set includes labeled samples while the testing set consists of unlabeled data. The result produced by the testing data represents the readiness of the classifiers to be used on actual real-world data. The learning models were trained several times and parameter-tuned to achieve high accuracy. The ensemble learning-based models that were built using random forest and gradient boosting have shown an overall better performance compared to individual learners that can be explained by the fact that ensemble learning algorithms incorporate techniques such as bagging and boosting to reduce the error rate. Our classifiers could produce different results if trained by other datasets. We admit that replications are needed to further enhance the external validity of our implemented models.

References

1. Elyassami, S., & Kaddour, A. (2021). Implementation of an incremental deep learning model for survival prediction of cardiovascular patients. *IAES International Journal of Artificial Intelligence. 10*(1), 101–109. ISSN 2252–8938
2. Elyassami, S., Hamid, Y., & Habuza, T.: Road crashes analysis and prediction using gradient boosted and random forest trees. In *2020 6th IEEE Congress on Information Science and Technology* (CiSt), Agadir—Essaouira, Morocco (pp. 520–525). https://doi.org/10.1109/CiSt49399.2021.9357298
3. Conroy, N. K., Rubin, V. L., & Chen, Y. (2015). Automatic deception detection: Methods for finding fake news. *Proceedings of the Association for Information Science and Technology, 52*(1), 1–4.
4. Pradhan, & Ajay, M. (2020). Fake news detection methods: Machine learning approach. *International Journal for Research in Applied Science and Engineering Technology, 8*(7), 971–975. https://doi.org/10.22214/ijraset.2020.29630
5. Maurice, V. (2018). Incorrect, fake, and false. journalists' perceived online source credibility and verification behavior. *Observatorio* (OBS*) *12.1* (2018): n. pag. Web.
6. Kuldeep, N. (2018). New social media and the impact of fake news on society. In *ICSSM Proceedings*, July (pp. 77–96).
7. Álvaro Ibrain, R., & Lloret Iglesias, L. (2019). Fake news detection using deep learning.
8. Federico, M et al. (2019). Fake news detection on social media using geometric deep learning.
9. Lyu, S., & Lo, D.C.-T. (2020). Fake news detection by decision tree. *SoutheastCon, 2020*, 1–2. https://doi.org/10.1109/SoutheastCon44009.2020.9249688
10. Natali, R et al. (2020). A hybrid deep model for fake news detection. *CSI, 4*(4). Accessed 27 Sept 2020.
11. Alao, A. (2020). How artificial intelligence tools are deployed in the fight against fake news. *The Nation 4*(4)
12. Nikhil, S. (2020). Fake news detection using machine learning. *International Journal Of Trend In Scientific Research And Development (IJTSRD),4*(4)
13. Kaggle. (2021). Fake news dataset 1. https://www.kaggle.com/clmentbisaillon/fake-and-real-news-dataset. Last Accessed 01 July 2021
14. Kaggle. (2021). Fake news dataset 2. https://www.kaggle.com/c/fake-news/data. Last Accessed 01 July 2021
15. Kaggle. (2021). Fake news dataset 3. https://www.kaggle.com/jruvika/fake-news-detection. Last Accessed 01 July 2021
16. Smelyakov, K., Karachevtsev, D., Kulemza, D., Samoilenko, Y., Patlan, O., & Chupryna, A. (2020). Effectiveness of preprocessing algorithms for natural language processing applications, In *2020 IEEE International Conference on Problems of Infocommunications. Science and Technology* (PIC S&T) (pp. 187–191). https://doi.org/10.1109/PICST51311.2020.9467919

17. Shah, F. P., & Patel, V. (2016) A review on feature selection and feature extraction for text classification. In *2016 International Conference on Wireless Communications, Signal Processing and Networking* (WiSPNET) (pp. 2264–2268). https://doi.org/10.1109/WiSPNET.2016.7566545
18. Shrivastava, P., & Shukla, M. (2015). Comparative analysis of bagging, stacking and random subspace algorithms. In *2015 International Conference on Green Computing and Internet of Things* (ICGCIoT) (pp. 511–516). https://doi.org/10.1109/ICGCIoT.2015.7380518
19. Kecman, V. (2005). *Support vector machines-an introduction in "Support vector machines: Theory and applications."* Springer.
20. Hofmann, T., Schölkopf, B., & Smola, A. J. (2008). Kernel methods in machine learning. *The Annals of Statistics, 36*(3), 1171–1220.
21. Saba Abdul-baqi, S. et al. (2018) A new model for iris classification based on naïve bayes grid parameters optimization. *International Journal of Sciences: Basic and Applied Research (IJSBAR) 40.2*, 150–155.
22. Singh, G., Kumar, B., Gaur, L., & Tyagi, A. Comparison between multinomial and bernoulli naïve bayes for text classification. In *2019 International Conference on Automation, Computational and Technology Management* (ICACTM) (pp. 593–596). https://doi.org/10.1109/ICACTM.2019.8776800.
23. Zhenhai, C., & Wei, L. (2012) Logistic regression model and its application. *Journal of Yanbian University(Natural Science Edition), 38*(01), 28–32.
24. Baldi, P. (1995). Gradient descent learning algorithm overview: A general dynamical systems perspective. *IEEE Transactions on Neural Networks, 6*(1), 182–195. https://doi.org/10.1109/72.363438
25. Friedman, J. H. (2001) Greedy function approximation: A gradient boosting machine. *Annual Statistics, 29* (5), 1189–1232.
26. Galdi, P., & Tagliaferri, R. (2018) Data mining: Accuracy and error measures for classification and prediction. *Encyclopedia of Bioinformatics and Computational Biology* 431–436
27. Powers, D. (2020). Evaluation: from precision, recall and Fmeasure to ROC, informedness, markedness and correlation. arXiv preprint arXiv:2010.16061
28. Lever, J., Krzywinski, M., & Altman, N. (2016). Classification evaluation. *Nature Methods, 13*, 603–604. https://doi.org/10.1038/nmeth.3945
29. Hofmann, M., & Klinkenberg, R. (2013) RapidMiner: Data mining use cases and business analytics applications.
30. Agarwal, A., Mittal, M., Pathak, A., et al. (2020). Fake news detection using a blend of neural networks: An application of deep learning. *SN Computer Science, 1*, 143.

Evaluation of Machine Learning Methods for Fake News Detection

Dimitrios Papakostas, George Stavropoulos, and Dimitrios Katsaros

Abstract In a cyber-connected world, fake information appears to be more enticing or interesting to the audience because of their limited attention spans and the plethora of content choices. Taking this into account, fake news detection/classification is definitely becoming of paramount importance in order to avoid the so-called reality vertigo, preclude misinformation and protect actual reality. This chapter presents a comprehensive performance evaluation of eight machine learning algorithms who perform fake news detection/classification based on regression, support vector machines, neural networks, decision trees and Bayes theorem. In every case, our study reaffirms that performance is governed by the nature of data, nevertheless, it sheds light and draws safe generic conclusions with respect to the dimensionality that each algorithm should have, the kind of training that should be performed beforehand for each one of them, and finally the method for generating vector representations of textual information.

Keywords Fake news · Misinformation · Reality vertigo · Machine learning · Algorithms

1 Introduction

Nowadays online information grows at unprecedented rates, and gradually more and more people consult online media, e.g., the Web, Online Social Networks (OSN) such as Facebook and Twitter, for satisfying their information needs. However, not

D. Papakostas (✉) · G. Stavropoulos · D. Katsaros
Department of Electrical and Computer Engineering, University of Thessaly, Volos, Greece
e-mail: papdimit@e-ce.uth.gr

G. Stavropoulos
e-mail: gstavropoulos@e-ce.uth.gr

D. Katsaros
e-mail: dkatsar@e-ce.uth.gr

M. Lahby et al. (eds.), *Combating Fake News with Computational Intelligence Techniques*, Studies in Computational Intelligence 1001,
https://doi.org/10.1007/978-3-030-90087-8_8

all information/knowledge producers are trustworthy, and the problem of fake news—fabricated stories presented as if they were originating from legitimate sources with an intention to deceive—and their spreading is getting more and more severe. It is speculated that in the current decade, people in developed countries will encounter more fake news than real news. This phenomenon is termed *reality vertigo*.[1]

This problem emerged as a major issue particularly during the 2016 US Presidential election, and it is even believed that fake news affected the final outcome. Unfortunately, this is not an isolated event; a study [1] shows that false medical information gets more views, likes, comments than true medical information. The COVID-19 pandemic is the most recent example of this; nearly 80% of consumers in the United States reported seeing misleading news about the coronavirus outbreak,[2] highlighting the extent of the issue and the reach fake news can achieve. Even worse, fake news not only is (more) popular, but spreads at a faster pace [2] than real news, too.

As a result, countermeasures against fake news began to emerge quickly. There are already fact-checking organizations, such as snopes.com, politifact.com, factcheck.com, truthorfiction.com. Efforts are taking place to deploy fact checking services in browsers; a notable effort which attracted media attention[3] is the development of a Google Chrome extension to combat fake news. Other anti-fake news techniques include adding publisher logos to the information items.

1.1 Motivations and Contributions

The need for detecting fake news—or classifying a news item as fake, true, or suspicious—is of paramount importance if we wish to avoid reality vertigo and protect our society, especially the less educated persons of our society. Even though manual or crowdsourced verification efforts could be a solid solution to the problem, scalability issues, due to the tremendous volume of items to be examined, would soon turn such efforts of limited applicability. Thus, algorithmic techniques are the only viable option for addressing the problem at its full scale.

Machine learning has been shown to be particularly effective in eliminating spam email, which is one type of disinformation. Consequently, algorithms in this category were among the first to be tested for efficacy. On the topic of detecting false news, the following machine learning paradigms have been investigated:

- Regression
  - L1 regularized logistic regression
- Support Vector Machines (SVM)

[1] https://www.nature.com/news/astronomers-explore-uses-for-ai-generated-images-1.21398.

[2] https://www.statista.com/statistics/1105067/coronavirus-fake-news-by-politics-us/.

[3] https://yaledailynews.com/blog/2018/01/22/yale-students-design-chrome-extension-to-combat-fake-news/.

  - C-support vector classification
- Bayesian methods
  - Gaussian naive Bayes
  - Multinomial naive Bayes
- Decision tree-based methods
  - Decision trees
  - Random forests
- Neural networks
  - Multi-layer perceptron (MLP)
  - Convolutional neural networks (CNNs)

The present chapter deals with the problem of detecting fake (or real) news from textual resources, and in particular it focuses on the exhaustive comparison of best performing algorithms from the most significant families of machine learning algorithms, i.e., those mentioned above, since their relative performance is unknown, and so is their generic behavior when tested against diverse datasets.

In that context, this chapter is going to answer these two broad questions, and make the corresponding contributions:

- It contrasts the effectiveness and efficiency of the competitors for several diverse datasets, and various performance measures.
- It contrasts the speed of the competitors for these datasets.

The rest of the chapter is organized as follows: Sect. 2 presents briefly the related work. Section 3 introduces the algorithms that will be evaluated. Section 4 describes the evaluation environment, i.e., competitors, datasets, performance measures, and on, and Sect. 5 presents the actual evaluation of the competing algorithms. Section 6 provides some future research directions, and finally Sect. 7 concludes the chapter.

2 Related Work

Machine learning and data mining algorithms have been considered as a very significant arsenal in the battle against fake news. Several supervised models have been proposed. For instance, a ranking model based on SVM and Pseudo-Relevance Feedback for tweet credibility has been developed in [3]. A credible news classifier based on regression was proposed in [4]. SVM on content-based features was utilized in [5] in order to detect fake, satirical and real news items. A comprehensive survey of data mining algorithms employed for fake news detection is contained in article [6].

A different line of research was taken by [7, 8] where the actual content was analyzed and news items were represented as multi-dimensional tensors. This is in contrast to aforementioned works which are based on feature extraction.

Some works investigated the issue of fake news detection following a credibility diffusion-based approach. These works [9] construct complex networks of heterogeneous entities (persons, tweets, events, message, etc.) and study the paths of fake news propagation in order to find non-credible sources of information, and thus infer fake news.

The authors in [10] investigated the characteristics that are more predictive for identifying social network accounts responsible for spreading fake news in the online environment, both from an automatic and human perspective. They conducted an offline analysis using deep learning techniques, as well as an online analysis involving real users in the classification of reliable/unreliable user profiles. The experimental results revealed the information that best enables machines and humans to detect rogue users effectively.

Interestingly, in [11] the authors proposed a fake news approach that included identifying potential fake news spreaders on social media as a first step toward preventing fake news from being spread among internet users. Thus, they investigated whether it is possible to distinguish credible authors from other authors who have shared fake news in the past. They conducted different learning experiments from a multilingual perspective (English and Spanish) and evaluated different textual features, handcrafted and automatically learned, that are primarily not tied to a specific language. The performance of their system achieved an overall accuracy of 78% and 87% on the English and Spanish corpus, respectively.

There are academic efforts to develop web services that will investigate how disinformation spreads and competes in online social networks. For instance, Hoaxy [1] is such a service for Twitter; it is actually a platform for the study of diffusion of misinformation in Twitter.

Less related areas are those concerning rumor classification, trust discovery, click-bait detection, spammer and bot detection, as well as related online services e.g., Botometer which checks Twitter accounts and assigns them a score based on how likely they are to be a bot. However, there are significant differences among those areas and fake news detection as explained in [6], and thus we do not consider them here. Finally, there are algorithms for detecting fake images [12, 13] and fake videos [14, 15], but these are beyond the scope of this chapter.

The interested reader may consult the following articles [16–22] for complete surveys on articles related to fake news detection.

3 Investigated Algorithms

In this section, we provide some background information which concerns the algorithms that are the focus of this chapter.

3.1 L1 Regularized Logistic Regression

Logistic Regression is basically a linear model accompanied by the sigmoid function which is being applied to the linear model in order to convert the output from any real number into the range of [0, 1]. Using the L1-regularization, we add the term w_1 to the cost function where $\|\cdot\|_1$ denotes the 1-norm and w values are the model's learned weights. So as an optimization problem is trying to minimize the following cost function:

$$\min_{w,c} \|w\|_1 + C \sum_{i=1}^{n} \log\big(\exp(-y_i (X_i^T w + c)) + 1\big),$$

with X_i, y_i being the input variables, c is the regularization parameter and C is the inverse of regularization strength.

3.2 C-Support Vector Classification

C-Support Vector Classification is one type of Support Vector Machines (SVM) that can incorporate different basic kernels. Given training vectors $x_i \in R^p$ $i = 1 \ldots, n$ in the two class case and the corresponding class labels decision $y_i \in \{-1, 1\}^n$, C-SVC solves the following problem [23, 24]:

$$\min_{w,b,\zeta} \frac{1}{2} w^T + C \sum_{i=1}^{n} \zeta_i$$

with constraints: $y_i\big(w^T \phi(x_i) + b\big) \geq 1 - \zeta_i$, and $\zeta_i \geq 0$, $i = 1 \ldots, n$ and $\phi()$ being the kernel function.

3.3 Gaussian and Multinomial Naive Bayes

Naive Bayes methods are a set of algorithms based on applying Bayes' theorem with the naive assumption of independence between every pair of features. For a given data point $x = \{x_1. \ldots x_n\}$ of n features and a class variable y, Bayes' theorem states the following relationship:

$$P(y|x_1, x_2, \ldots, x_n) = P(y) \frac{P(x_1, x_2, \ldots, x_n|y)}{P(x_1, x_2, \ldots, x_n)}.$$

Using the naive independence assumption that

$$P(x_i|y, x_1, x_2, \ldots, x_{i-1}, x_{i+1}, \ldots, x_n) = P(x_i|y)$$

and since $P(x_1, x_2, \ldots, x_n)$ is constant given the input, this can be formulated as:

$$P(y|x_1, x_2, \ldots, x_n) \propto P(y) \prod_{i=1}^{n} P(x_i|y).$$

Thus, the most likely class assignment for a data point $x = x_1, x_2, \ldots, x_n$ can be found by assigning the class for which the above value is largest. In mathematical notation, this is defined as:

$$\hat{y} = \arg\max_{y} P(y) \prod_{i=1}^{n} P(x_i|y).$$

Gaussian Naive Bayes The likelihood of the features is assumed to be Gaussian:

$$P(x_i|y) = \frac{1}{\sqrt{2\pi\sigma_y^2}} \exp\left(-\frac{(x_i - \mu_y)^2}{2\sigma_y^2}\right).$$

Multinomial Naive Bayes Multinomial Naive Bayes adapts the naive Bayes algorithm for multinomially distributed data. The distribution is parameterized by vectors $\theta_y = (\theta_{y1} \ldots \theta_{yn})$ for each class y, where n is the number of features and θ_{yi} is the probability $P(x_i|y)$ of feature i appearing in a sample of class y. The parameter θ_y is estimated by relative frequency counting:

$$\hat{\theta}_{yi} = \frac{N_{yi} + \alpha}{N_{y+an}}$$

where $\boldsymbol{N}_{yi} = \sum_{x \in T} \boldsymbol{x}_i$ is the number of times feature *i* appears in a sample of class *y* in the training set $\boldsymbol{T}$, and $\boldsymbol{N}_y = \sum_{i=1}^{n} \boldsymbol{N}_{yi}$ is the total count of all features for class *y*. In our tests we used Laplace smoothing by setting $\alpha = 1$.

3.4 Decision Trees

Despite the various decision tree algorithms, the type of decision tree that we used was first discussed by Breiman [25] and is known as CART (Classification And Regression Trees). The decision tree begins with a root node *t* derived from whichever variable in the feature space minimizes a measure of the impurity of the two sibling nodes. Let $p(w_j|t)$ be the proportion of patterns x_i allocated to class w_j at node *t*. Then, the measure of the impurity (in our case we chose Gini) at node *t*, denoted by

$i(t)$ is computed by:

$$i(t) = \sum_{k} p(w_j|t)\big(1 - p(w_j|t)\big).$$

Each non-terminal node is then divided into two further nodes, t_L and t_R, such that p_L, p_R are the proportions of entities passed to the new nodes t_L, t_R respectively. The best division is that which maximizes the difference given in the equation below:

$$\Delta_i(s, t) = i(t) - p_L i(t_L) p_R i(t_R).$$

The decision tree grows by means of the successive sub-divisions until a stage is reached in which there is no significant decrease in the measure of impurity when a further additional division s is implemented. When this stage is reached, the node t is not subdivided further, and automatically becomes a terminal node. The class w_j associated with the terminal node t is that which maximizes the conditional probability $p(w_j|t)$.

3.5 Random Forests

The Random forests algorithm belongs to the family of ensemble methods. It was introduced by Breiman [26]. During training, the algorithm creates multiple trees using the CART [25] methodology with each tree trained on a bootstrapped sample of the original training data. In contrast to the original publication, the scikit-learn implementation combines classifiers by averaging their probabilistic prediction, instead of letting each classifier vote for a single class.

3.6 Multi-Layer Perceptron (MLP)

A Multi-Layer Perceptron belongs to the class of feed-forward neural networks, and it includes at least three layers of nodes; an input layer, an output layer, and an arbitrary number of hidden layers. A fully connected MLP with three input neurons, a single hidden layer, and an output layer with two-output neurons can be represented graphically as shown in Figure 1. MLPs can be trained using first-order methods, such as classical backpropagation [27], Stochastic Gradient Descent [28], Adam [29] or second-order methods, such as L-BFGS [30].

A one-hidden-layer MLP is a function $f : R^D \rightarrow R^L$, where D is the size of input vector x, L is the size of the output vector such that:

$$f(\mathrm{x}) = softmax(\mathbf{W}_1^T logsig(\mathbf{W}_2^T \mathrm{x} + \boldsymbol{b}_1) + \boldsymbol{b}_2$$

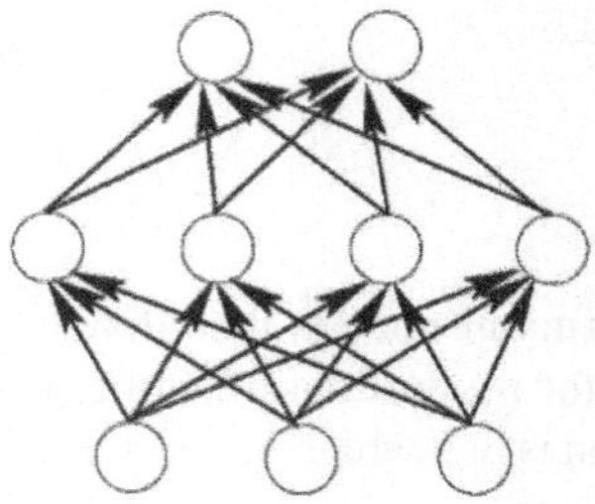

Fig. 1 The topology of a multi-layer perceptron

with $\boldsymbol{b_1}, \boldsymbol{b_2}$ being the bias vectors of the two layers, $\boldsymbol{W_1}, \boldsymbol{W_2}$ being the weight matrices of the two layers, $logsig$ being the logistic sigmoid function, and the $softmax$ function being defined as $softmax(z_i) = \frac{\exp(z_i)}{\sum_{l=1}^{k} \exp(z_l)}$ (where z_i represents the ith element of the input to $softmax$, which corresponds to class i, and K is the number of classes). To train the MLP, in order to learn the set of parameters $\boldsymbol{b_1}$, $\boldsymbol{b_2}$, $\boldsymbol{W_1}$, $\boldsymbol{W_2}$ the L-BFGS quasi-Newton optimization algorithm is used in our experiments.

3.7 Convolutional Neural Networks (CNNs)

A Convolutional Neural Network (CNN) is a type of multi-layer feed-forward neural network with a grid-like topology, which uses the mathematical operation of convolution in place of general matrix multiplication in at least one of its layers. Usually, convolution used in CNNs does not correspond precisely to the convolution employed in other engineering fields and mathematics; almost all CNNs use the so-called pooling operation.

A typical layer in a CNN consists of three stages; in the first stage the layer performs several convolutions in order to produce a set of linear activations; each one of them—in the second or detector stage—passes through nonlinear activation function, and finally, in the third stage a pooling function modifies the output. This pooling function is usually an aggregation (summary) statistic, e.g., max, over the nearby outputs. So, pooling makes the representation invariant to small translations. Training of CNNs can be performed with standard backpropagation methods [32, Ch. 8].

Figure 2 presents a traditional Deep CNN network [31]. The fully connected layer is a standard MPL that uses a softmax activation function in the output layer. The term "fully connected" implies that every neuron in the previous layer is connected to every neuron on the next layer. The output of the feature extraction stage represents the input image's high-level features. The fully connected layer's goal is to use these features to classify the input image into several classes based on the training dataset.

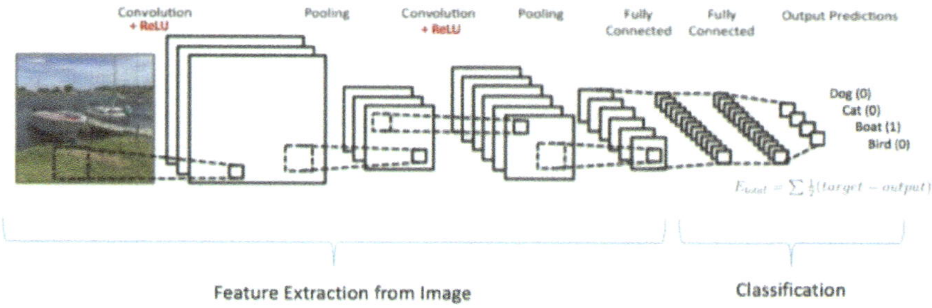

Fig. 2 A typical CNN [31]

4 Evaluation Environment and Settings

This section presents the performance evaluation of the algorithms for fake news classification. We will briefly present the competitors, the datasets, and the performance measures, whereas the actual evaluation will be presented in the next section.

4.1 Competing Algorithms

The competitors are the eight algorithms presented in Sect. 3, namely L1 Regularized Logistic Regression, C-Support Vector Classification, Gaussian Naive Bayes, Multinomial Naive Bayes, Decision Trees, Random Forests, Multi-Layer Perceptron, and Convolutional Neural Networks. The version of the first seven algorithms is that provided by scikit-learn [33], whereas for the last one we developed our own code according to [34].

4.2 Execution Environment

Our tests were executed in two different servers, the first one was used for training the CNNs on a Tesla K20x GPU, and the second one for the rest of the algorithms. This is due to the fact that CNN training is a highly CPU-intensive task. The following Table 1 has the detailed specifications of the machines used in our experiments.

4.3 Datasets

We strived for using freely available datasets that have been used in earlier studies, to ease reproducibility. The datasets are described in Table 2.

Table 1 Servers specifications

	Server 1	Server 2
CPU architecture	Haswell	Ivy Bridge
Model No.	Xeon E5-2695V3	Xeon E5-2620V2
# of cores	14	6
Core frequency (GHz)	2.30	2.10
Main memory (GB)	128	128
GPU	Nvidia Tesla K20x	None

Table 2 Datasets used in the evaluation

	Dataset properties		
Dataset name	Size	Property	Source
"Liar, liar pants on fire": a new benchmark dataset for fake news detection	Training set size of 10,269 articles	Two labels for the truthfulness ratings (real/fake) were used instead of the original six	[36]
The signal media one-million news articles dataset	1 million articles	13,000 articles were selected at random and marked as real news	Signalmedia[4]
Getting real about fake news	13,000 articles	All 13,000 articles were marked as fake news	Kaggle[5]

Before using any of our datasets, firstly we subjected them to some refinements like stop-word, punctuation and non-letters removal and finally we used the Porter2 English Stemmer algorithm for stemming, due to its improvements over the widely used Porter stemmer [35]. This was done in order to avoid noise in our data and make classification faster and more efficient.

Using the datasets from Table 2, we created three input datasets (experiments) on which we evaluated the algorithms. For the first experiment we used the Wang's training dataset [36] which contains various statements from PolitiFact,[6] a Pulitzer Prize-winning Website. From this dataset we used only the headline of each news story and two labels for the truthfulness ratings (real/fake).

Using the two remaining datasets, we created two new datasets which contained a mix of true/fake headlines and a mix of true/fake body texts respectively. For the newly created datasets we chose to keep a balance between the true and fake news using the same number for them from the original datasets. The headlines dataset finally contained 25000 news stories titles that were selected at random from both original datasets and about the body text dataset, using the fact that the average length of stories from five of the top sites that were shared on social media on December

[4] http://research.signalmedia.co/newsir16/signal-dataset.html.

[5] https://www.kaggle.com/mrisdal/fake-news.

[6] http://www.politifact.com/.

2016 was between 200 and 1000 words7, we collected 10000 body texts of a length between 150 and 4000 words. We will call these three datasets as Dataset1, Dataset2 and Dataset3.

4.4 Performance Measures

Since we consider the fake news detection problem as a binary classification task, we evaluated the competitors in terms of the following commonly used measures, namely F1-measure and accuracy whose precise definition are as follows:

- *Accuracy* is the fraction of predictions that are correctly classified as either fake or real news by the model.
- *F1-measure* is the harmonic mean of precision and recall, where precision and recall are defined as follows:
  - *Recall* is the percentage of all fake news that are correctly classified as fake by the model.
  - *Precision* is the percentage of news items being actually fake out of all news items returned as fake by the model.

Moreover, we consider the execution time as another significant quantity to measure; it is comprised by the time to complete two tasks, namely training and classification. So, we measured the following two quantities:

- *Training time*, which indicates the total time (in seconds) needed for training the model.
- *Classification time*, which indicates the total time (in seconds) needed for providing the classification decision.

5 Performance Evaluation

In this section we will present the details of the evaluation setting and illustrate the results.

5.1 Text-To-Vector Transformation

First of all, we needed to transform the text into some numeric or vector representation. This numeric representation should depict significant characteristics of the text. There are many such techniques, for example, occurrence, term-frequency, TF-IDF, word co-occurrence matrix, word2vec and GloVe. In our tests, we used the following two techniques:

- *Word Embeddings.* A word embedding is a parameterized function mapping words of some language to high-dimensional vectors $W : words \rightarrow R^n$. In our tests two different techniques were used:
  - *Pre-trained Word Vectors.* We use the publicly available Glove vectors [37] trained on 6 billion tokens of Wikipedia 2014 + Gigaword 5. The vectors have dimensionality of 50, 100, and 300.[7]
  - *Trained Word Vectors Based on our datasets.* We use word2vec from genism library to train our own vectors based on the selected datasets. The vectors have dimensionality of 50, 100, 300 and were trained using the continuous bag-of-words model. In order to get a single vector representation within each headline/article we averaged the corresponding word vectors.
- *Term Frequency-Inverse Document Frequency (TF-IDF).* TF-IDF weighting scheme is the combination of two terms, the Term Frequency (TF) and Inverse Document Frequency (IDF). Term Frequency, measures how frequently a term t occurs in a document and Inverse Document Frequency, measures the importance of this term t in the whole collection, i.e., its rareness. Even though there are exist many variation of the scheme in literature [38], we use a simple formula; more specifically, we define TF-IDF as follows:

$$tf_{t,d} = \frac{\text{number of times term } t \text{ appears in a document}}{\text{total number of terms in the document}}$$

$$idf_t = \log \frac{\text{total number of documents}}{\text{number of documents with term } t \text{ in it}}$$

So, the final TF-IDF weight of the term t is given by the following product:

$$tf - idf_{t.d} = tf_{t,d} \times idf_t$$

As a result, every document can be interpreted as a vector with one component corresponding to each term in the dictionary together with its weight. For any other term that does not occur in the document, we assign this weight equal to zero.

So, each competitor has seven variants, i.e., three variants due to the three different dimensions of the pre-training, three variants due to the three different dimensions of the training based on our datasets, and one variant based on TF-IDF. So our first step is to discover which of the six former variants is the best one for each competitor.

5.2 How Many Dimensions Are Necessary?

We ask the following two questions:

---

[7] https://nlp.stanford.edu/projects/glove/

- *How many dimensions are preferable for our algorithms? and*
- *Is it training based on the examined dataset or on benchmark datasets a better solution?*

We present the average accuracy of the six variants of each algorithm in Fig. 3. Deviation is small, so average is quite a good measure for all algorithms with the exception of DT for Dataset1. We observe that CNN is the best performing algorithm, with a significant gap from the next best performing which are Multi-Layer Perceptron and Random Forest. For Dataset1 all algorithms achieve the same performance.

We present the average F1-measure of the six variants of each algorithm in Fig. 4. The obtained results are similar to those observed for average accuracy, with CNN being again the champion method.

We present the average precision of the six variants of each algorithm in Fig. 5. Deviation is small, so average is quite a good measure for all algorithms with the exception of DT for any Dataset and MNB for Dataset2. The relative performance of the methods remains the same as for the accuracy measure.

We present the average recall of the six variants of each algorithm in Fig. 6. Here we see that CNN is again the champion algorithm, but the differentiation of the rest is not very sound.

It is expected that no choice on the number of dimensions and/or training on any kind of data can generate a variant of an algorithm that will be the champion one; such

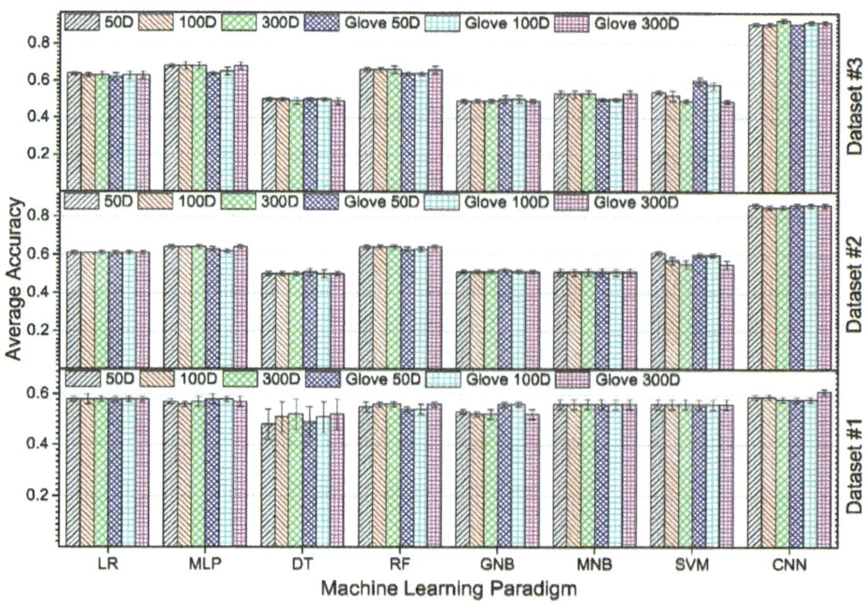

Fig. 3 Average accuracies

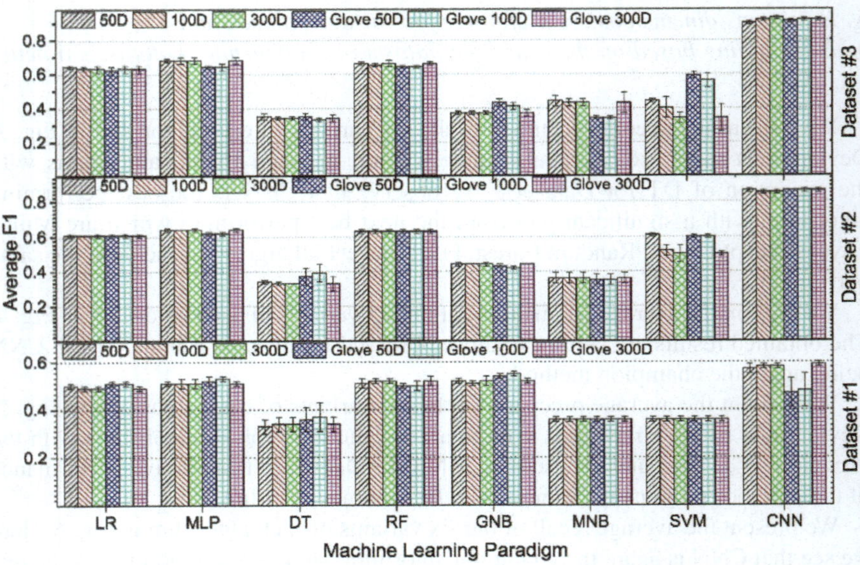

Fig. 4 Average F1-measure

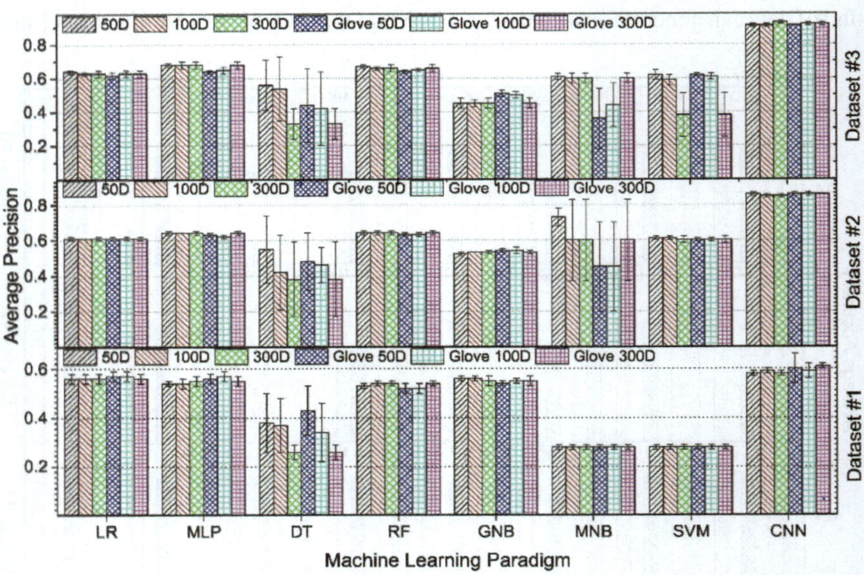

Fig. 5 Average precision

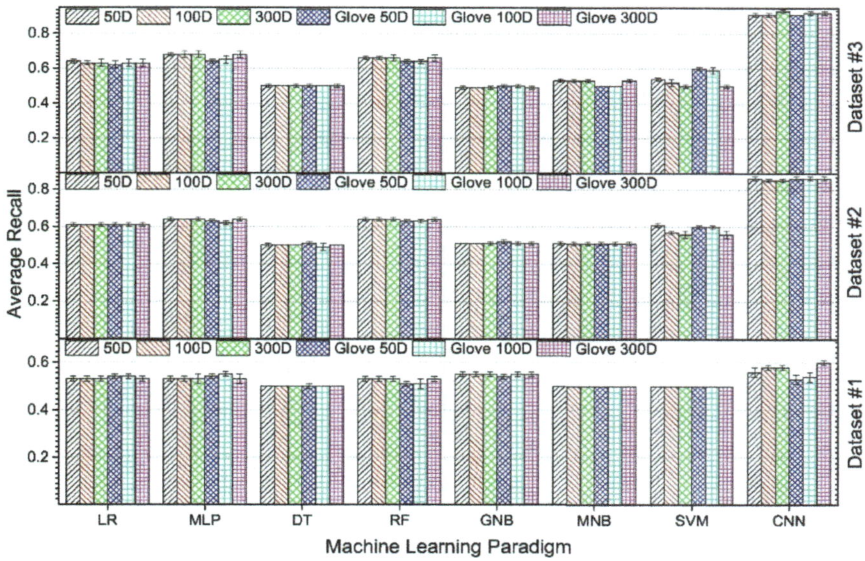

Fig. 6 Average recall

problems and the associated algorithms are highly dependent on data distributions. In Table 3 we present the variant of each algorithm that showed the best performance.

We can draw two quite evident conclusions from Table 3. The first observation is that a small or moderate number of dimensions is preferable because they do not create overfitted models. Secondly, pretraining based on benchmark datasets can be quite effective, meaning that such kind of pretraining is able to create models beating those generated on the specific data that are the target of investigation; this is a quite encouraging result.

Table 3 Champion variant of each algorithm with respect to the number of dimensions and type of training

Algorithm	Dataset1	Dataset2	Dataset3
LR	100D glove	100D	50D
MLP	100D glove	100D	50D
DT	100D glove	100D glove	50D glove
RF	100D	300D glove	50D
GNB	100D Glove	300D glove	50D glove
MNB	Any variant	300D glove	50D
SVM	Any variant	50D	50D glove
CNN	300D glove	100D glove	300D glove

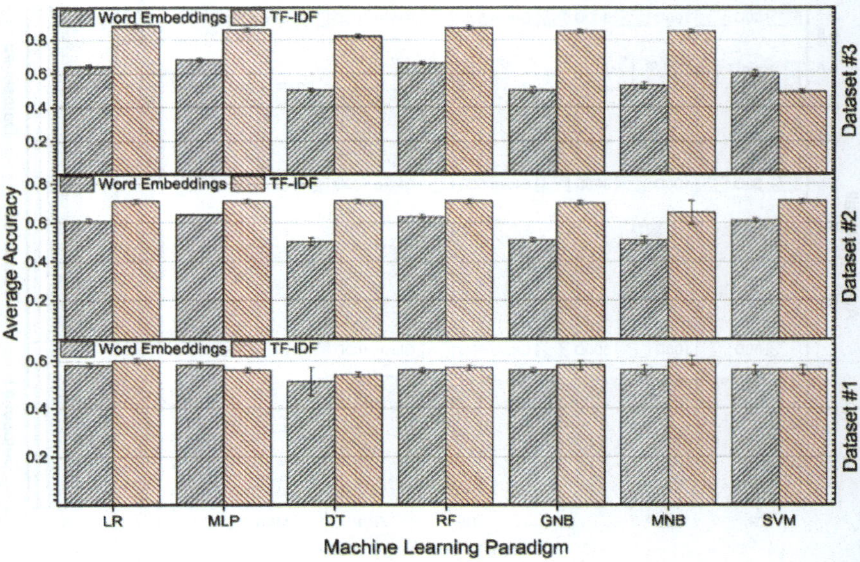

Fig. 7 Average accuracies

5.3 Method of Choice to Generate Vector Representations

Based on the identified "champion" variant of each algorithm from the previous section, we ask the following question:

Is it preferable to use a TF-IDF scheme or word embeddings to generate vector representations of textual information?

The answer to this question is illustrated in Figs. 7 and 8. The first three plots compare the performance of the champion word embedding variant against the TF-IDF variant of each algorithm from the perspective of average accuracy; whereas the other three plots contain the results from the perspective of average F1-measure.

The results show clearly that the TF-IDF representation is a better alternative for the great majority of cases and algorithms. In particular, this representation achieves a 10% better performance in almost all cases, in some cases this gap widens to reach a 30%. The only exception is for SVM in the case of Dataset3.

5.4 Execution Time

As far as the execution time is concerned, Table 4 shows the execution time—training and classification time—of all variants of the algorithms for Dataset1. In general, SVM and the neural network-based algorithms are the most time-consuming during the training phase, which is expected.

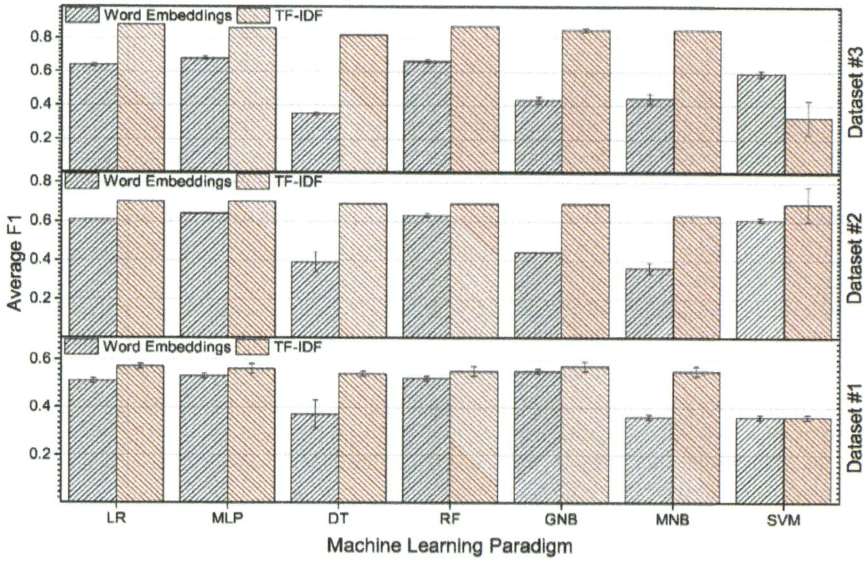

Fig. 8 Average F1-measure

Table 4 Training/classification times (in seconds) for Dataset1

				Glove vectors			
Model	50D	100D	300D	50D	100D	300D	TF/DT
LR	0.69–0.01	0.97–0.01	0.58–0.01	5.75–0.01	7.36–0.00	3.13–0.01	0.04–0.00
MLP	8.37–0.00	7.40–0.00	11.45–0.00	8.12–0.00	6.45–0.00	10.74–0.0	8.46–0.00
DT	1.10–0.00	2.01–0.00	6.39–0.00	1.10–0.00	1.76–0.00	5.44–0.00	0.58–0.00
RF	1.02–0.01	1.39–0.01	2.31–0.01	0.96–0.01	1.33–0.01	2.26–0.01	0.84–0.01
GNB	0.01–0.00	0.01–0.00	0.03–0.01	0.01–0.00	0.01–0.00	0.03–0.01	0.05–0.01
MNB	0.01–0.00	0.01–0.00	0.03–0.00	0.01–0.00	0.01–0.00	0.02–0.00	0.00–0.00
SVM	14.44–1.08	19.08–1.68	54.86–4.81	13.04–1.09	19.09–1.72	53.44–4.78	10.39–0.91
CNN	9.88–0.24	12.28–0.27	16.99–0.27	12.11–0.29	14.72–0.28	17.15–0.29	

5.5 *Summary of findings*

In summary, our experimentation showed that a small or moderate number of dimensions is adequate, and that pre-trained models based on benchmark datasets can achieve steadily good performance. As far as the method to generate vector representation of textual information is concerned we found out that the TF-IDF method is the clear winner. Finally, among all examined methods and their variants, convolutional neural networks can be considered as the champion algorithm.

6 Future Directions

Although significant developments have been made in recent years with regards to combating the spread of fake news, the lack of standard datasets and benchmarks generates uncertainty, basically due to the lack of authoritative benchmarks within the respective IT community, that precludes more robust achievements. In future research, standard datasets and practical evaluation metrics are needed for comparing various fake news algorithms and promoting the development of more efficient methods.

Another significant problem is the increase in the number of users that spread or share information and the quality of the disseminated data that might be potentially uncertain due to inconsistencies, incompleteness, noise and unstructured nature. This complexity probably jeopardizes the legitimacy of the results of any standard analytic processes and decisions that would be based on them. Designing tailor-made advanced analytical techniques that could conclude on future courses of action with efficacy remains very challenging.

A worthwhile future research point is to investigate the cognitive mechanisms of false information. To elaborate, if we manage to perceive the cognitive mechanisms that fabricated information dissemination is built upon, then more effort can be focused on the respective counter measures/tactics, thus, making them more efficient.

Taking into account that the battle against fake news never ends, counter measures/tactics generality or adaptability is quite important in order to improve their robustness. Fake news identification methods should be able to track unseen, newly coming events, even if the internal system data may differ from contents of emerging events. An insightful research direction to be explored and then adapted accordingly on the fake news detection domain concerns web security, virus/spam detection methods, which also suffer from similar issues such as early detection and model generalization.

Future work is also required in areas of social bots and troll detection, which often act as a catalyst in generating and spreading fake news. The main problem in this particular example is not the fake news rather it is the magnitude of sharing and speed of spreading of the fake news that is causing more harm.

The process of detecting fake information is by nature the learning of a classifier to identify the credibility of some distributed material information. Embracing of novel machine learning models, combining the characteristics of different machine learning models to provide adaptability and improve system efficacy, or even further exploring and extending the potential capabilities of readily available machine learning models signify that there are still more that can be explored.

Providing explainable results should improve fake news detection system efficacy since it is increasing user trust in the detection models. In this research direction the respective experience investigated in other related domains, such as recommender systems, could be very useful.

Moreover, developing counter measures to confront adversarial attacks that target the fake information detection systems is also an area of interest. These adversarial attacks might impede the robustness of the fake news identification models which means that adding some small perturbations to input vectors could make these models get wrong results.

7 Conclusions

The fast spreading of fake news and the impact they are having on our society, along with the in scalability of manually detecting them, have created a surge of research and development in machine learning algorithms to battle them. In this article, we evaluated representatives from eight well-known families of algorithms, namely regression, support vector classification, multi-layer perceptron, Gaussian and multinomial naive Bayes, random forests, decision trees and convolutional neural networks against three publicly available datasets. We tested the efficiency and training speed of these algorithms. We concluded that a space with a hundred dimensions is of adequate dimensionality to capture the needed text features and get high accuracy of detection. Moreover, we established that the TF-IDF method for generating vectors from the text is a better alternative relative to word embeddings, and finally that pretraining based on benchmark datasets is able to reap performance benefits similar to that when training is performed based on the data under study. As far as the champion algorithm is concerned, we have shown that convolutional neural networks is the best performing algorithm with the downside of requiring significantly higher training time.

Remarks This chapter is an extended version of [39]. We are making this version available in order to have more clear results and discussions in comparison to its short version.

Referesnces

1. Liu, X., Zhang, B., Susarla, A., & Padman, R. Go to YouTube and see me tomorrow: The role of social media in managing chronic conditions. Available at https://ssrn.com/abstract=3061149
2. Vosoughi, S., Roy, D., & Aral, S. (2018). The spread of true and false news online. *Science, 359*(6380), 1146–1151.
3. Gupta, A., & Kumaraguru, P. (2012). Credibility ranking of tweets during high impact events. In *Proceedings of the workshop on privacy and security in online social media.*
4. Hardalov, M., Koychev, I., & Nakov, P. (2016). In search of credible news. In *Proceedings of the artificial intelligence: Methodology, systems and applications* (pp. 172–180).
5. Horne, B. D., & Adali, S. *This just in: Fake news packs a lot in title, uses simpler, repetitive content in text body, more similar to satire than real news.* Technical Report. Available at http://arxiv.org/abs/1703.09398
6. Shu, K., Sliva, A., Wang, S., Tang, J., & Liu, H. (2017). Fake news detection on social media: A data mining perspective. *ACM SIGKDD Explorations, 19*(1), 22–36.

7. Guacho, G. B., Abdali, S., Shah, N., & Papalexakis, E. (2018). *Semi-supervised content-based detection of misinformation via tensor embeddings*. Technical Report. Available at https://arxiv.org/abs/1804.09088
8. Hosseinimotlagh, S., & Papalexakis, E. (2018). Unsupervised content-based identification of fake news articles with tensor decomposition ensembles. In *Proceedings of the workshop on misinformation and misbehavior mining on the web (MIS2)*.
9. Gupta, M., Zhao, P., & Han, J. (2012). Evaluating event credibility on Twitter. In *Proceedings of the SIAM international conference on data mining (SDM)* (pp. 153–164).
10. Sansonetti, G., Gasparetti, F., D'aniello, G., & Micarelli, A. (2020). Unreliable users detection in social media: Deep learning techniques for automatic detection. *IEEE Access*, 213154–213167.
11. Vogel, I., & Meghana, M. (2020). Detecting fake news spreaders on Twitter from a multilingual perspective. In *Proceedings of the 2020 IEEE 7th international conference on data science and advanced analytics (DSAA)* (pp. 599–606).
12. Gupta, A., Lamba, H., Kumaraguru, P., & Joshi, A. (2010). Faking sandy: Characterizing and identifying fake images on Twitter during hurricane sandy. In *Proceedings of the ACM international conference on world wide web (WWW)* (pp. 729–736).
13. Galbally, J., Marcel, S., & Fierrez, J. (2014). Image quality assessment for fake biometric detection: Application to iris, fingerprint, and face recognition. *IEEE Transactions on Image Processing*, 710–724.
14. Güera, D., & Delp, E. J. (2018). Deep fake video detection using recurrent neural networks. In *15th IEEE international conference on advanced video and signal based surveillance (AVSS)* (pp. 1–6).
15. Agrawal, R., & Sharma, D. K. (2021). A survey on video-based fake news detection techniques. In *8th International conference on computing for sustainable global development (INDIACom)* (pp. 663–669).
16. Guo, B., Ding, Y., Yao, L., Liang, Y., & Yu, Z. (2020). The future of false information detection on social media: New perspectives and trends. *ACM Computing Surveys*.
17. Zhou, X., & Zafarani, R. (2020). A survey of fake news: Fundamental theories, detection methods, and opportunities. *ACM Computing Surveys*, 1–40.
18. Ghafari, S. M., Beheshti, A., Joshi, A., Paris, C., Mahmood, A., Yakhchi, S., & Orgun, M. A. (2020). A survey on trust prediction in online social networks. *IEEE Access*, 144292–144309.
19. Collins, B., Hoang, D. T., Nguyen, N. T., & Hwang, D. (2021). Trends in combating fake news on social media—A survey. *Journal of Information and Telecommunication*, 247–266.
20. Kumar, S., Kumar, S., Yadav, P., & Bagri, M. (2021). A survey on analysis of fake news detection techniques. In *Proceedings of the 2021 international conference on artificial intelligence and smart systems (ICAIS)* (pp. 894–899).
21. Choudhary, M., Jha, S., Prashant, Saxena, D., & Singh, A. K. (2021). A review of fake news detection methods using machine learning. In *Proceedings of the 2021 2nd international conference for emerging technology (INCET)* (pp. 1–5).
22. Kumar, P. J. S., Devi, P. R., Kumar, S. S., & Benarji, T. (2021). Battling fake news: A survey on mitigation techniques and identification. In *Proceedings of the 2021 5th international conference on trends in electronics and informatics (ICOEI)* (pp. 829–835).
23. Cortes, C., & Vapnik, V. (1995). Support-vector network. *Machine Learning, 20*, 273–297.
24. Vapnik, V. (1998). *Statistical learning theory*. Wiley.
25. Breiman, L., Friedman, J. H., Olshen, R. A., & Stone, C. J. (1984). *Classification and regression trees*. Brooks/Cole Publishing.
26. Breiman, L. (2001). Random forests. *Machine Learning, 45*(1), 5–32.
27. Goodfellow, I., Bengio, Y., & Courville, A. (2016). *Deep learning*. The MIT Press.
28. Robbins, H., & Monro, S. (1951). A stochastic approximation method. *The Annals of Mathematical Statistics, 22*(3).
29. Kingma, D. P., & Ba, J. L. (2015). ADAM: A method for stochastic optimization. In *Proceedings of the international conference on learning representations (ICLR)*.

30. Nocedal, J. (1980). Updating quasi-Newton matrices with limited storage. *Mathematics of Computation, 35*, 773–782.
31. Selva. CNN classifier using 1D, 2D and 3D feature vectors. MATLAB central file exchange. https://www.mathworks.com/matlabcentral/fileexchange/68882-cnn-classifier-using-1d-2d-and-3d-feature-vectors. Retrieved August 4, 2021.
32. Aggarwal, C. C. (2018). *Neural networks and deep learning: A textbook*. Springer.
33. Pedregosa, F., Varoquaux, G., Gramfort, A., Michel, V., Thirion, B., Grisel, O., Blondel, M., Prettenhofer, P., Weiss, R., Dubourg, V., Vanderplas, J., Passos, A., Cournapeau, D., Brucher, M., Perrot, M., & Duchesnay, E. (2011). Scikit-learn: Machine learning in python. *Journal of Machine Learning Research, 12*, 2825–2830.
34. Kim, Y. (2014). Convolutional neural networks for sentence classification. In *Proceedings of the empirical methods in natural language processing (EMNLP)*.
35. Porter, M. F. (1980). An algorithm for suffix stripping. *Program, 14*(3), 130–137.
36. Wang, W. Y. (2017). "Liar, liar pants on fire: A new benchmark dataset for fake news detection. In *Proceedings of the annual meeting of the association for computational linguistics* (pp. 422–426).
37. Pennington, J., Socher, R., & Manning, C. D. (2014). Glove: Global vectors for word representation. In *Proceedings of the empirical methods in natural language processing (EMNLP)*.
38. Manning, C. D., Ragnavan, P., & Schütze, H. (2008). *Introduction to information retrieval*. Cambridge University Press.
39. Katsaros, D., Stavropoulos, G., & Papakostas, D. (2019). Which machine learning paradigm for fake news detection? In *Proceedings of the 2019 IEEE/WIC/ACM international conference on web intelligence (WI)* (pp. 383–387).

Credibility and Reliability News Evaluation Based on Artificial Intelligent Service with Feature Segmentation Searching and Dynamic Clustering

Ming-Shen Jian

Abstract Recently, fake news are spread through Internet, social media, specific organizations or parties. Considering the affection of the fake news to the credibility and reliability, to check and be aware of the news is needed. Based on the artificial intelligence and suitable k-means grouping method, the existed and proved fake news could be used to train the proposed system. The features of the fake news could be classified and identified according to the proposed system in this research. In addition, according to these found features, the news announced or spread by the specific person, organizations, or group, could be classified as the doubtful news.

Keywords Fake news · Artificial intelligence · Big data · K-means · Segment

1 Introduction

Today, online media or society or community is very popular. People would exchange the information, photos, or news between each other. Furthermore, people could share the information or news from any other social media user who they may not understand. During the sharing and broadcasting between each other, the information may be misrepresented [1]. Corresponding to real life related to the news spreading, some news are made by the none-recognition or unidentified social media user for the specific purposes. These news may involve some true and fake information, or be the completely fictitious news, to confuse the readers or information receivers. If the news are spread with the specific purpose by the malicious organizations or groups, the fake news would be flooded through the social media and Internet [2, 3]. Although there are various search engines for searching information and it sources, to find the correct information instead of the flooded fake news is still difficult.

Some different methods for fake news detection are proposed based on the supervision [3, 4]. By supervising all the news spreading online, the fake news and be

M.-S. Jian (✉)
Cloud Computing and Intelligent System Laboratory, Department of CSIE, National Formosa University, Huwei, Taiwan
e-mail: jianms@nfu.edu.tw

M. Lahby et al. (eds.), *Combating Fake News with Computational Intelligence Techniques*, Studies in Computational Intelligence 1001,
https://doi.org/10.1007/978-3-030-90087-8_9

found and marked as the doubtful news. However, to supervise all the information and news through Internet is difficult and the workload is high. To develop tools for un-supervising all news should be the possible solution [5]. In addition, since some fake news with specific purposes would come from the malicious person, organizations, or groups intentionally, to find the possible fake news based on the background of the news sources could be also the available solution [6, 7]. Some social media or websites would verify and recognize the fake news based on the crowd-sourcing approach which depends on the public annotations or replies. Due to the difficulty of news verification, especially in some specific field or polities, the intentionally spread news are not easy to confirm [8].

Considering the language structures, the articles or information written by some languages are more difficult to verify. For example, in Asia, Japanese, Korean, or Chinese vocabularies are the complex languages which may be completely different in meanings due to different combination of vocabularies or words [9]. By matching the keywords on demand defined in the database, the algorithm could find the possible and available information or features according to the statics of the past searching according to these keywords. In other words, to find the suitable keywords or sentences included in these articles or news are important. Considering the language structures, different vocabularies could be combined as the new extended words or sentence [9–11]. Therefore, if the extended potential words or vocabularies could be found, the features of the news can be defined and described with higher accuracy.

However, due to the rapid change of the information and news broadcasted through the Internet and various social media users, partial content of the information and news would be changed. Some features or keywords would also change. If the malicious fake news is spread intentionally, the total amount of the fake news would be spread and created would be very huge online by social media. Then, to deal with the big data of these information and news will require the huge computing resource and storage spaces. In other words, the continuous data collection for fake news evaluation and training, and the fast computing for news evaluation and artificial intelligent service training, are both needed.

First, corresponding to each language, the structure of sentence should be defined and divided into several segments. Since most information and news are spread through the network and represented via web browser or social media applications, using the web crawler for information searching and collection is the available method [12].

Then, according to the meaningful vocabularies matching and searching, the features of the specific content could be found. Considering the amount of found vocabularies would be huge, using the Hadoop/MapReduce algorithm would be the possible solution to reduce the data size [11, 13, 14]. The analysed data could be represented as the key-value format. Each key indicates the found vocabulary and value means the appearance time of this specific vocabulary in this analysed data.

Currently, the malicious fake news spread through the Internet would not be the incident event. There could be various types and content of the fake news related to many people or purposes. Therefore, to clustering and grouping these data is needed for separating different topics of these fake news. Suppose that the features of these fake news could be found individually, then the fake news with the similar or the same features could be grouped. Since that the total amount of features corresponding to each fake news group could be different, the dynamic clustering method is needed.

In addition, too many features would also cause the grouping and grouping more difficult. Therefore, to reduce the amount of features should be considered. In opposition to the short vocabularies found as the features of fake news, to extend the length of the words could find the features of the specific topic with higher accuracy [11]. Therefore, how to extend and evaluate the extended vocabularies becomes an important issue.

The main contributions of this research are listed as follows:

1. By collecting the verified fake news through web crawler, the various fake news with the same or similar topic could be automatically obtained through Internet search engine or social media. These collected data could be used as the artificial intelligent service training data for fake news clustering and prediction. Less manual operation and supervision is needed.
2. According to the proposed procedure and system, languages such as English or Chinese with different sentence structures could be analysed and managed. In addition, based on the proposed method, the keywords as the features of the fake news are extendable. The feature words or sentences for finding fake news could be developed with higher accuracy.
3. Different topics of fake news could be clustered and grouped by the features. After evaluation and statistical methods, even some malicious people or groups could be found. In other words, the fake news creators or spreading chain would be found possibly.
4. After clustering the fake news with the found features, the warning could be given to the reader when the new unverified news is announced from the possible malicious groups or organizations.

This research is organized as follows. Section 2 presents the related works. The proposed procedure and method for fake news classification and prediction is given in Sect. 3. The verification and results are given in Sect. 4. Then, the discussion is presented in Sect. 5. Finally, the conclusion is provided in Sect. 6.

2 Preparation of Your Paper

Recently, through the online services such as social media, multimedia streaming, or web-forum, the pure-text or multimedia content is spread quickly and easily. Various information is shared or provided actively to the readers. Internet fraud and phishing are the common types of network crime. In addition, due to some malicious purposes including malicious political purposes, malicious business purposes, or even bullying, fake news are spread through the social media or online forum [6].

Many researches were proposed to separate and differentiate the true and fake news [1–3]. By supervising the news and information, the news could be announced as the fake news or the true one. However, the news would be various and the total amount the news would be huge. The workload of the news supervisor could be huge. Therefore, by the responses from crowd to detect and reduce the spread of fake news and misinformation was proposed [10]. However, if the malicious fake news is spread by the specific organizations or groups, the responses from crowd would be not trustable. In other words, the credibility of the responses from crowd is doubtful. To avoid the possibility of misjudgement related to the news and information, unsupervised fake news detection or classification with less manual operation was proposed [5]. Based on the proposed algorithm, the content of the news could be differentiated and classified. However, the fake news changes every day. To judge the news is true or fake is difficult due to the total amount and the frequent changing. A fake news detecting system should keep colleting the data from the social media or search engine and update the rules of judgement as soon as possible. Along with the huge unverified news spread through Internet every day, to automatically collect, cluster, and identify the information is needed. However, different languages with various structures and segments of the sentences or articles would be independent. To find the suitable way for finding the key features and segments of the analysed news is required.

Various languages would have different structures and segments for presentation by sentences. For example, some languages cannot directly analysed because of missing spaces between words. Different segments combined with vocabularies would result in different meanings. Furthermore, the names of people or stores would not be the common used words. Some new developed words by the social media would be excluded in the dictionary or database. Therefore, how to find the segments with different combination and extension of vocabularies should be considered [11].

To deal with the Chinese language, different algorithms and methods were proposed to merge the words for the unknown word extraction [9, 15, 16]. Based on the corpus or database, different vocabularies could be identified and recognized as the specific text-pattern or segments. By extending the size of the segments or combining with different continuous segments, the possible vocabularies could be found [9, 14]. However, the new found vocabularies for article analysis cannot provide 100% accuracy [16]. To recursively verify the possible vocabularies and find the new words becomes an important issue for the various fake news spread through Internet.

Since the information and Data is spread through the Internet, to collect the news spread by the social media and online forum automatically [12], the web crawler could be the possible and available solution [11, 14, 17]. By using the web crawler, the online information could be continuously and automatically collected from the given website hyperlink address or search engine according to the specific keywords. In addition, based on the web crawler application, the keep updating information and news could be collected as the training data for the fake news recognition. With more similar news or web content, the feature could be found with higher accuracy [11].

The length of the obtained news or data by the web crawler is various. To segment the vocabularies is one thing, to calculate and count the frequency of occurrence related to individual found vocabulary is another. The total amount of the raw data after vocabulary segment could be huge, to identify and classify individual word requires computing resource Therefore, to parallel process these raw data-through the distributed system is the available solution [18, 19]. Today, Hadoop or MapReduce which is proposed based on the distributed system for big data processing is the popular tool for data analysis [11, 13, 14]. Since the length of the collected data from the Internet would be various, using the fixed length and size database is meaningless. Hence, the noSQL type of database which accepts the various length of data should be used. The research based on noSQL database, HBase, was proposed for storing the analysed big data [20]. To support the enough computing resource and storage spaces for the parallel processing and distributed computing in these big data management, the cloud computing platform is needed [21]. Different collected data from various news sources could be divided into several partitions for parallel article text analysing. According to the Hadoop/MapReduce structure, these divided partitions of the original article or data are called Tasks. In opposition, each article is called a Job. In other words, one Job could be merged by multiple Tasks. Based on Map procedure of the Hadoop/MapReduce, these Tasks will be delivered and distributed to multiple computing nodes which could be established as the multiple virtual machines [21]. After counting at computing nodes, each found vocabulary will be defined as the 'Key'. Hence, there will be huge amount of Keys with different frequency of occurrence, which is defined as 'Value'. Therefore, the found vocabulary with its frequency of occurrence cab be defined as the set{Key, Value}. Some Keys may be repeatedly found from different partitions. To summarize the frequency values of occurrence of the same Key vocabulary is needed. In other words, based on Reduce procedure of the Hadoop/MapReduce, the multiple Values of the same Key vocabulary will be summarized and added. If the workload of summarizing is huge, the summarising work will be divided as multiple tasks for distributed computing and parallel processing. After several times or iterations of the Hadoop/MapReduce procedure, finally, the report related to the independent Key vocabularies with individual frequency value of occurrence in the whole data or article found from the Internet can be given. Considering the possible extension of the Key vocabularies, to integrate the Key vocabularies counting and the suitable vocabulary extension method is needed.

To provide huge computing resource, cloud computing platform is the popular technology for various services. Through network and virtualization technology, most hardware are virtualized such as central processing unit (CPU), memory, storage spaces. These virtualized resources are merged as the resource pool. According to the on demand given requirements, the corresponding virtual CPU, virtual memory, and virtual storage spaces could be integrated and represented as the physical single computer as possible. This computing environment similar to the physical computer is called virtual machine. Users could rent the required virtual hardware infrastructure to establish the temporary virtual machine, called Infrastructure as a Service (IaaS). In addition, the online operation system or platform could be also available when user requires to do the online operation, called Platform as a Service (PaaS). If the software or application is packaged and available through the network, users could directly obtain the software service without installation at the local computer, called Software as a Service (SaaS). When no longer using the rented services, the computing resource is returned into the computing resource pool for rapidly using. In other words, if there is the template to establish the virtual machine for specific purpose, the multiple virtual machines which all follow the configurations of the template could be quickly implemented. Therefore, by including the required services and applications into the template of virtual machine, then multiple virtual machines with the same services or applications can be rapidly established and provided. In other words, if the processes or procedures are recursively executed, cloud computing platform could provide the suitable computing resource management for the computing.

Recent years, the artificial intelligent for services are popular. Most services should be trained in advance for higher accuracy of prediction or classification. Corresponding to the fake news, the content changes quickly and always new fake news spread through Internet. To train the artificial intelligent system for service, providing the known and identified fake news is the first step. In addition, different recognized fake news should be divided into different groups. In other words, to classify all collected fake news should be done for training the artificial intelligent system. Since the features or keywords corresponding to various fake news could be different, to classify these news is possible. K-mean algorithm for classification is generally used for artificial intelligent services. Based on the vector quantization after the features analysing, the targets for classification which is near to others would be grouped together. With enough group centre for the vector quantization the similar analysed targets would be more closer to each other. Assume that total k randomly group centres are given in the beginning that represented as S_1 to S_k. In addition, there are total f features which indicates the f dimensions data for measuring the distance from the group centre. After calculating the distances of all f dimension data between the target x and the group k, the group with the shortest distance from all f features of x is the group that the x belongs to. Therefore, all the classification targets which included in the same group G would be called the members of group. The K-mean clustering function is shown as Eq. 1.

$$\arg\min_{S} = \sum_{i=1}^{k} \sum_{x \in S_i} (x - \mu_i)^2 \tag{1}$$

3 Additional Information Required from Authors

For identify and recognize the fake news, in this research, the artificial intelligent service is proposed. There are several partitions included in the proposed services: Web Crawler procedure, Text Segmentation and Extension Method for Chinese language, Classification and Clustering Procedure, and Fake New Prediction and Warning Procedure.

3.1 Web Crawler Procedure

Since the proposed services for fake news identification and recognition is based on the artificial intelligence, to provide the training data for the artificial algorithm is needed.

First, the training data should be verified and proved as the true or fake news. Some organizations or online social media provide the corresponding information and proof of the news, such as Taiwan Fact Check Center [22]. Most of these proved news are based on the supervising method which would need the assistance of manual operation. Therefore, the fake news proved by these organizations are identified with high accuracy. However, The longer processing time by manual operation is needed. Therefore, these fake news proved with high accuracy could be used as the training data for the proposed artificial intelligent service. To continuously collect the proved fake news from these organizations or online websites, the web crawler is implemented in this research.

In this research, these organizations or social media websites are the available training data sources which provide the text content related to the proved fake news. Hence, by on demand given the source hyperlink addresses, the web crawler could search the corresponding web pages and download the text content. In addition, since the news is already proved as the fake news, using the titles of these fake news as the searching keywords would find more similar news. In other words, based on the proved fake news, web crawler could collect more similar or the same fake news through the web search engine.

Hence, to continuously collect the proved fake news, reliable organizations or online social media, and online search engines could be the training data sources. Through the web crawler with on demand given networked web sources and related keyword or titles, the training data corresponding to the independent group of proved fake news could be continuous obtained. Therefore, based on the proved training data, the proposed artificial intelligent service could enhance the performance with higher accuracy. In addition, every time a new fake news is proved, the proposed service could automatically obtain the proved fake news for further training.

3.2 Text Segmentation and Extension Method

In this research, the fake news are presented in Chinese language (Traditional Chinese in Taiwan). Hence, to segment the vocabularies of the article content written in Chinese is needed.

To segment the Chinese sentences in the text article, the ambiguity resolution rules for dividing the sentences and words should be considered. Maximum Matching segment method was proposed for Chinese vocabulary or segment finding [23]. By moving a chunk for including the neighbour words or vocabularies, the possible extended vocabularies with largest summation degree corresponding to morphemic freedom about one-character word, with minimum variance word length and maximum average word length, and the largest matching corresponding to the known vocabularies, can be found as the new keywords or extended vocabularies.

After recursively segment and extended the possible vocabularies, original text content of the fake news could be divided into various independent segments which indicate the various key vocabularies included in the fake news. However, most segment extensions are developed based on the largest word length and matching. Therefore, to find the new created vocabularies without including in the database is difficult. In this research, jieba algorithm is implemented.

Jieba algorithm is the method to find the vocabulary segmentation in Chinese [24]. To segment the words, two possible methods are used: Rule Segmentation and Statistical Segmentation. Rule Segmentation would match the available vocabulary based on the Trie-structure, also called dictionary tree. However, some new words may not be included in the dictionary. Hence, Statistical Segmentation was proposed. According to the appearance frequency of the connected words in the various text content, if the frequency is large, these connected words would be recognized as the new word or vocabulary. In other words, by on demand giving the threshold, the potential new vocabulary could be found.

All the possible words and vocabularies are described and structured as the directed acyclic graph according to the word graph scanned by prefix dictionary. Then, the path with maximum probability according to the dynamic programming could be found. In opposition to the existed vocabulary included in the dictionary, jieba also provides the method to find the new vocabulary. Based on the Hidden Markov Models (HMM), the vocabulary without included in the dictionary could be learn and found. In other words, based on the statistical results from various text data, the new created vocabulary could be found.

Therefore, by recursively execute the jieba algorithm, the text segmentation can be done. By using the Statistical Segmentation and Hidden Markov Models, the hidden new created vocabulary which is merged by several continuously connected words cal also defined recognized. In other words, the key segments of the text content in Chinese can be extended.

In addition to jieba algorithm, this research also uses the MapReduce algorithm for the segmentation statistics. Due to the requirements of Statistical Segmentation, huge data should be counted for the statistical results of individual vocabulary. According to the MapReduce, the various fake news key segments could be counted via the {Key, Value} format. Since the Hidden Markov Models is used to find the potential new vocabulary according to the statistical results, to recursively recognize and identify the appearance frequency of the connected words is needed. Considering the fake news spread today, personal names, some homophonic vocabularies or metaphor words developed by social media would be completely new and excluded in the dictionary. When the new found potential key vocabularies are various, to count the values of all found potential key vocabularies would require huge computing resource. Therefore, following the same MapReduce procedure, the recursive calculation and computing can be enhanced by the cloud computing with enough computing resource.

For example, the appearance frequency of the vocabularies which merged as the complete new vocabulary, such as a name, must be very closer to each other. Take Chinese word and vocabulary, "蔡" and "英文" as an example. "蔡" is the popular and normal last name in Chinese. "英文" is the vocabulary and proper nouns which the meaning is English. When combing these two words, a completely new vocabulary "蔡英文" appears. This new vocabulary points to a personal name instead of the original meaning, English. Suppose that the news are related to the specific person, then the name should be frequently mentioned. In other words, these two initial vocabularies would be mentioned together as many times as possible. Therefore, according to the jieba algorithm and Hidden Markov Models, the hidden or potential new vocabularies could be found. In other words, the new possible features of the fake news could be recognized.

3.3 Classification and Clustering Procedure

In addition to Hidden Markov Models and MapReduce, term frequency-inverse document frequency (TF-IDF) method is also used in this research. Term frequency-inverse document frequency is also the statistical method in weighting techniques for information exploration and text mining [25, 26]. Individual word or vocabulary can be evaluated the importance to a file set, text content, or the data in a corpus. If a word or vocabulary is more important to the file or data, the number of appearance related to this word or vocabulary, called term frequency (TF), will increase proportionally. In opposition, the number of appearance related to this specific word or vocabulary in the corpus will decrease in inverse proportion, called inverse document frequency (IDF). Therefore, the term frequency could be defined as Eq. 2.

$$\mathrm{tf}_{i,j} = \frac{n_{i,j}}{\sum_k n_{k,j}} \tag{2}$$

where $\mathrm{tf}_{i,j}$ indicate the importance of the ith vocabulary included in the jth data file. $n_{i,j}$ means the appearance numbers (times) of the ith vocabulary in the jth data file. The denominator is the sum of the number of occurrences of all words in the file. In addition, the inverse document frequency of the ith vocabulary can be counted according to Eq. 3.

$$\mathrm{idf}_i = \log \frac{|D|}{1 + \left|\{j : t_i \in d_j\}\right|} \tag{3}$$

where |D| indicates the total amount of the files in the corpus. Then, the denominator is the total amount of the files which include the ith vocabulary. In other words, if the $n_{i,j}$ is not zero, this jth data file which includes the ith vocabulary will be added and counted. Finally, the term frequency-inverse document frequency method can be obtained by Eq. 4.

$$\text{tfidf}_{\text{ij}} = \text{tf}_{\text{i,j}} \times \text{idf}_{\text{i}} \quad (4)$$

Therefore, term frequency-inverse document frequency method often filters out common words and keeps important vocabularies. In other words, some vocabularies are important in the specific files or fake news. These specific vocabularies can be found according to the by the proposed Text Segmentation and Extension Method. Since various vocabularies found by the term frequency-inverse document frequency method are important to different fake news, the relative vocabularies corresponding to the specific fake news could be used as the features.

During the artificial intelligent service training, the types or kinds of the fake news could be on demand defined. In other words, for the purpose of training the proposed service, the total group numbers of the fake news can be decided in advance. Therefore, by giving the suitable value of k in the Eq. 1, the proved fake news could be clustered or grouped according to the features found by the proposed method.

To integrate the term frequency-inverse document frequency (TF-IDF) method and the K-means clustering procedure, the Scikit-learn [27, 28] with the classification accuracy is used in this research. Scikit-learn is one of the machine learning method. It uses Numpy to perform the linear algebra and array operations efficiently. Scikit-learn provides various classification, regression and clustering algorithms, such as support vector machine, gradient boosting, term frequency-inverse document frequency, k-means clustering, DBSCAN, etc. Therefore, based on the Scikit-learn, the features of the fake news clustered in each independent group could be found and defined. In other words, the feature text segmentation and be used as the feature to define and describe the group of specific fake news. Therefore, by recursively executing the artificial intelligent training based on Scikit-learn, the existed and proved fake news could be recognized and clustered automatically without manual operation.

3.4 Fake New Prediction and Warning Procedure

According to the training data the identified fake news could be recognized and clustered according to the feature key vocabularies. In other words, based on the multiple similar fake news clustered in the same group, the feature key vocabulary segmentations could be found. These feature key vocabulary segmentations could be used as the definition or description of the corresponding fake news. Considering the intentional malicious fake news sources, the people or organizations would rapidly spread the news. Some intentional malicious news related to specific purposes or person from the personal social media, organization, or forums are fake and proved before. Since the fake news are spread though the Internet, to collect these existed and proved fake news could be easy. Hence, the personal social media, organization, or forum could be possibly included in the feature vocabulary segmentations of the segmentations.

When an unrecognized and unidentified new news is sent to the proposed artificial intelligent service, the Text Segmentation and Extension Method, and the Classification and Clustering Procedure would be executed recursively. Then, the feature key vocabulary segmentations of this unrecognized and unidentified news can be found. According to the Classification and Clustering Procedure, this news could be clustered into one proved group.

However, this news could be completely new and not belong to any proved fake news group. There are two possible warning methods proposed in this research: (1) the probability by evaluating the distance between nearest group and the unidentified news, and (2) cross feature vocabulary segmentations matching between multiple groups.

First, sometimes, an unrecognized news would be forcedly clustered into one fake news group just due to the K-means and DBSCAN classification/clustering algorithm. However, the detail of the content would not really match or completely the same as the fake news group it belongs to. Therefore, according to the distance between this news and its corresponding group centre, the probability can be evaluated as the Eq. 5:

$$p_{1new} = \frac{d_{boundary} - d_{new}}{d_{boundary}} \times 100\% \quad (5)$$

where the $d_{boundary}$ indicates the distance from the group centre to the group boundary after proposed Classification and Clustering Procedure. The value of d_n presents distance from the group centre to this new news. According to the probability p_{1new}, the similarity between the new news and the corresponding group could be evaluated.

Second, suppose that the new collected unrecognized news is a new type or content of the fake news that no group could cover it. It means that this unrecognized news could be a true news or undefined new fake news. According to the found feature vocabulary segmentations of all clustered groups, the potential possibility of fake news related to this new news could be evaluated. Although the content of the new news is different from all recognized fake news groups, the organization, social media, forums with malicious intention may be the same. Therefore, according to the cross feature vocabulary segmentations matching, one new news with many feature vocabulary segmentations matching with various fake news groups could be recognized as the potential fake news. The evaluation function can be shown as Eq. 6.

$$p_{2new} = \frac{m_i}{|V|} \times 100\% \quad (6)$$

where m_i means the matching numbers (times) of the ith feature vocabulary segmentation in the new news with all the feature vocabulary segmentations found in all fake news groups. $|V|$ shows the total amount of the feature vocabulary segmentations found corresponding to all fake news groups. Finally, by on demand giving the threshold values, if the probabilities, p_{1new} and p_{2new}, are larger than the threshold values, the fake news warning will be triggered.

4 Verification and Results

A prefilled copyright form is usually available from the conference website. Please send your signed copyright form to your conference publication contact, either as a scanned PDF or by fax or by courier. One author may sign on behalf of all of the other authors of a particular paper, providing permission has been given to do so. In this case, the author signs for and accepts responsibility for releasing this material on behalf of any and all co-authors. Digital signatures are not acceptable.

To verify the proposed artificial intelligent service with feature vocabulary segmentation searching and clustering, open source cloud computing platform is implemented. Based on the Linux type cloud computing, the virtual machines could be established according to the on demand configured templates. The proposed Web Crawler, Text Segmentation and Extension Method for Chinese language, Classification and Clustering Procedure are distributed implemented on different virtual machines. The data exchanging between virtual machines could be done though the internal network.

The artificial intelligent service is created based on the Scikit-learn machine learning method. Though the python programming language, the scikit-learn for Classification and Clustering Procedure and the jieba for Text Segmentation and Extension Method can be connected. In addition, the software, pandas, is included in the proposed service for data analysis. In this research, the various data are stored in the noSQL HBase. To be more easily presented by pandas, the data from the HBase will be represented as the multiple Excel files. Furthermore, to provide the visual operation interface, the python matplotlib which includes the NumPy is used to present graphic results.

Figure 1 presents the collected fake news from the Internet by the web crawler. These fake news are already proved and recognized by the Taiwan Fact Check Center [22]. According to the published content online, the fake news collected are divided into three partitions: News Title, News Article, and News Type. News Title is the title of this fake news spread through network or social media. It also could be a short abstract of this fake news. News Article is the content of the fake news written in Chinese. The name of the news reporters, organization or group which publishes or releases news, even the name of the news photographer are also included in the article. News Type indicates that the news is fake or true after proving. Since the articles are all proved by the Taiwan Fact Check Center, these online announced news are all recognized as the fake news. Therefore, by web crawler, these fake news could be collected as the proved training data for the proposed artificial intelligent service. In Fig. 1, the collected data is recorded in the json file.

Since the fake news collected from the Taiwan Fact Check Center are various, the feature vocabulary segmentations found by the proposed term frequency-inverse document frequency (TF-IDF) method of Text Segmentation and Extension Method would be also various. In other words, the total amount of the feature vocabulary segmentations would be huge. To clustering all fake news into multiple groups, K-means of Scikit-learn is used. However, the numbers of the groups should be

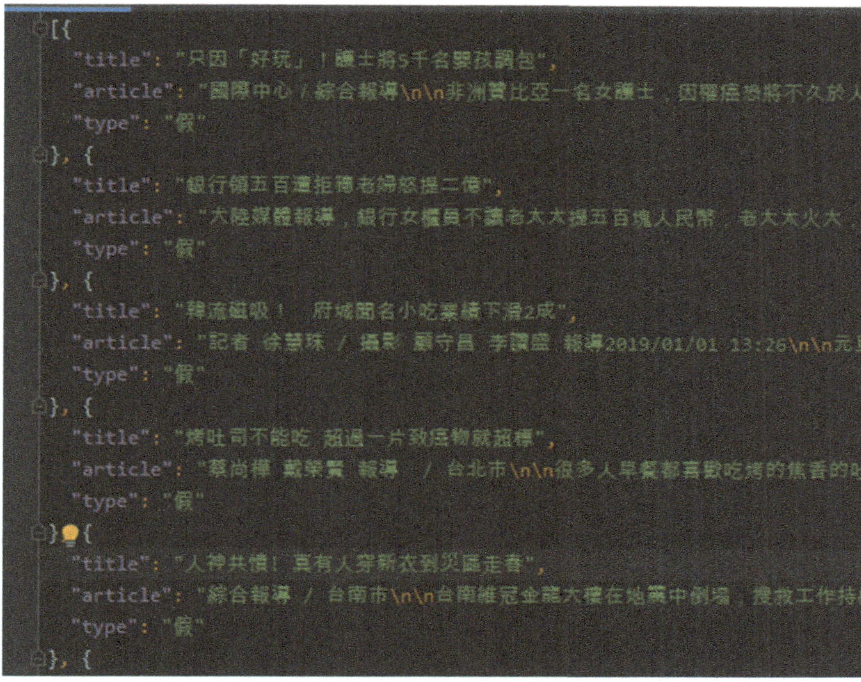

Fig. 1 The fake news collected by web crawler for artificial intelligent service training shown as a json file

decided. To find the suitable values of k for total k clustering groups, the Elbow Method is used [29]. Elbow method is heuristic in determining the total number of clusters. By recursively executing the K-means integrated with the Elbow methods, the best numbers of the clustering groups could be automatically found for the K-means algorithm. In this research, the collected fake news are chosen related to 4 topics. When more related fake news are clustered recursively, the required numbers of the clusters or groups are increased. Figure 2 presents the training based on the Elbow Method. The elbow point is located at k-value 4. The summation of the squared distance corresponding to the fake news in the related cluster or group can be evaluated with little differences. In other words, by using the Elbow Method, the minimum k-value for clustering the fake news into k groups could be found after recursively training. In other words, when any new topic of the fake news is given for training and recognition, the K-means algorithm with Elbow Method could provide and define the suitable clusters according to the feature vocabulary segmentations from the Text Segmentation and Extension Method.

However, considering the presentation, the multiple dimensions of each fake news group is not easy for representation as the two dimensional graphics. Therefore, Principal Component Analysis [30], a popular method for dimensionality reduction is

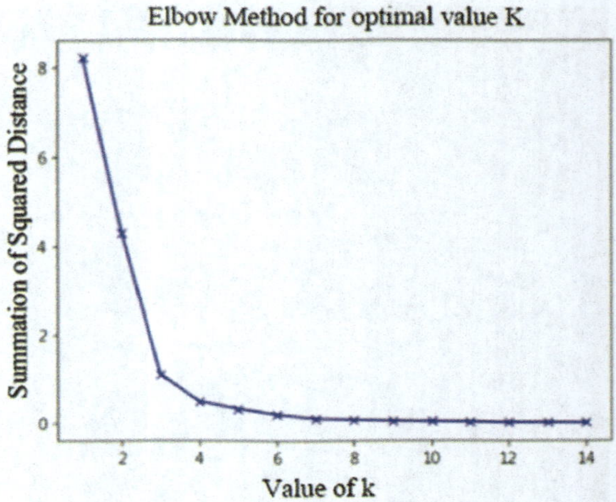

Fig. 2 The minimum k-value with little differences summation of the squared distance of for clustering the fake news into k groups could be found after recursively training

used in this research. Principal Component Analysis method reduces the number of features and minimizes the information loss during dimensionality reduction. Therefore, multiple dimensional fake news groups are represented as only two dimensional data. Note that these two dimensional data of all the fake news are only used for graphic representation. Figure 3 presents the fake news with two dimensional coordinates: pca_1 and pca_2 as the X-coordinate and Y-coordinate. After Classification and Clustering Procedure with multiple feature vocabulary segmentations, each fake news are clustered into corresponding groups. The number of the cluster in Fig. 3 indicates the fake news group number it belongs to. Comparing the cluster numbers with the X-coordinate and Y-coordinate based on Principal Component Analysis method, the most original features related to each fake news are saved. In addition, the coordinates of the fake news in the same group are closed to each other. In other words, the graphic based on the Principal Component Analysis method still correctly reflect the original state of the multiple dimensional fake news groups.

In this research, according to the fake news proved by the Taiwan Fact Check Center, four topics are selected for training the proposed artificial intelligent service. Based on these four topics, the web crawler found the related news from the search engine via the partial title of the selected fake news. To train the artificial intelligent service with the collected news from Internet as less as possible, only thirty additional news related to these four fake news topics are chosen. After Text Segmentation and Extension Method and Classification and Clustering Procedure, these total thirty-four fake news are successfully classified and clustered into corresponding four groups. Figure 4 presents the clustering training results of these fake news. In other words, the proposed Text Segmentation and Extension Method and Classification and Clustering Procedure are reliable.

Furthermore, these 34 fake news are also represented as the two dimensional coordinates data by the Principal Component Analysis method. Figure 5 presents the two dimensional coordinates clustering results of total 34 fake news. Obviously, 34

Fig. 3 Two dimensional coordinates of the multiple dimensional fake news after principal component analysis method

	A	B	C	D
1		pca_1	pca_2	cluster
2	0	-0.0368	-0.0274	1
3	1	-0.0382	-0.0274	1
4	2	-0.0335	-0.0199	1
5	3	-0.021	-0.0027	1
6	4	-0.0156	-0.0183	1
7	5	0.6327	0.0361	2
8	6	0.5629	0.025	2
9	7	0.6327	0.0361	2
10	8	0.5752	0.0444	2
11	9	0.5764	0.0246	2
12	10	0.6552	0.027	2
13	11	0.4363	0.034	2
14	12	0.2566	-0.0063	2
15	13	0.4403	0.0272	2
16	14	0.5476	0.0313	2
17	15	-0.2785	-0.5601	4
18	16	-0.294	-0.5579	4
19	17	-0.2886	-0.5712	4
20	18	-0.1801	-0.2795	1
21	19	-0.1856	-0.2906	1
22	20	-0.2994	-0.5675	4
23	21	-0.3038	-0.5781	4
24	22	-0.2082	-0.4033	4
25	23	-0.0842	-0.12	1
26	24	-0.146	-0.2283	1
27	25	-0.3494	0.6038	3
28	26	-0.2771	0.4151	3

Training Results of Total 34 Fake News

Group	
1	News #1. #2. #3, #4, #5, #19, #20, #24, #25
2	News #6. #7. #8, #9, #10, #11, #12, #13, #14, #15
3	News #26. #27. #28, #29, #30, #31, #32, #33, #34
4	News #16. #17. #18, #21, #22, #23

Fig. 4 The clustering training results of these 34 fake news

Fig. 5 Total 34 fake news are successfully divided into four independent clusters

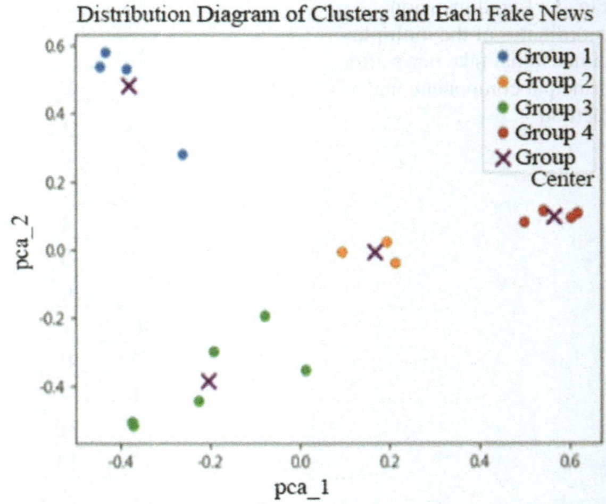

fake news are successfully divided into four independent clusters shown as Group1 to Group 4. Since the fake news would be rewritten or represented, some fake news in one cluster would be a little far away from the group center. However, due to the same topic of the fake news in the same group, the distance of the specific fake news from the group center is still short enough for clustering in the same group.

5 Discussion

According to the proposed artificial intelligent service, the verification shows that fake news already proved can be used as the training data. Then, another similar news with the same topic could be recognized and classified as the fake news corresponding to the correct cluster. Therefore, the potential fake news could be found and classified by the proposed artificial intelligent service.

However, to definitely recognize one news as the fake news is still difficult. First, the news may be announced as the fake news but finally recognized as the true one. Or in opposition, a news would be announced as the true news but at last recognized as the fake one. As the time goes on, the state of the news related to the true or fake would be changed. In other words, to predict the news based on the past recognized fake news still exists the risk of wrong recognition. However, this wrong recognition is due to the wrong training data for the artificial intelligent service but not the procedure of the proposed system.

In addition, social media or some online forums spread huge amount of news every day. There are some fake news spread on these media. Due to that the name of the social media or forum is repeatedly found as one of the feature vocabulary segmentation according to the proposed Text Segmentation and Extension Method, after across matching by Fake New Prediction and Warning Procedure, these news with the repeatedly appeared name would be easily classified as the fake news. In opposition, only the media always with the ability to verify all news could be excluded as the fake news media.

Moreover, judging a new news according to the history of the people, organization, even media companies may not be fair. Without enough proof and time, to provide the prediction of fake news could only depend on the percentage of the credibility and reliability corresponding to the multimedia, people, organizations, etc., but not depend on news itself.

6 Conclusion

In this research, the proposed Artificial Intelligent Service with Feature Segmentation Searching and Dynamic Clustering could find the feature vocabulary segmentations of each collected news though the web crawler and the proposed Text Segmentation and Extension Method. Based on the proposed Classification and Clustering Procedure, different topics of the fake news are identified and clustered into the corresponding fake news groups. In addition, the total number of the groups could be dynamically increased by recursively training in Classification and Clustering Procedure. Following the Fake New Prediction and Warning Procedure, the probability of the predicted fake news could be presented by comparing with the distance from the nearest fake news group center or the cross matching of the found feature vocabulary segmentations of the proved fake news.

Acknowledgements Thanks for the support of Cloud Computing and Intelligent System Lad. (CCIS Lab.) of National Formosa University.

References

1. Vosoughi, S., Roy, D., & Aral, S. (2018). The spread of true and false news online. *Science, 359*(6380), 1146–1151. https://doi.org/10.1126/science.aap9559
2. Gaozhao, D. (2020). Flagging fake news on social media: An experimental study of media consumers' identification of fake news. *SSRN Journal*. https://doi.org/10.2139/ssrn.3669375
3. Albahar, M. (2021). A hybrid model for fake news detection: Leveraging news content and user comments in fake news. *IET Information Security, 15*, 169–177. https://doi.org/10.1049/ise2.12021
4. Pivoda, K. (2019). Review of news literacy: Helping students and teachers decode fake news. *Education Review, 26*. https://doi.org/10.14507/er.v26.2459
5. Yang, S., Shu, K., Wang, S., Gu, R., Wu, F., & Liu, H. (2019). Unsupervised fake news detection on social media: A generative approach. In *Proceedings of the 33rd AAAI conference on artificial intelligence*. https://doi.org/10.1609/aaai.v33i01.33015644
6. Castillo, C., Mendoza, M., & Poblete, B. (2011). Information credibility on twitter. In *Proceedings of the 20th international conference on world wide web ACM* (pp. 675–684).

7. Wang, W. Y. (2017). Liar, liar pants on fire: A new benchmark dataset for fake news detection. *arXiv preprint*. arXiv:1705.00648
8. Kedar, H. E. (2019). Fake news in media art: Fake news as a media art practice versus fake news in politics. *Postdigital Science and Education, 2*, 132–146. https://doi.org/10.1007/s42438-019-00053-y.
9. Chen, K. J., & Bai, M.-H. (1998). Unknown word detection for Chinese by a corpus-based learning method. *International Journal of Computational linguistics and Chinese Language Processing, 3*(1), 27–44.
10. Kim, J., Tabibian, B., Oh, A., Scholkopf, B., & Gomez Rodriguez, M. (2018). Leveraging the crowd to detect and reduce the spread of fake news and misinformation. In *Proceedings of the 11th ACM international conference on web search and data mining* (pp. 324–332).
11. Jian, M.-S., Hong, W.-C., Tsai, S.-C., Chen, Y.-W., & Fan, C.-L. (2020). Based on automatic correlation keyword grouping and combination based deep information search corresponding to specific language big data—Case of leisure recreation. In *Proceedings of the IEEE ICACT* (pp. 372–377). https://doi.org/10.23919/ICACT48636.2020.9061481
12. Erlandsson, F., Nia, R., Boldt, M., Johnson, H., & Wu, S. (2015). Crawling online social networks. In *Proceedings of the 2015 2nd ENIC* (pp. 9–16).
13. Hadoop—Introduction. Available online : https://www.tutorialspoint.com/hadoop/hadoop_introduction.htm
14. Jian, M.-S., Fang, Y.-C., Wang, Y.-K., & Cheng, C. (2017). Big data analysis in hotel customer response and evaluation based on cloud. In *Proceedings of the IEEE ICACT* (pp. 791–795). https://doi.org/10.23919/ICACT.2017.7890201
15. Ma, W.-Y., & Chen, K. J. (2003). A bottom-up merging algorithm for Chinese unknown word extraction. In *Proceedings of the ACL workshop on Chinese language processing* (pp. 31–38).
16. Yu, C., Pengyu, M., Bessmertny, I. A., Platonov, A. V., & Poleschuk, E. A. (2017). Term extraction from Chinese texts without word segmentation. In *IEEE 11th international conference on application of information and communication technologies (AICT)* (pp. 1–4). https://doi.org/10.1109/ICAICT.2017.8687047
17. Saini, C., & Arora, V. (2016). Information retrieval in web crawling: A survey. In *International conference on advances in computing, communications and informatics (ICACCI)* (pp. 2635–2643). https://doi.org/10.1109/ICACCI.2016.7732456
18. Varghese, M., & Jose, V. (2018). Big data and cloud computing review and future trends. *International Journal of Computer Sciences and Engineering, 6*(12), 361–365. https://doi.org/10.26438/ijcse/v6i12.361365
19. Verma, J. P., Patel, B., & Patel, A. (2015). Big data analysis: Recommendation system with Hadoop framework. In *IEEE international conference on computational intelligence and communication technology* (pp. 92–97). https://doi.org/10.1109/CICT.2015.86
20. Hassan, M. U., Yaqoob, I., Zulfiqar, S., & Hameed, I. A. (2021). A comprehensive study of HBase storage architecture—A systematic literature review. *Symmetry, 13*(109). https://doi.org/10.3390/sym13010109
21. Jian, M.-S., & You, M.-S. (2016). Cloud based hybrid evolution algorithm for NP-complete pattern in nurse scheduling problem. *International Journal of Innovation, Management and Technology, 7*(5), 234–237.
22. https://tfc-taiwan.org.tw/articles/report
23. Tsai, C.-H. MMSEG: A word identification system for Mandarin Chinese text based on two variants of the maximum matching algorithm. http://technology.chtsai.org/mmseg/
24. Zhang, X., Wu, P., Cai, J., & Wang, K. (2010). A contrastive study of Chinese text segmentation tools in marketing notification texts. *Journal of Physics: Conference Series, 1302*(2). https://doi.org/10.1088/1742-6596/1302/2/022010
25. Hakim, A. A., Erwin, A., Eng, K. I., Galinium, M., & Muliady, W. (2014). Automated document classification for news article in Bahasa Indonesia based on term frequency inverse document frequency (TF-IDF) approach. In *6th International conference on information technology and electrical engineering (ICITEE)* (pp. 1–4). https://doi.org/10.1109/ICITEED.2014.7007894

26. Havrlant, L., & Kreinovich, V. (2017). A simple probabilistic explanation of term frequency-inverse document frequency (tf-idf) heuristic (and variations motivated by this explanation). *International Journal of General Systems, 46*(1), 27–36. https://doi.org/10.1080/03081079.2017.1291635
27. Hishamuddin, M. N. F., Hassan, M. F., Tran, D. C., & Mokhtar, A. A. (2020). Improving classification accuracy of Scikit-learn classifiers with discrete fuzzy interval values. In *International conference on computational intelligence (ICCI)* (pp. 163–166). https://doi.org/10.1109/ICCI51257.2020.9247696
28. Brites, D., & Wei, M. (2019). PhishFry—A proactive approach to classify phishing sites using SCIKIT learn. In *IEEE Globecom workshops (GC Wkshps)* (pp. 1–6). https://doi.org/10.1109/GCWkshps45667.2019.9024428
29. Syakur, M. A., Khotimah, B. K., Rochman, E. M. S., & Satoto, B. D. (2017). Integration K-means clustering method and elbow method for identification of the best customer profile cluster. *IOP Conference Series: Materials Science and Engineering, 336*.
30. Wold, S., Esbensen, K., & Geladi, P. (1987). Principal component analysis. *Chemometrics and Intelligent Laboratory Systems, 2*, 37–52.

Deep Learning with Self-Attention Mechanism for Fake News Detection

Ivana Cvitanović and Marina Bagić Babac

Abstract Nowadays, fake news is one of major concerns in our society, that is a form of news consisting of deliberate disinformation or hoaxes spread via traditional news media or online social media. Thus, this study aims to explore state-of-the-art methods for detecting fake news in order to design and implement classification models. Four different classification models based on deep learning with self-attention mechanism were trained and evaluated using current datasets that are available for this purpose. Three models explored traditional supervised learning, while the fourth model explored transfer learning by fine-tuning the pre-trained language model for the same task. All four models yield comparable results with the fourth model achieving the best classification accuracy.

Keywords Natural language processing · Fake news · Self-attention · Deep learning · Transfer learning

1 Introduction

With the rapid growth and the diversification of its usage, the Internet is becoming a huge part of daily life for most of the population. Social media and news platforms are especially becoming involved in every aspect of the daily routine for many people [1]. One of those aspects is news consumption regarding every possible domain like politics, pop culture, weather, etc. that is shared online every day [2]. More traditional news sources like TV, newspapers, etc. are getting significantly dominated by social media and online media platforms. The reason for this is the low cost and easy access to those news sources [3].

Often the news about the event is posted by those who are the witnesses, and the public is informed about it even before the traditional news outlets get to report on it.

I. Cvitanović · M. B. Babac (✉)
Faculty of Electrical Engineering and Computing, University of Zagreb, Unska 3, 10000 Zagreb, Croatia
e-mail: marina.bagic@fer.hr

M. Lahby et al. (eds.), *Combating Fake News with Computational Intelligence Techniques*, Studies in Computational Intelligence 1001,
https://doi.org/10.1007/978-3-030-90087-8_10

Furthermore, traditional news sources are recognizing this, and they are now sharing news online as well [4].

The positives of news accessibility and the speed of sharing can quickly become overshadowed by misinformation and fake news spreading. This is a serious threat that has arisen in recent years and already many studies are trying to solve this problem [5].

One of the phenomena that have additionally supported the spreading of fake news is software-controlled robots often called bots. These fake accounts are often used to produce or spread false information online and significant fraction of them do become viral [6].

There are many examples from recent history where fake news greatly impacted real-life events in a variety of different fields. One of the fields where fake news can be the most impactful is politics. People commonly read about politicians online and shape their opinion using the information that they are provided with [1]. Therefore, online social platforms are suitable for political opponents and their followers to try and manipulate the mass into shaping a bad opinion about a specific politician or a party.

Economic gain is also often an incentive behind creating and spreading fake news [3]. Posts on business news sites can go viral and manipulate the direction of the stock market [7]. Additionally, important global events like pandemics, terrorist attacks, and natural disasters oftentimes cause a great number of news articles and posts to be produced [8]. This is a usual occurrence because publishers want to get attention by exploiting a viral subject and sometimes, they share unreliable information [9].

While the best solution for detecting and stopping the fake news from spreading might be employing professionals to determine the authenticity of articles, for which there are websites that do just that, such as *Politifact* and *FactCheck*, the amount of news being produced on an everyday basis simply cannot be processed by humans based on how time and money consuming it would be [10]. Therefore, automatic fake news detection is introduced as a practical NLP problem useful to all online content providers and sets the goal of automatic fake news detection to reduce human time and effort while trying to stop the spreading of fake news [11].

This study aims to explore state-of-the-art methods for detecting fake news to design and implement classification models. Four different classification models based on deep learning with self-attention mechanism were trained and evaluated. Three models explored traditional supervised learning, while the fourth model explored transfer learning by fine-tuning the pre-trained language model for the same task. All four models yielded comparable results with the fourth model achieving the best classification accuracy. For this task, a novel dataset was constructed combining different existing public datasets.

2 Related Work

To understand and distinguish fake news, a solid definition of what is considered fake news is needed. There are various types of fake news such as hoaxes, false news, and disinformation. According to Zhou and Zafarani [12], these categories are distinguished based on the following three characteristics: authenticity, intention, and whether the observed information is news.

The definition of what is considered a piece of fake news varies in previous research. Kshetri and Voas [2] described the difference between misinformation and disinformation. They define a misinformation as just an incorrect piece of information, while disinformation has the intention of deceiving the public. They also classify fake news as a form of disinformation. Moreover, Allcott and Gentzkow [13] described a fake news article as an article that has been verified as false and intended to deceive a reader.

In addition, there are different perspectives to fake news detection, such as [12]: knowledge-based, style-based, propagation-based, and source-based detection.

In this work, the focus is on the style-based approach to detecting fake news, where the main task is exploiting the text and its characteristic to understand the nature of the text and use those characteristics to classify a piece of news as fake or not fake. Previous work regarding automatic style-based fake news detection either utilizes traditional machine learning or deep learning techniques. Usually, the earlier work focuses more on the first category, while newer works exploit deep learning for the same task [14].

As Shu et al. [15] stated, fake news publishers usually have an intent to mislead the reader and influence a larger group of people, and for that purpose, a specific style of writing is used, thus it is reasonable to make use of linguistic features that can capture writing styles that are characteristic for deception. These features capture the style of the text on four levels: lexicon, syntax, discourse, and semantic [16]. Moreover, Afroz et al. [17] argue that it is possible to use this linguistic and contextual information to achieve high accuracy in distinguishing deception from regular writing.

While using hand-crafted linguistic features and traditional machine learning techniques might help capture linguistic and psychological causes, these features fail to classify text well and thus limit the performance [18]. One of the arguments on why deep learning models are better for fake news detection is that hand-crafted linguistic features of fake news are still not fully understood and can vary across different topics and types of fake news [19].

On the other hand, some argue that deep learning-based models sometimes fail to interpret and explain why a certain news article is predicted to contain fake news [20]. While it might take more effort to interpret deep learning models, some of the recent works have been focusing on investigating and providing interpretations of their models and explanations on why something is classified as fake or not [21].

Qiao et al. [22] used a bidirectional neural network and train the classifier on interpretable language-based features for the task of detecting fake news. Language-based features make this model more white-box, therefore more explainable, than the other approaches that use similar deep learning techniques which use word embeddings.

Deep learning models have shown to be very efficient in detecting fake news [23]. Bajaj [24] used pre-trained embeddings for text representation and classified fake news using eight different models based on machine learning techniques such as RNNs, LSTMs, GRUs, and as well as an attention-like mechanism, however the best results are achieved when using an RNN architecture.

After the invention of a Transformer, an architecture that employs a self-attention mechanism [25] for sequence modelling, these kinds of architectures have been widely used in the field of text classification. Shu et al. [20] exploited both the text from the news article and comments for that article to develop a model that provides explainable classifications using the attention mechanism to extract latent features of a comment.

Moreover, Fang et al. [26] combined CNN and attention mechanism into their architecture that achieved 95.5% accuracy. They used multi-headed attention that allowed them to extract different relations in the same sentence. Their idea is that the attention mechanism can make up for the CNN not being able to obtain connections of distant words in the text.

Recent trends in NLP show an increasing interest in Transfer learning. Large pre-trained models trained on unsupervised textual data capture linguistic knowledge that can be reused and fine-tuned for downstream NLP tasks using supervised learning [27].

After the invention of BERT [28], a Bidirectional Encoder Representation from Transformers, that is an attention-based bidirectional network and other pre-trained models such as RoBERTa [29], ALBERT [30], these large-scale models that are pre-trained on large data corpora, became state-of-the-art models for NLP. Their main advantage is their ability to be fine-tuned for many different sequence model-ling tasks.

Gundapu and Mamid [31] constructed several different models to detect fake news regarding the coronavirus disease pandemic. They compared traditional supervised machine learning models such as Support Vector Machines (SVM), Passive Aggressive (PA) Classifier, and Multi-Layer-Perceptron (MLP) against deep learning models such as LSTM, CNN, and pre-trained transformer models such as BERT, ALBERT, and XLNet that are fine-tuned for this task. Transformer-based models showed best results.

Based on the previous work explained in this section, this work targets at using a Transformer-like model that employs a self-attention mechanism with some additional changes to better fit the task. Furthermore, to compare traditional learning methods with knowledge transfer methods, the BERT model will be fine-tuned for the same task.

3 Theoretical Framework

3.1 Word Embeddings

Text representation models such as *Bag-of-Words* (BoW) and *Term Frequency-Inverse Document-Frequency* (TF-IDF) are commonly used together with traditional machine learning approaches in many text classification tasks such as fake news detection [32].

The main drawback for these word representations is sparsity, that is the dimension of the vectors that represent each word is as big as the number of words in the vocabulary, and as a result, a small amount of information is held in a large representation space. Another problematic aspect of BoW models is that they commonly ignore semantic relationships between words in the vocabulary, which leads to even larger representation vectors that include synonymous and polysemous words [33].

Word embeddings [34] on the other hand, use dense word representations. Each word is represented by a fixed-size vector. These vectors contain real values that correspond to several different aspects of the word, placing each word as a point in a vector space. This also allows the representation of these vectors to have a lot smaller dimensions than the size of the vocabulary [35]. Word embeddings are learned by how words are used in the text, which results in words that are used in a similar way having similar representation. Word embeddings can be looked at as a model's understanding of the word meaning.

Almeida and Xexéo [36] described two method types for obtaining pre-trained word embeddings: prediction-based and count-based methods. These methods are established on the distributional hypothesis [37, 38] indicating that words used in a similar context have similar semantics.

Prediction-based methods, usually exploit language models to learn word embeddings through a process of unsupervised learning. This method of learning word representations was first demonstrated on training a neural language model to learn distributed word representations [35]. Count-based methods use word-context co-occurrences represented by word-context matrices. These two methods produce generic word-embeddings that are not specific to a certain task. Task-specific word embeddings can be obtained through a supervised learning process, in this case, word embedding vectors are learned jointly with all other neural network parameters.

While unsupervised text embedding methods are scalable and effective, creating word embeddings that are general and not specialized for a specific task does not compare. Tang et al. [33] argue that these methods result in inferior results when used in deep neural networks, proposing that this happens because unsupervised methods are not making use of the labels when learning word representations.

Our study exploits pre-trained word embeddings obtained using a count-based method, as well as task-specific embeddings. The aim is to establish if text classification models benefit from using pre-trained word representations.

The pre-trained embeddings that we use are Glove word embeddings [39]. In contrast to other similar models that use sparse word count matrices, Glove embeddings are based on ratios of co-occurrence probabilities. These embeddings are learned on five large online corpora. Each corpus is tokenized and lowercased, and the final vocabulary includes 400,000 frequent tokens. Task-specific word embeddings are obtained by randomly initializing parameters of the embedding layer and allowing them to be optimized through the process of training the network for fake news classification.

Bogoychev [40] distinguishes three different types of parameters of Transformers: embedding, attention, and feed-forward neural network parameters to determine how important each category of parameters is compared to the other, by freezing certain parameter categories during the training.

It is shown that the embedding layer parameters are the least important for text translation tasks but very important in language modelling.

Our study examines how important are the embedding layer parameters in the text classification tasks, by freezing the embedding layer when using pre-trained word embeddings.

3.2 Self-Attention Mechanism

Recurrent Neural Networks (RNNs), Long Short-Term Memory (LSTM), and Gated Recurrent neural networks (GRUs) have been a state-of-the-art approach for most of the NLP sequence modelling problems [25].

Encoder-decoder-based architectures [41] have been widely used for understanding sequential data in a large variety of NLP tasks primarily translation, but also classification, question answering, etc.

Encoder creates an internal representation. This representation is then passed to the decoder to create an output sequence; therefore, encoder-decoder structures are often used for sequence-to-sequence tasks. For sequence-to-label tasks such as text classification usually only the encoder component is used. The encoder produces the output sequence and additional layers are used on top of the encoder to map the output sequence to a label [42].

While initially encoder-decoder architectures mostly relied on using RNNs, these networks suffer from information getting wiped out after its propagated over multiple time steps [43]. This problem is caused by a vanishing gradient [44]. LSTM and GRU networks attempt to solve this problem by introducing gates that control the flow of information [45]. While they do manage to store longer-term information, they are still hard to train.

Another solution for the stated shortcomings is described as a Transformer [25], that is an encoder-decoder architecture that relies on the self-attention mechanism to compute sequence representations. Both the encoder and decoder use self-attention as the primary mechanism for capturing the semantic meaning of the text. Another advantage of the self-attention compared to RNNs and CNNs is that it is less computationally expensive [46].

Since our study is focusing on building a sequence-to-sequence model for fake news detection, only the encoder component will be explored. The encoder consists of a self-attention layer and a feed-forward network. The encoder receives a sequence of tokens, each represented by a vector. Self-attention associates each token representation in the sequence with all other input token representation in the sequence. This allows the model to encode the words based on how they interact with other words in the sequence [47].

A common way to transform a sequence-to-sequence architecture into a sequence-to-label architecture is to apply a global average pooling. Global average pooling is applied to the final output sequence, followed by a linear layer to map the results to a vector where each dimension represents one of the classes [25]. Using the global average pooling method for classification, instead of fully connected layers helps to prevent overfitting [48].

Self-attention is a sequence-to-sequence transformation that maps the input sequence into a vector representation, picking up its contextual information. Tokens from an input sequence are initially represented by embeddings (vectors). These input vectors $(x_1, x_2, x_3, \ldots, x_n)$ are used to calculate the weights like dot products (1).

$$w'_{ij} = x_i^T x_j \tag{1}$$

Since the value of these weights is bigger when using embedding that has a greater dimension, this dot product is scaled. In this work, the dot product is scaled by dividing the dot product with the square root of embedding dimension (2), in this case d = 100, so the weights are scaled by 10.

$$w'_{ij} = \frac{x_i^{\top} x_j}{\sqrt{d}} \tag{2}$$

The *softmax* function is applied to normalize the weights to values between zero and one (3), finally, the output vectors are calculated as a weighted average over all input vectors.

$$w_{ij} = \frac{\exp w'_{ij}}{\sum_j \exp w'_{ij}} \tag{3}$$

Input token representations are transformed using linear operations (4, 5, 6) into three different vectors (*query, key, value*), This introduces additional controllable parameters (matrices W_q, W_k and W_v) in the self-attention layer which allows it to modify incoming vectors to suit the roles of *key, query*, and *value* derived from the same input vector [25].

$$q_i = W_q x_i \tag{4}$$

$$k_i = W_k x_i \tag{5}$$

$$v_i = W_v x_i \tag{6}$$

Then the weights are calculated:

$$w'_{ij} = \frac{q_i^\top k_j}{\sqrt{d}} \tag{7}$$

$$w_{ij} = \frac{\exp w'_{ij}}{\sum_j \exp w'_{ij}} \tag{8}$$

And the outputs of the attention layer are calculated as a weighted average off all input vectors v_j:

$$y_i = \Sigma_j w_{ij} v_j \tag{9}$$

Another idea that is built into this mechanism is multi-headed attention. Input token representations (embeddings) are split into several same-sized vectors and on each of those vectors, the attention function is applied in parallel. This results in the model being able to learn information from different subspaces of the input representation space [25].

To inject a sensitivity to word order, input tokens should have information about the position of the token in the input sequence [25]. Some of the options that can be used are position embeddings, position encodings, and relative positions. When using position embedding, the problem could occur when the model faces sequences longer than the ones it came across in training [49]. This is the reason why in this work position encodings are used.

Position encodings are obtained by passing the position of the token into a sine or a cosine function depending on the position and the model dimension [50]. The model dimension is the dimension of the word embeddings that the model uses. Final embedding vectors are obtained by summing word embeddings and positional encodings.

3.3 Sliding Window Self-Attention

While self-attention is a powerful tool, one problem that transformer-based models face is the inability to process long sentences. The cause of this problem is the self-attention mechanism that scales quadratically with the length of the input sequence [51].

In theory, self-attention can be applied to very long input sequences. However, because of the computational usage and memory consumption that scales quadratically with the input sequence length, applying regular self-attention to very long input sequences can be very inefficient and costly.

Motivated by this limitation several researchers attempt to develop new methods to overcome the stated problem in a way that does not decrease the performance of the model. Some of the naive approaches might be cutting the input or segmenting only the most important parts which can be no longer than the predefined fixed length. For example, the BERT transformer that usually has an input sequence length of 512 tokens, was adapted [52]. The long input sequence was segmented into smaller overlapping pieces and each piece was passed through the model. In the end, an additional transformer was employed with an LSTM network to perform the classification.

Another way to reduce memory and computational cost is to sparse out the attention mechanism, by not allowing every element to attend to every other element in the input. Child et al. [53] called transformers that use sparse self-attention Sparse Transformers, where they managed to scale down the time and memory cost of their model from $O(n^2)$ to $O(n\sqrt{n})$ by factorizing the self-attention operation. This work also follows this lead, as the dataset established for this task has a significant amount of input sequences that have more than 1000 tokens. Performing the full attention matrix computations on this dataset could lead to unnecessary complexity. This is especially unnecessary since models that employ similar changes to sparse out the attention mechanism show promising results [54, 55].

Our study refers to the sparse self-attention [51], where the sliding window attention incorporates local context into the model, and the dilated sliding window is used to additionally expand the receptive field. Another related concept is the global attention, which takes care of times when the models fuse the representation of the entire sequence into just one token, these tokens are called special tokens and they are often used in language modelling networks such as BERT. The global attention mechanism makes it so that these special tokens attend to all other tokens in the input [56].

In our work, only the sliding attention mechanism is explored. Sliding window [51] allows all tokens in the sequence to attend only to a pre-defined fixed number of tokens to the left and right (Fig. 1b), as opposed to full self-attention where every token attend to every other token in the sequence (Fig. 1a).

The overall number of tokens that each token attends to is called the window size. Beltagy et al. [51] also described the occurrence when the multiple layers of self-attention, with the fixed window size, are stacked, the receptive field widens. If l is the number of layers of self-attention and w is the window size (in case all layers

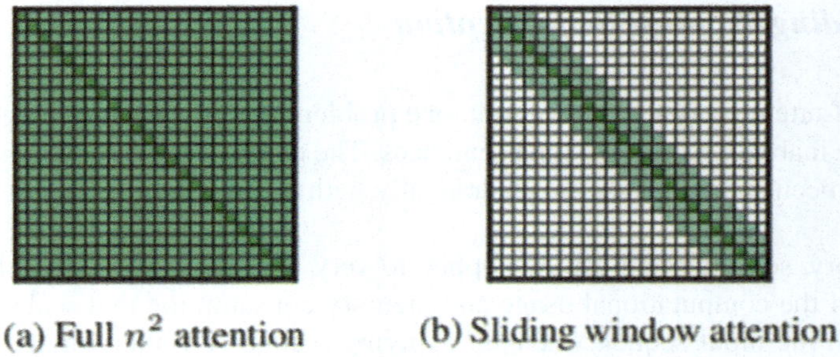

Fig. 1 **a** Full self-attention, **b** sliding window self-attention [51]

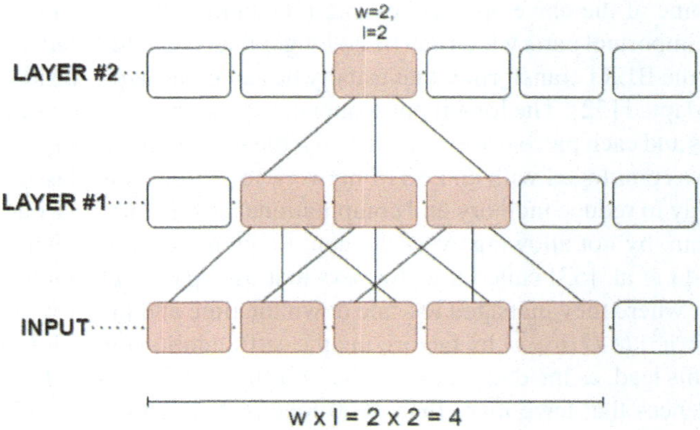

Fig. 2 Receptive field for sliding window attention [51]

use same window size), this creates a receptive field of size $l \times w$ in the top layer (Fig. 2).

In this implementation, the fixed sized window is used, the same sized window is used in all the attention layers. Each token attends to itself and 256 tokens on each side. And by using 4 self-attention layers, the effective receptive field on the top layer is 2048 tokens.

3.4 Bidirectional Encoder Representations from Transformers—BERT

In recent years, traditional supervised learning that requires defining a specialized model, strictly defined task and a large corpus of training data is commonly replaced with transfer learning methods. This is especially the case in sequential transfer learning, where transfer learning models achieve state-of-the-art results when fine-tuned for different NLP tasks [57].

Transfer learning or knowledge transfer methods follow the idea that knowledge learned previously can be reused in different tasks even if the domains and distributions are different [58].

Traditional machine learning methods to train the model on the specific domain and tasks, naturally use data labelled within the same domain and task. On the other hand, with transfer learning knowledge acquired from training the model on a specific data and task can be reused in models with a different domain and task [59].

Based on the taxonomy of transfer learning [60], sequential transfer learning is a subtype of transfer learning where source (pre-training) and target (fine-tuning) models are trained for different tasks, also labelled data is used only in the fine-tuning part. Additionally, the two different tasks are learned sequentially.

A paper published by the researchers working for Google AI Language [28] describes a model for language representations called BERT, that is Bidirectional Encoder Representations from Transformers. This model is designed with the purpose of pre-training on unlabeled text to obtain text representations, that can later be fine-tuned for different NLP tasks.

BERT architecture is based on the standard transformer architecture [25]. Transformer has two components, the encoder that encodes the input text and the decoder that generates prediction which is depended on the task. The goal of BERT is to generate a language model so only the encoder component is employed [61], where bidirectional language representations are important. While the other models process the text from left-to-right or right-to-left, some even process the text in both ways and then concatenate the embeddings to obtain final language representations [62], BERT takes in the whole input sequence at once.

In this work, the BERT model is fine-tuned for text classification in the domain of fake news detection. This model is used to analyze how transfer learning methods compare to traditional supervised learning methods within this specific task of fake news detection.

4 Experimental Setup and Results

4.1 Dataset Collection

There are many public fake news datasets that can be used to train a fake news detection classifier, such as FakeNewsNet datasets [63], LIAR dataset [64], ISOT dataset [65], etc. Structures of these datasets vary. Some contain only the text and the label while others have additional data about the speaker, timestamp, etc. While this additional information can be useful for different classification models, this work focuses mainly on the language used in the news articles.

Since the task is formulated as a binary classification, the aim is to create a dataset that contains the text and the label for each of the data points. The text component contains the full text of the article. The label is a numeric value, eighter *zero* or *one*, where *one* indicates that the article contains fake news and *zero* indicates that the article does not include fake news.

For this purpose, public fake news datasets are adjusted and reused. Combining multiple datasets could help avoid bias in the network as well as include articles from different domains and subjects [61].

Oshikawa et al. [5] determined three different categories of public fake news datasets: claims, entire articles, and Social Networking Services (SNS) data. With the aim for the final model to be able to classify news articles with different number of tokens, both short claims and full articles are incorporated into the dataset.

For the purposes of this study, three public datasets were combined to build a final dataset:

1. Kaggle Fake News Dataset (https://www.kaggle.com/c/fake-news/data)
   This dataset was created for developing machine learning programs that can identify when an article might be fake. The data is split into two different files, train data and test data. Train file consists of labelled data. Labelled data has 5 fields (*id, title, author, text, label*) but only the text and the label fields were reused in the new dataset.
2. ISOT Fake News Dataset [65]
   This dataset consists of 44,898 articles, where real news articles are collected by crawling Reuters.com and fake news are obtained from unreliable sources flagged by Politifact (https://www.politifact.com/). The data is split into two files one containing fake and one containing real news. Both files have 4 fields: *title, text, subject*, and *date*, again only the text is used and given a label based on the file it belongs to. This dataset is also labelled by the subject it belongs to, where subjects used in this dataset are Government-news, Middle East, US news, Left-news, Politics and News.
3. LIAR Dataset [64]
   This dataset uses short claims that were collected from the Politifact website. The dataset uses 6 different labels: *pants-fire*, *false*, *barely-true*, *half-true*, *mostly-true*, and *true*. Since in this case the model preforms a binary classification, the data is relabeled, such that all the claims that fall into first three categories are labelled as *fake* and claims that fall into last three categories are labelled as *true*. This dataset mostly uses short claims. Each claim also has additional fields like the name of the speaker, his job title, party affiliation etc., but only the text and the label are exploit for this task.

Table 1 shows number of data points in the dataset per category for each of the datasets explored. The last row shows the distribution of articles in the final dataset.

All the datasets used are balanced, making the final dataset also balanced. The average number of tokens is 443, and the maximum number of tokens is 24,234. Most of the sequences have less than 500 tokens, with 54,113 sequences in that category. However, the number of sequences in other categories is not to be disregarded.

Table 1 Final dataset

Datasets	Articles labelled as fake news	Articles labelled as true news	Total articles
Kaggle dataset	10,374	10,387	20,761
ISOT dataset	23,481	21,417	44,898
LIAR dataset	5041	6466	11,507
Final dataset	38,896	38,270	77,166

4.2 Models Design and Implementation

For the purposes of classifying the news article based on its credibility, four models were designed, implemented, and evaluated. First three of the four models use supervised learning and a transformer-based architecture. This is used to try exploring as many tokens in the input sequence as possible, without significantly increasing the complexity of the model. For these models, the sliding window self-attention is used.

The final of the four models is built by fine-tuning the pre-trained BERT model. The pre-trained model is trained on unlabeled data and is designed to be fine-tuned for a variety of different down-stream tasks. All models are trained and evaluated using the same dataset, as described in previous section.

Sparse self-attention models. The *PyTorch* machine learning library [66] is utilized to build a transformer model. This model is modified to behave as a classifier. The classifier is trained using the prepared dataset to distinguish fake news articles from the true news articles.

The neural network model is constructed following the general transformer structure [25], but instead of the regular self-attention, sliding window self-attention is used.

The encoder component of the transformer architecture is used to build a language model. To adjust the model to preform binary classification the output of the encoder component is passed through additional linear classification layer. The classification layer generates the probability distribution over the two classes.

The architecture of the model consists of the embedding layer, set of transformer blocks and linear layer that maps the output sequence to a probability distribution as shown in Fig. 3.

The first layer of the network is the embedding layer, this layer maps input token indexes to their corresponding embedding vectors. Values in the embedding vectors are parameters of the network and can be configured as trainable or not.

In this part of the network, positional encoding is used. Final embedding vector is a sum of the original embedding vector and the positional encoding. Obtained embeddings are then passed through a set of 4 transformer blocks. These blocks are the main part of the network since they carry out multi-headed self-attention operations. The number of heads used in the multi-headed attention layers is 5.

Figure 4 shows the architecture of the transformer block. Referring to the original architecture of the self-attention encoder [25], the transformer block is designed with

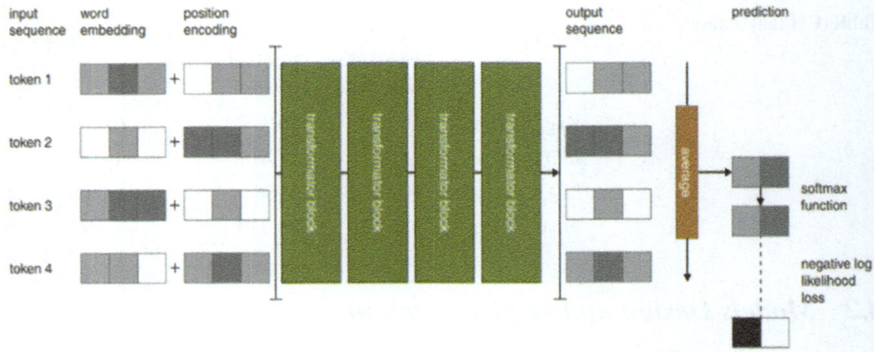

Fig. 3 Model architecture

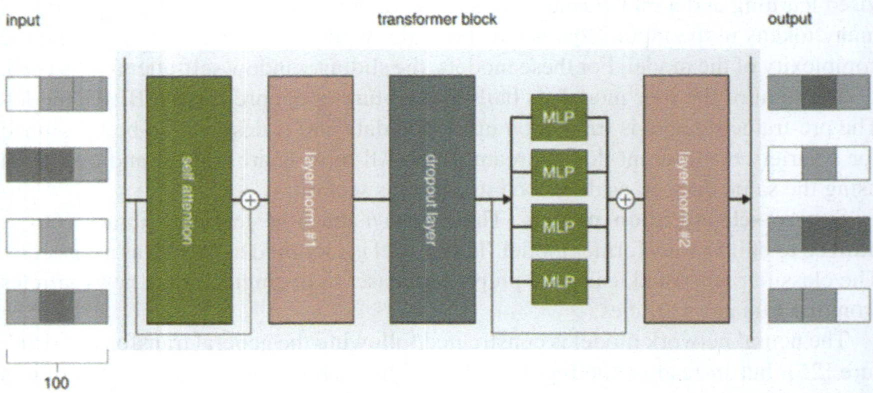

Fig. 4 Transformer block architecture

two sublayers. The first sublayer is a layer of self-attention and the second one is a fully connected feed-forward network, each followed by a layer of normalization. Residual connections are added around both layers. These connections allow easier training of deep neural networks while still allowing it to appropriately map inputs to the desired outputs [67]. This is accomplished by employing the actual layers to learn residual functions ($F(x) := H(x) - x$), where H(x) is the actual desired output and x is the input. Since both input and output are of the same dimension, the residual connection propagates the input to the output allowing the final output to be $F(x) + x = H(x)$ as desired.

Additionally, a dropout layer between two of the sublayers is added to prevent the model from overfitting [68]. The dropout rate used in all the models is 0.25.

All self-attention layers use sparse, sliding window self-attention as described in previous section. Window size is 256, meaning each token is attending to 256 tokens on the left, 256 tokens to the right and itself. This creates receptive field of 2048 tokens in the top transformer.

The global average pool is applied to the final output sequence, and the linear layer with the two-dimensional output is added to represent the probabilities of the input sequence residing to each of the classes. Final output of the network is obtained by applying a *softmax* function, that normalizes probabilities to values between zero and one.

Using the architecture from previous section, three models are trained, with the difference between these three models in the embedding layer. The neural network models can either use pretrained embeddings or employ the embedding layer to learn embedding parameters jointly with the rest of the model's parameters. By using different embedding layer configurations, the aim is to explore how important embedding layer is when using the self-attention based neural network for text classification.

Model 1: The first model uses fully trainable embedding layer, meaning word embeddings are initialized to random values and learned together with other network parameters.

Model 2: The second model also uses a trainable embedding layer, but the embedding layer is initialized with pre-trained Glove embeddings.

These two networks make use of supervised learning for word-representations. The aim is for the embedding layer to hold semantic relationships between words as well as their correlation with the two categories that the models are designed to distinguish.

Model 3: The third model also utilizes pre-trained Glove embedding, but this time the parameters of the embedding layer are frozen. This would mean that all word embedding vectors will stay the same throughout the training of the network and will only hold general semantic relationships between words and not their correlation to classes the model is set to recognize.

Table 2 shows numbers of trainable parameters for each of the models. One can conclude that the number of parameters that needs to be trained when freezing the embedding layer is significantly smaller then when optimizing the embedding layer as well. This is because the number of parameters in the embedding layer is somewhat larger than the number of all other parameters in the network.

For all models, the same dimension of the embedding vectors is used ($d = 100$). Other than embedding layer configurations, the setup for all the three models is the same.

Table 2 Number of trainable parameters per model

Model	Number of trainable parameters
1	31,897,802
2	31,640,702
3	483,802

Since the Transformer based models work with raw input sequences, the text does not require a lot of pre-processing. Pre-processing steps include tokenization and padding or truncating if needed.

The first step in the training process is randomly splitting the data into a train and test set. The dataset is split dedicating 80% of data for training and the remaining 20% for validation.

Moreover, tokenized train set is used to construct a vocabulary. All the words that occur in the train set are given an index. The vocabulary maps all tokens used in the training data to an index, since the model works with tokens represented as indexes.

All the model's trainable parameters are optimized using a gradient decent optimization. More specifically mini-batch gradient decent is used. Using mini-batch gradient decent allows a stable convergence by lowering the variance of parameter updates [59]. Batch size is set to 4. After each batch, the gradients are calculated and all the parameters in the network are updated.

Since the *softmax* activation function is used on the output layer of the network, to calculate the loss function *negative log-likelihood* function (NLL) is used [69].

Adaptive Moment Estimation (Adam) is used as a gradient decent optimization algorithm. This optimization algorithm generates adaptive learning rates that are specific for each parameter in the network [19].

Learning rate is set to 0.0001 and a learning rate scheduler is used to adapt the learning rate during the training. Furthermore, to prevent exploding gradients while training the network, gradient scaling is used. This means that if the vector norm for a gradient is higher than 1, then all values in in the vector will be scaled so that the norm of the vector is 1.

Each model is trained for total of 5 epochs. After each epoch, the model is evaluated using the validation set. Table 3 shows how accuracy and loss changed during training and validation across all five epochs for the first model that does not use pre-trained embeddings. The model achieves 91.6% accuracy after five epochs of training. Each epoch takes around 21 min to train on the Tesla T4 GPU that is used for this model.

When analyzing how training and validations loss change over five epochs, one can notice (Fig. 5) that the validation loss starts to increase after the third epoch and is higher than the training loss, this indicates that the model is overfitting. This means that the model is becoming more specialized to the data in the train dataset. When testing this model on the validation set that was not used to adjust the parameters

Table 3 Training and validation loss and accuracy for the model that uses custom word embeddings

epoch	Training loss	Valid. loss	Valid. Accur.	Training time	Validation time
1	0.406	0.273	0.890	0:21:47	0:01:55
2	0.235	0.310	0.892	0:21:43	0:01:54
3	0.206	0.226	0.909	0:21:46	0.01:54
4	0.194	0.229	0.914	0:21:45	0:01:53
5	0.185	0.231	0.916	0:21:37	0:01:54

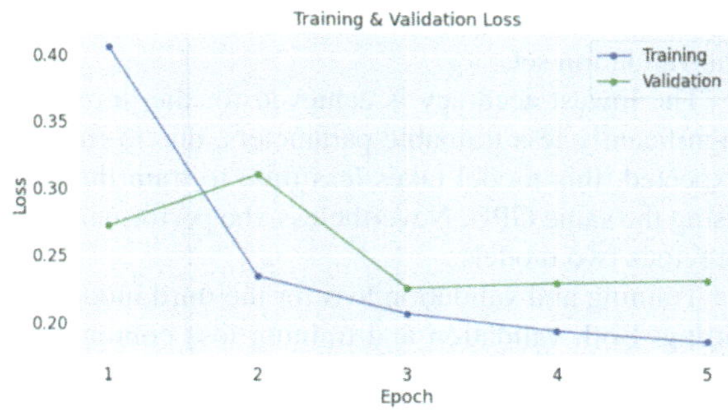

Fig. 5 Training and validation loss for the model that uses custom word embeddings

of the network the loss is bigger, because model learned information specific to the train set that cannot be applied on the validation set.

The second model that utilizes the pre-trained word embeddings shows better results than the previous model (Table 4). Final accuracy achieved is 93%. For the training of this model the same GPU Tesla T4 is used, but training time is slightly higher than in the previous model.

The second model also shows better loss then the first model, the validation loss is lower than the training loss (Fig. 6), indicating the model is picking up on the information that is general for all articles and not just ones from the training set. After the fourth epoch of training, the model starts to show signs of overfitting as the

Table 4 Training and validation loss and accuracy for the model that uses pre-trained word embeddings

epoch	Training loss	Valid. loss	Valid. Accur.	Training time	Validation time
1	0.378	0.272	0.908	0:23:23	0:02:15
2	0.226	0.223	0.909	0:23:34	0:02:16
3	0.199	0.171	0.926	0:23:41	0:02:17
4	0.184	0.168	0.928	0:23:31	0:02:18
5	0.174	0.183	0.930	0:23:36	0:02:17

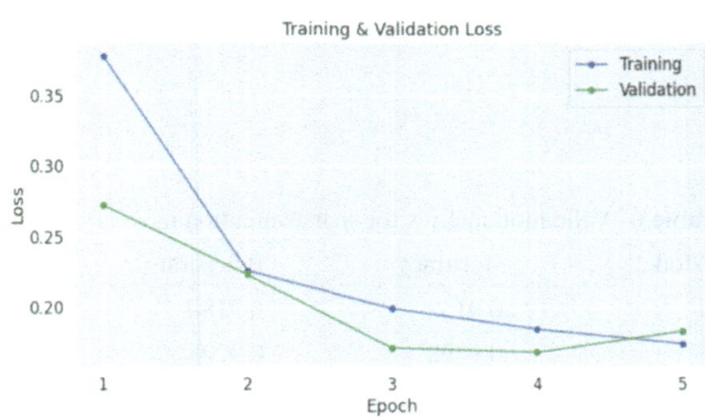

Fig. 6 Training and validation loss for the model that uses pre-trained word-embeddings

validation loss starts to increase. Nevertheless, the model achieves good accuracy on the validation set.

The lowest accuracy is achieved for the third model (Table 5). This model has significantly less trainable parameters, due to freezing of the embedding layer. As expected, this model takes less time to train then the previous two models, when using the same GPU. Nevertheless, the performance of this model is not far from the previous two models.

Training and validation loss for the third model (Fig. 7) also show signs of overfitting. Both validation and training loss continue to decrease, indicating that this model could be trained on more epochs to achieve better accuracy.

All three models are compared based on the accuracy, precision, recall and F1 score (Table 6). The first row shows metrics for the model that uses custom embeddings. The second row in the table shows metrics for the model that uses pre-trained embeddings and keeps the embedding layer trainable. Finally, the third row shows metrics for the model that uses pre-trained word embeddings, but the embedding

Table 5 Training and validation loss and accuracy for the model that uses pre-trained word embeddings and does not update embedding layer parameters

epoch	Training loss	Valid. loss	Valid. Accur.	Training time	Validation time
1	0.409	0.376	0.871	0:20:06	0:02:02
2	0.276	0.243	0.905	0:20:09	0:02:03
3	0.232	0.238	0.913	0:19:56	0:02:01
4	0.212	0.236	0.912	0:20:09	0:02:03
5	0.203	0.223	0.913	0:20:07	0:02:01

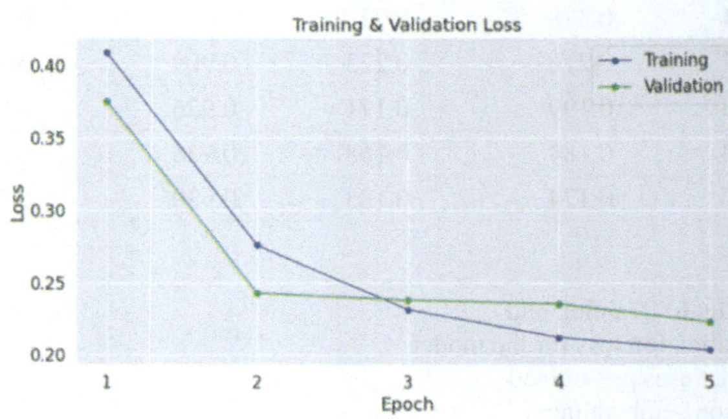

Fig. 7 Training and validation loss for the model that uses pre-trained word embeddings and does not update embedding layer parameters

Table 6 Validation metrics for sparse attention models

Model	Accuracy	Precision	Recall	F1
1	0.9158	0.9180	0.9162	0.9171
2	0.9298	0.9298	0.9298	0.9298
3	0.9129	0.9162	0.9134	0.9148

layer is not trainable. The results for all the three models are relatively close in comparison.

The model that uses pre-trained word-embeddings and keeps the embedding layer trainable preforms the best across all the forementioned metrics. The model that uses pre-trained embeddings and does not optimize the embedding layer parameters has the lowest score for all the metrics.

Even though the third model is significantly smaller than the other two models in terms of the number of trainable parameters, the performance is not far off from the first model.

Fine-tuning BERT model. Following the fine-tuning process described in [70], the model is fine-tuned for fake news detection using our dataset. To fine-tune a BERT model for a text classification task, *Hugging Face Transformer* library is used. The library provides several different BERT modifications. All the different configurations are based on two general BERT models; BERT base and BERT large.

For this experiment, BERT-base-uncased model was used. BERT-base uses 12 transformer blocks as well as 12 self-attention heads in each of the transformer blocks. Hidden layer is 768-dimensional, and the model has total of 110 M parameters. Maximum sequence length of the input is 512 tokens. Uncased version converts all tokens to lower case. For the text classification task, BERT for text classification model was used. To adjust the BERT model for text classification, this implementation used a linear layer that is placed on top of the regular BERT architecture to yield probabilities for each class.

For this experiment, the dataset is also split using 80%-20% ratio for the train and test set, respectively. The pre-trained model is trained for additional three epochs. To optimize the parameters of the network, the mini-batch gradient decent method was used.

Like the setup described for the previous models, the mini-batch gradient decent method is used. Batch size used for this experiment is five. The optimizer used to fine-tune BERT is the AdamW optimizer [71], which is the improved version of Adam optimizer. The learning rate is set to 0.00002 and the learning rate scheduler is used.

Table 7 shows how loss and accuracy change throughout the training and validation, as well as the accuracy after each epoch. Losses for the training and validation are also visualized in Fig. 8.

Since the validation loss was growing after the first epoch and was higher than the training loss, this showed that model was starting to overfit after the first epoch. The model achieved 94.1% accuracy after 3 epochs of training. Each epoch took a

Table 7 Training and validation loss and accuracy during fine-tuning the BERT model

epoch	Training loss	Valid. loss	Valid. Accur.	Training time	Validation time
1	0.183	0.146	0.939	1:54:12	0:10:02
2	0.144	0.191	0.940	1:54:02	0:10:01
3	0.102	0.270	0.941	1:53:57	0:10:01

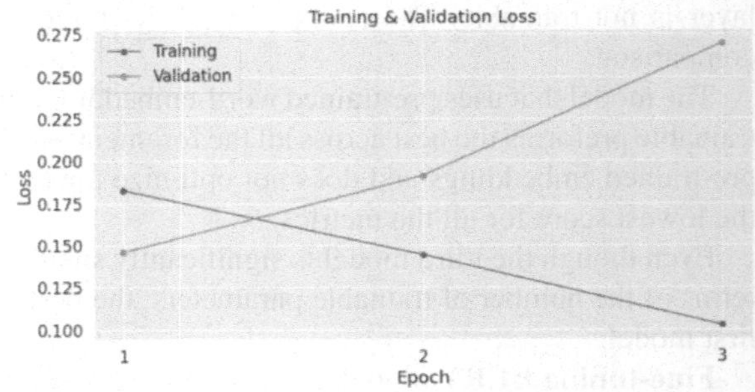

Fig. 8 Training and validation loss for BERT fine-tuning

little less than 2 h on the same GPU Tesla T4 as previous 3 models. Thus, fine-tuned BERT model showed better results than the previously described models, proving that transfer learning is a promising approach for the detection of fake news.

5 Conclusion

While fake news is not a new phenomenon [12], recently it attracts increasingly more public attention. The possible cause is that fake news can be created and published online faster and cheaper when compared to traditional news media [20].

From the results presented in this study, it can be concluded that language models that use self-attention are suitable for text classification and fake news detection. Additionally, using a sliding window self-attention can be useful when trying to reduce memory and computational usage.

Our main results show high overall accuracy performance in fake news detection. The best result was obtained when utilizing a pre-trained model, showing that transfer learning methods hold enough general knowledge about the language and can be fine-tuned for fake news detection. The pre-trained model achieved the best results even after just one epoch of fine-tuning, proving pre-trained BERT model can be fine-tuned easily using the labelled data [72]. The second-best result was achieved when exploiting pre-trained word embeddings and making them more task specific through the process of supervised learning, proving that deep learning methods do profit from using labels when creating language representations.

While the best results are achieved when specializing the embedding layer for a task, using pre-trained word embeddings, and freezing the embedding layer also shows decent results when compared to all others. Therefore, it can be concluded that the task-specific word embeddings can help a language model to classify text but are not crucial to achieve good results. The experiment also showed that pre-trained word embeddings are a powerful tool and can be used as a static part of the neural network, which can be helpful when trying to reduce the number of trainable parameters.

The results from our models outperformed the results presented in [64], which explored fake news detection using LIAR dataset based on integrated text with metadata and hybrid convolutional neural network. Our approach has also outperformed results based on Fast-TransE model [73], which combined machine learning algorithms and knowledge graphs, as well as the CNN based architecture from [61].

However, certain limitations of our approach should also be addressed. It is worth noting, that even with the given dataset, only textual part of the information was used. Our study did not include domain knowledge related features, leaving a room for future work to explore the possibility of using metadata, such as the source, age of the news, author of the article, and user response.

In addition, social and psychological factors play an important role in fake news gaining public trust and further facilitate the spread of fake news [74], thus more research should combine different aspects of the phenomenon. For example, user connections and their historical activity on social media should be analyzed where the user can reflect how they relate to the spread of fake news. It is important to incorporate the human mental condition with the user's historical data, which can better analyze the user's activity [75].

Another limitation of this study is that the overall training process is computationally intensive due to large amount of data and model parameters, thus requiring many computational and memory resources to shorten the training process. In addition, this study limits to currently publicly available datasets, so the novel datasets would confirm if a model working well on a specific dataset, generalize well.

The future research should experiment with the network by freezing some of the other parts of it, such as the feed forward layer parameters to establish the importance of other parameters in the network compared to the embedding layer parameters. Then, incorporating dilated sliding window and global self-attention in addition to the already explored sliding window self-attention with the aim to achieve even better results. In addition, fine-tune other existing pre-trained models for the same task would determine how they perform compared to the BERT model.

Regarding the future applications of this work, it should be also used for detection of fake URLs, fake blogs, fake printed papers, etc. In addition, it can be extended for other multimedia data like video, or blogs, and applied to the detection of multi-source and multi-class fake news.

References

1. Bian, T., Xiao, X., Xu, T., Zhao, P., Huang, W., Rong, Y., & Huang, J. (2020). Rumor detection on social media with bi-directional graph convolutional networks. *Proceedings of the AAAI Conference on Artificial Intelligence, 34*(1), 549–556.
2. Kshetri, N., & Voas, J. (2017). The economics of "fake news." *IT Professional, 19*(6), 8–12.
3. Vosoughi, S., Roy, D., & Aral, S. (2018). The spread of true and false news online. *Science, 359*(6380), 1146–1151.
4. Przybyla, P. (2020). Capturing the style of fake news. *Proceedings of the AAAI Conference on Artificial Intelligence, 34*(1), 490–497.

5. Oshikawa, R., Qian, J., & Wang, W. Y. (2018). A survey on natural language processing for fake news detection. arXiv:1811.00770
6. Shao, C., Ciampaglia, G. L., Varol, O., Flammini, A., & Menczer, F. (2017). The spread of fake news by social bots. arXiv:1707.07592
7. Ruchansky, N., Seo, S., & Liu, Y. (2017). Csi: A hybrid deep model for fake news detection. In *Proceedings of the 2017 ACM on conference on information and knowledge management* (pp. 797–806).
8. Tenney, I., Xia, P., Chen, B., Wang, A., Poliak, A., McCoy, R. T., Pavlick, E., et al. (2019). What do you learn from context? Probing for sentence structure in contextualized word representations. In *International conference on learning representations.*
9. Starbird, K., Maddock, J., Orand, M., Achterman, P., & Mason, R. M. (2014). Rumors, false flags, and digital vigilantes: Misinformation on Twitter after the 2013 Boston marathon bombing. In *Conference 2014 Proceedings.*
10. Chung, J., Gulcehre, C., Cho, K., & Bengio, Y. (2014). Empirical evaluation of gated recurrent neural networks on sequence modeling. arXiv:1412.3555
11. Pérez-Rosas, V., Kleinberg, B., Lefevre, A., & Mihalcea, R. (2017). Automatic detection of fake news. arXiv:1708.07104
12. Zhou, X., & Zafarani, R. (2020). A survey of fake news: Fundamental theories, detection methods, and opportunities. *ACM Computing Surveys (CSUR), 53*(5), 1–40. https://doi.org/10.1145/3395046
13. Allcott, H., & Gentzkow, M. (2017). Social media and fake news in the 2016 election. *Journal of Economic Perspectives, 31*(2), 211–236.
14. Chernyavskiy, A., Ilvovsky, D., & Nakov, P. (2021). Transformers: "The end of history" for NLP?. arXiv:2105.00813
15. Shu, K., Sliva, A., Wang, S., Tang, J., & Liu, H. (2017). Fake news detection on social media: A data mining perspective. *ACM SIGKDD Explorations Newsletter, 19*(1), 22–36.
16. Conroy, N. K., Rubin, V. L., & Chen, Y. (2015). Automatic deception detection: Methods for finding fake news. *Proceedings of the Association for Information Science and Technology, 52*(1), 1–4.
17. Afroz, S., Brennan, M., & Greenstadt, R. (2012). Detecting hoaxes, frauds, and deception in writing style online. In *IEEE symposium on security and privacy* (pp. 461–475). IEEE.
18. Girgis, S., Amer, E., & Gadallah, M. (2018). Deep learning algorithms for detecting fake news in online text. In *13th International conference on computer engineering and systems (ICCES)* (pp. 93–97). IEEE.
19. Kingma, D. P., & Ba, J. (2014). Adam: A method for stochastic optimization. arXiv:1412.6980
20. Shu, K., Cui, L., Wang, S., Lee, D., & Liu, H. (2019). Defend: Explainable fake news detection. In *Proceedings of the 25th ACM SIGKDD international conference on knowledge discovery and data mining* (pp. 395–405).
21. Zhang, J., Cui, L., Fu, Y., & Gouza, F. B. (2018). Fake news detection with deep diffusive network model. arXiv:1805.08751
22. Qiao, Y., Wiechmann, D., & Kerz, E. (2020). A language-based approach to fake news detection through interpretable features and BRNN. In *Proceedings of the 3rd international workshop on rumours and deception in social media (RDSM)* (pp. 14–31).
23. Agarwal, A., Mittal, M., Pathak, A., & Goyal, L. M. (2020). Fake news detection using a blend of neural networks: An application of deep learning. *SN Computer Science, 143*(3), 1–9. https://doi.org/10.1007/s42979-020-00165-4
24. Bajaj, S. (2017). The pope has a new baby! Fake news detection using deep learning. CS 224N (pp. 1–8).
25. Vaswani, A., Shazeer, N., Parmar, N., Uszkoreit, J., Jones, L., Gomez, A. N., & Polosukhin, I. (2017). Attention is all you need. arXiv:1706.03762
26. Fang, Y., Gao, J., Huang, C., Peng, H., & Wu, R. (2019). Self multi-head attention-based convolutional neural networks for fake news detection. *PloS One, 14*(9), e0222713.
27. Durrani, N., Sajjad, H., & Dalvi, F. (2021). How transfer learning impacts linguistic knowledge in deep NLP models?. arXiv:2105.15179

28. Devlin, J., Chang, M. W., Lee, K., & Toutanova, K. (2018). Bert: Pre-training of deep bidirectional transformers for language understanding. arXiv:1810.04805
29. Liu, Y., Ott, M., Goyal, N., Du, J., Joshi, M., Chen, D., Stoyanov, V., et al. (2019). RoBERTa: A robustly optimized BERT pretraining approach. arXiv:1907.11692
30. Lan, Z., Chen, M., Goodman, S., Gimpel, K., Sharma, P., & Soricut, R. (2019). Albert: A lite bert for self-supervised learning of language representations. arXiv:1909.11942
31. Gundapu, S., & Mamid, R. (2021). Transformer based automatic COVID-19 fake news detection system. arXiv:2101.00180
32. Al Asaad, B., & Erascu, M. (2018). A tool for fake news detection. In *20th International symposium on symbolic and numeric algorithms for scientific computing (SYNASC)* (pp. 379–386). IEEE.
33. Tang, J., Qu, M., & Mei, Q. (2015). Predictive text embedding through large-scale heterogeneous text networks. In *Proceedings of the 21th ACM SIGKDD international conference on knowledge discovery and data mining* (pp. 1165–1174).
34. Mikolov, T., Sutskever, I., Chen, K., Corrado, G., & Dean, J. (2013). Distributed representations of words and phrases and their compositionality. In *Advances in neural information processing systems, 26*. NIPS.
35. Bengio, Y., Ducharme, R., Vincent, P., & Janvin, C. (2003). A neural probabilistic language model. *The Journal of Machine Learning Research, 3*, 1137–1155.
36. Almeida, F., & Xexéo, G. (2019). Word embeddings: A survey. arXiv:1901.09069
37. Firth, J. R. (1957). A synopsis of linguistic theory, 1930–1955. *Studies in Linguistic Analysis*.
38. Harris, Z. S. (1954). Distributional structure. *Word, 10*(2–3), 146–162.
39. Pennington, J., Socher, R., & Manning, C. D. (2014). Glove: Global vectors for word representation. In *Proceedings of the 2014 conference on empirical methods in natural language processing (EMNLP)* (pp. 1532–1543).
40. Bogoychev, N. (2020). Not all parameters are born equal: Attention is mostly what you need. arXiv:2010.11859
41. Cho, K., Van Merriënboer, B., Bahdanau, D., & Bengio, Y. (2014). On the properties of neural machine translation: Encoder-decoder approaches. arXiv:1409.1259
42. Bahdanau, D., Cho, K., & Bengio, Y. (2014). Neural machine translation by jointly learning to align and translate. arXiv:1409.0473
43. Hu, D. (2019). An introductory survey on attention mechanisms in NLP problems. In *Proceedings of SAI intelligent systems conference* (pp. 432–448). Springer.
44. Bengio, Y., Simard, P., & Frasconi, P. (1994). Learning long-term dependencies with gradient descent is difficult. *IEEE Transactions on Neural Networks, 5*(2), 157–166. https://doi.org/10.1109/72.279181
45. Thota, A., Tilak, P., Ahluwalia, S., & Lohia, N. (2018). Fake news detection: A deep learning approach. *SMU Data Science Review, 1*(3), 10.
46. Vaswani, A., & Huang, A. (2020). *Self-attention for generative models*. Presentation slides at Stanford University. https://web.stanford.edu/class/cs224n/slides/cs224n-2019-lecture14-transformers.pdf
47. Alammar, J. (2018). The illustrated transformer. https://jalammar.github.io/illustrated-transformer/
48. Lin, M., Chen, Q., & Yan, S. (2013). Network in network. arXiv:1312.4400
49. Nasir, J. A., Khan, O. S., & Varlamis, I. (2021). Fake news detection: A hybrid CNN-RNN based deep learning approach. *International Journal of Information Management Data Insights, 1*(1), 100007. https://doi.org/10.1016/j.jjimei.2020.100007
50. Singhania, S., Fernandez, N., & Rao, S. (2017). 3HAN: A deep neural network for fake news detection. In: D. Liu, S. Xie, Y. Li, D. Zhao, & E. S. El-Alfy (Eds.), *Neural information processing*. ICONIP 2017. Lecture notes in computer science (vol. 10635). Springer. https://doi.org/10.1007/978-3-319-70096-0_59
51. Beltagy, I., Peters, M. E., & Cohan, A. (2020). Longformer: The long-document transformer. arXiv:2004.05150

52. Pappagari, R., Zelasko, P., Villalba, J., Carmiel, Y., & Dehak, N. (2019). Hierarchical transformers for long document classification. In *IEEE automatic speech recognition and understanding workshop (ASRU)* (pp. 838–844). IEEE. https://doi.org/10.1109/ASRU46091.2019.9003958
53. Child, R., Gray, S., Radford, A., & Sutskever, I. (2019). Generating long sequences with sparse transformers. arXiv:1904.10509
54. Cui, B., Li, Y., Chen, M., & Zhang, Z. (2019). Fine-tune BERT with sparse self-attention mechanism. In *Proceedings of the 2019 conference on empirical methods in natural language processing and the 9th international joint conference on natural language processing (EMNLP-IJCNLP)* (pp. 3539–3544).
55. Huang, L., Yuan, Y., Guo, J., Zhang, C., Chen, X., & Wang, J. (2019). Interlaced sparse self-attention for semantic segmentation. arXiv:190712273
56. Rogers, A., Kovaleva, O., & Rumshisky, A. (2020). A primer in BERTology: What we know about how BERT works. *Transactions of the Association for Computational Linguistics, 8*, 842–866. https://doi.org/10.1162/tacl_a_00349
57. Ruder, S. (2019). The state of transfer learning in NLP. Sebastian ruder. https://ruder.io/state-of-transfer-learning-in-nlp/
58. Pan, S. J., & Yang, Q. (2009). A survey on transfer learning. *IEEE Transactions on Knowledge and Data Engineering, 22*(10), 1345–1359. https://doi.org/10.1109/TKDE.2009.191
59. Ruder S. (2021). Transfer learning—machine learning's next frontier. Sebastian ruder. https://ruder.io/transfer-learning/index.html#whatistransferlearning
60. Ruder, S., Peters, M. E., Swayamdipta, S., & Wolf, T. (2019). Transfer learning in natural language processing. In *Proceedings of the 2019 conference of the North American chapter of the association for computational linguistics: Tutorials* (pp. 15–18).
61. Mouratidis, D., Nikiforos, M. N., & Kermanidis, K. L. (2021). Deep learning for fake news detection in a pairwise textual input schema. *Computation, 9*(20). https://doi.org/10.3390/computation9020020
62. Peters, M. E., Ammar, W., Bhagavatula, C., & Power, R. (2017). Semi-supervised sequence tagging with bidirectional language models. arXiv:1705.00108
63. Shu, K., Mahudeswaran, D., Wang, S., Lee, D., & Liu, H. (2020). FakeNewsNet: A data repository with news content, social context, and spatiotemporal information for studying fake news on social media. *Big Data, 8*(3), 171–188. https://doi.org/10.1089/big.2020.0062
64. Wang, W. Y. (2017). "Liar, liar pants on fire": A new benchmark dataset for fake news detection. In *Proceedings of the 55th annual meeting of the association for computational linguistics* (vol. 2, pp. 422–426). https://doi.org/10.18653/v1/P17-2067
65. Ahmed, H., Traore, I., & Saad, S. (2018). Detecting opinion spams and fake news using text classification. *Journal of Security and Privacy, 1*(1).
66. Paszke, A., Gross, S., Massa, F., Lerer, A., Bradbury, J., Chanan, G., & Chintala, S. (2019). PyTorch: An imperative style, high-performance deep learning library. In *Advances in Neural Information Processing Systems* (vol. 32, pp. 8024–8035). Curran Associates, Inc.
67. He, K., Zhang, X., Ren, S., & Sun, J. (2016). Deep residual learning for image recognition. In *Proceedings of the IEEE conference on computer vision and pattern recognition* (pp. 770–778).
68. Srivastava, N., Hinton, G., Krizhevsky, A., Sutskever, I., & Salakhutdinov, R. (2014). Dropout: A simple way to prevent neural networks from overfitting. *The Journal of Machine Learning Research, 15*(1), 1929–1958.
69. Miranda, L. J. (2017). Understanding softmax and the negative log-likelihood. https://ljvmiranda921.github.io/notebook/2017/08/13/softmax-and-the-negative-log-likelihood/
70. McCormick, C., & Ryan, N. (2020). BERT fine-tuning tutorial with Pytorch. mccormickml.com. https://mccormickml.com/2019/07/22/BERT-fine-tuning/
71. Loshchilov, I., & Hutter, F. (2017). Decoupled weight decay regularization. arXiv:1711.05101
72. Sun, C., Qiu, X., Xu, Y., & Huang, X. (2019). How to fine-tune BERT for text classification?. In *China national conference on Chinese computational linguistics* (pp. 194–206). Springer.
73. Shakeel, D., & Jain, N. Fake news detection and fact verification using knowledge graphs and machine learning. https://doi.org/10.13140/RG.2.2.18349.41448

74. Deepak, S., & Chitturi, B. (2020). Deep neural approach to fake-news identification. *Procedia Computer Science, 167*, 2236–2243.
75. Liu, N. F., Gardner, M., Belinkov, Y., Peters, M. E., & Smith, N. A. (2019). Linguistic knowledge and transferability of contextual representations. arXiv:1903.08855

Modeling and Solving the Fake News Detection Scheduling Problem

Said Aqil and Mohamed Lahby

Abstract The large amount of information on social networks makes it difficult to distinguish fake news from good news. Society and individuals are often affected by the extent of the dissemination of false information. Currently the fake news detection (FND) is becoming a concern for scientific researchers. In this article, we propose a new approach to modeling the FND document processing process using a flow shop scheduling problem (FSSP). Our approach consists in modeling the FND data processing in the FSSP. In fact, in this model the process is composed of several phases arranged in series forming a FSSP in the flow work. The documents are considered as tasks to be processed available in front of the first machine of the model will be processed by all the machines in the same order. In order, to solve this problem we propose three metaheuristics: the iterated greedy (IG), the genetic (GA) and the artificial bee colony (ABC) algorithms. The goal is to minimize the maximum completion time of the set of documents called the makespan ($C_{\max}$). Various instances are tested by varying the number of documents in front of the queue in the all machines. The simulation results showed that the IG algorithm performs the best compared to other algorithms.

Keywords Fake news detection · Flow shop scheduling problem · Metaheuristic · Makespan

1 Introduction

Technological development all over the world has provided instant access to information for large numbers of people. This, easily makes it possible to share false information by a general public. Social media are often forced to disseminate false

S. Aqil (✉)
Faculty of Sciences and Technology of Mohammedia, University Hassan II, Casablanca, Morocco

M. Lahby
University Hassan II, Casablanca, Morocco

M. Lahby et al. (eds.), *Combating Fake News with Computational Intelligence Techniques*, Studies in Computational Intelligence 1001,
https://doi.org/10.1007/978-3-030-90087-8_11

information without bothering to verify it. The overriding objective is to increase the number of viewers for televisions, or subscribers for users of internet social networks, etc. The problem arises when information turns out not to be true, which makes it possible to lose the credibility of the dissemination resource. Ensuring the accuracy of information has become a concern for the industrial in this field. Often a research team is responsible for ensuring the accuracy of the information before releasing it. The use of computer models to classify information [1] and distinguish variation from false is a task entrusted to specialties in this field. The models developed are often software allowing to convert the documents in multidimensional vector form in order to facilitate the manipulation of the data set.

The information is grouped by similarity and classes of equivalence. Classification models are primarily based on scientific approaches such as support machine vector (SVM). The news is therefore studied according to the history of each news group, for example the BBC, the news group, etc. The contradiction detection procedure is based on mathematical techniques using powerful data manipulation algorithms. Indeed, the approaches resulting from the regression technique and scheduling algorithms constitute powerful platforms for detecting anomalies in the processing information. Various methods are applied to classify and identify FNDs problem. We distinguish more particularly artificial intelligence [2, 3] algorithms such as neural networks [4, 5] or machine learning [6] algorithms, stochastic descending gradient algorithms, etc. They are also put in hybrid form with other optimization techniques and procedures. In [7–9], we find nature-inspired metaheuristic approaches also being implemented in solving the FND problem. The studies carried out are tested on a set of real databases. Especially on social networks like Facebook, Twitter, etc.

The data processing process is forced to detect a fairly large amount of data. An ordering technique is necessary to find the best sequencing of documents in the different phases of the process [10]. The techniques generally use scheduling algorithms to process the total number of documents in a shorter time. These approaches are implemented with hybrid techniques in nested procedure in order to reduce the computation time as much as possible. Indeed, meta-heuristics [11] are often an approach favored by developers of data analysis software. Recently, resolution approaches based on meta-heuristics[12] have experienced a great expansion. Industrial applications are diverse covering different fields of optimization. We find more particularly the problems related to the production industry and the service industry. For this reason, we have opted to model a new problem recently frequented as FND. This modeling is based on the FSSP [13, 14] model often treated in the manufacturing industry, but rarely treated in the service industry such as the information industry. This contribution therefore constitutes a new contribution and new application of the FSSP manufacturing process.

Our paper is organized as follows: in Sect. 2, we propose the modeling of the FND problem by a FSSP. In Sect. 3, we develop the resolution approaches, in Sect. 4 we present the simulation study of metaheuristic tests. In Sect. 5, we conclude our work by highlighting the strengths of our contribution as well as the possible perspectives.

2 Modeling of the FND Problem with FSSP

The process of handling data in the FND problem consists of a series successive phases. Generally the first phase represents the classification and read of the document, the second phase concerns for tokenization, the third phase is analyzes the document, the fourth phase is intended to stemming the content, the five phase completes the processing process by digitizing the encoding of the document. The proposed problem thus consists in treating a set of document represented by a set of task $\mathcal{T} = \{T_1, T_2, \ldots, T_n\}$ designating the documents to be studied on a set of processors $\mathcal{P} = \{P_1, P_2, \ldots, P_m\}$ designating the different processing phases. Each document T_j requires a processing time t_{ji} on the processors P_i. The objective is to find the best sequence $\tau = [\tau_1, \tau_2, \ldots, \tau_n]$ to run in the processing process modeling the flow work of documents. Knowing that each document T_j has an finish time FT_{ji} on the processor P_i. Figure 1 indicates the FND model in FSSP of five processors.

The model will be described by the following equations

$$FT_{\tau_1 1} = t_{\tau_1 1} \tag{1}$$

Equation (1) denotes the finish time of the first task on the first processor.

$$FT_{\tau_j 1} = FT_{\tau_{(j-1)} 1} + p_{\tau_j 1};\ j = 2, \ldots, n \tag{2}$$

Equation (2) represents the finish time of the task T_j in the first processor.

$$FT_{\tau_1 i} = C_{\tau_1 (i-1)} + p_{\pi_1 i};\ i = 2, \ldots, m \tag{3}$$

Equation (3) represents the finish time of the first task on the other processors.

$$FT_{\tau_j i} = \max\left\{FT_{\tau_{(j-1)} i}, FT_{\tau_j (i-1)}\right\} + p_{\tau_j i}; \tag{4}$$

$$i = 2, \ldots, m$$

$$j = 2, \ldots, n$$

Equation (4) designates the finish time of the task T_j in the others processors.

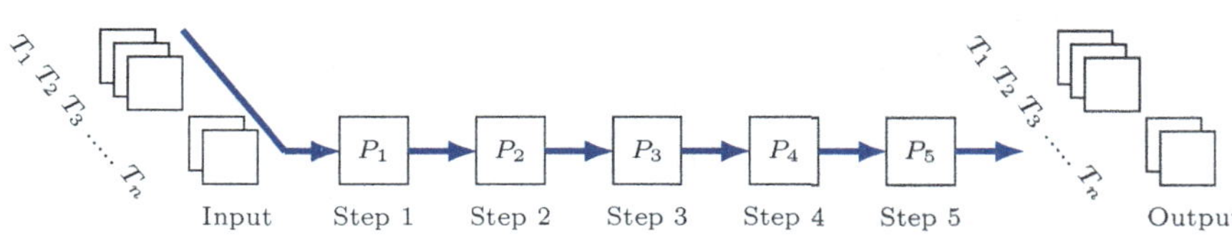

Fig. 1 Manufacturing process for FND problem

$$\text{Makespan} : C_{\max} = \max\{FT_{\tau_j m}\}, \quad j = 1, \ldots, n \tag{5}$$

Equation (5) denotes the maximum finish time called the makespan and is denoted $C_{\max}$.

In this model the objective is to find the best sequence τ^* in the space of neighboring solutions Π such that $C_{\max}(\tau^*) < C_{\max}(\tau)$ with $\tau \in \Pi$. We also retain the scheduling hypothesis in the FSSP workshop. Each T_j document admits m successive operations on all processors. A single processor processes only one activity at a time and interruption of processing is not allowed. Each activity can only be processed by one processor. The buffer stock between the processor is of unlimited capacity.

We present here the illustration numerical for scheduling problem in FND manufacturing process. The processing times is given by Table 1.

The sequence to be scheduled in the production system is $\tau = [4, 3, 1, 5, 2]$. We respect the constraints imposed by the FSSP model. We note that the maximum duration is given by the finish time of the last task T_2 in this sequence. In the flowchart given by Gantt diagram of the Fig. 2, the makespan value is $C_{\max} = 43$ unit of time. In the next section, we describe the solution approach based on three inspired nature metaheuristics: IG, ABC and GA. The principle consists in disturbing the current solution by a neighborhood exploration technique in order to improve the current solution. This technique is repeated until a stop criterion is reached that allows us to stabilize and converge our metaheuristics towards their best solutions.

Table 1 The processing time of five jobs on five processors

Task	T_1	T_2	T_3	T_4	T_5
P_1	4	6	6	3	5
P_2	6	6	3	5	7
P_3	7	5	4	5	5
P_4	4	2	5	5	4
P_5	5	4	6	4	3

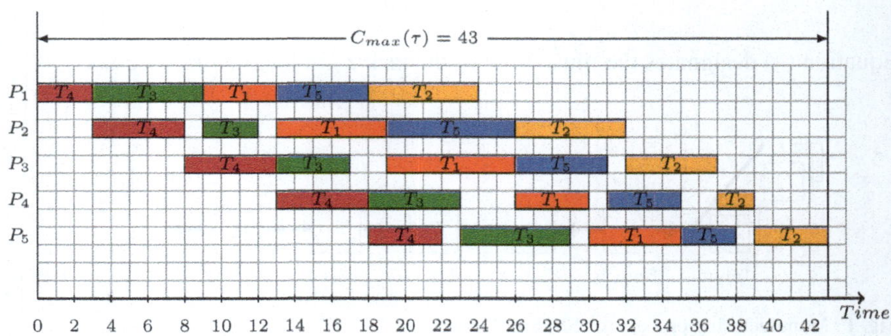

Fig. 2 The flowchart for a sequence τ

3 Resolution Approach

The resolution of the kind problem based on approximate methods using metaheuristics. In the different metaheuristic algorithms, we start from an initial solution, then a neighborhood exploration procedure allows the generation of one or a set of neighboring solutions. If the criterion is improved, i.e. the makespan is reduced, the new solution will be adopted. In the areas optimization scheduling problem, the IG, AG and ABC metaheuristics are among the powerful algorithms. We adapt this algorithms for solving the FND problem modeled as a FSSP industry.

3.1 *The IG Algorithm*

The IG algorithm is a procedure based on two fundamental steps. The first is that of destruction and construction. The second is based on local research involving the simulated annealing model.

Algorithm 1: IG algorithm for FND problem.

---

input : $\tau = [\tau_1, \tau_2, ..., \tau_n]$ Given by priority rule, $\tau^* \leftarrow \tau$, $C^*_{\max} \leftarrow C_{\max}(\tau)$

while *Stopping criterion is not satisfied* **do**

    $\Gamma \leftarrow \emptyset,\ \tau' \leftarrow \tau$

    for $r = 1$ *to* d **do**

        Extract a random task τ_r *from the* τ' *and add it to* Γ *subset*

    for $r = 1$ *to* $|\Gamma|$ **do**

        Extract the task τ'_r *from the subset* Γ *and test the task on the different positions in the* τ' *and choose the best position giving the best makespan,* $\tilde{\tau}$ *is the best sequence*

    if $C_{\max}(\tilde{\tau}) \leqslant C_{\max}(\tau)$ **then**

        $\tau \leftarrow \tilde{\tau}$

        if $C_{\max}(\tilde{\tau}) \leqslant C_{\max}(\tau^*)$ **then**

            $\tau^* \leftarrow \tilde{\tau}$

    else

        if $rand() \leqslant e^{-\left(\frac{C_{\max}(\tilde{\tau}) - C_{\max}(\tau)}{T_e}\right)}$ **then**

            $\tau \leftarrow \tilde{\tau}$

output: τ^*

The IG (see Algorithm 1) is applied in solving a fairly large number of optimization problems. In [15–17], we find different implementation approaches for solving FSSP optimization problems. We apply the same approach for solving the scheduling FND problem. The implementation of the IG algorithm requires knowledge of the number d to be extracted from the sequence during the destruction sub-phase. Subsequently, the insertion procedure in the different positions is applied to determine the best constructed sub-sequence. Also the parameter T_e of the temperature used in the

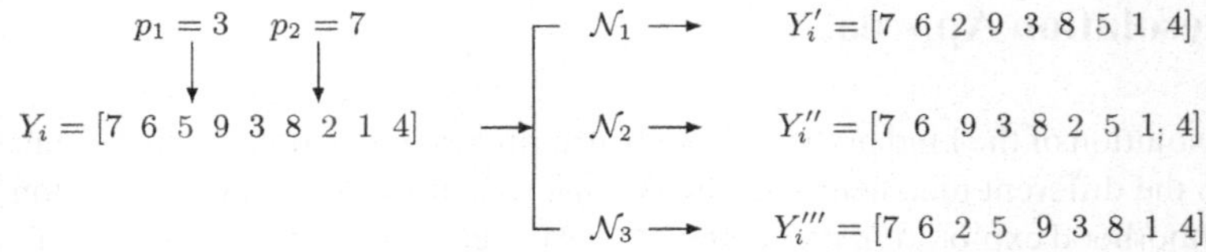

Fig. 3 An example of neighborhood structure $\mathcal{N}_i$

simulated annealing model which allows a good exploration of the neighborhood system. We have considered the same model used in [18] for the resolution of the FSSP optimization problem.

3.2 The ABC Algorithm

The ABC algorithm is an optimization approach that simulates the behavior of the bee colony in food search. This metaheuristic is based on three phases: the employer phase, the onlooker phase and the scout phase. During the first phase a group of employees work for the search for food. In the onlooker phase, if the search is unsuccessful or if the bees do not find better areas than the previous phase, the search area is updated. In the third phase, the scout bees fly for the last time for a larger exploration in the search for food.

Algorithm 2: ABC algorithm for FND problem.

input : Generate the population $\mathcal{Y} = \{Y_1, Y_2, ..., Y_K\}$ the of sequences.

Evaluate each sequence Y_k : $k = 1, ..., K$ *in the* $\mathcal{Y}$

$Y_{ABC} = argmin_{1 \leqslant k \leqslant |\mathcal{Y}|}\{C_{\max}(Y_k), Y_k \in \mathcal{Y}\}$

while *Stopping criteria not satisfied* **do**

  for $k = 1$ *to* K **do**

    Employed phase: generate the neighborhood $\mathcal{N}_1$

    Onlooker phase: generate the neighborhood $\mathcal{N}_2$

    Scout phase: generate the neighborhood $\mathcal{N}_3$

    Update the population $\mathcal{Y}$ *and the best solution* Y_{ABC}.

output: Y_{ABC}

In the Algorithm 2, we describe the model adopted in our approach for solving the studied problem. By simulating this process the ABC algorithm is implemented with three neighborhood system exploration phases $\mathcal{N}_1$, $\mathcal{N}_2$ and $\mathcal{N}_3$. Each neighborhood system represents a phase of the food (the solution) search in the initial algorithm.

Figure 3, shows an illustrative example of a neighborhood system used in the ABC algorithm. Considering a current solution $Y_i = [7, 6, 5, 9, 3, 8, 2, 1, 4]$, we choose two random positions in the current sequence, such that the position $p_1 = 3$ occupied by the task T_5 and the position $p_2 = 7$ occupied by the task T_2. In the first neighborhood system $\mathcal{N}_1$, we consider a permutation of position p_1 and p_2, that is to say the new sequence $Y_i' = [7, 6, 2, 9, 3, 8, 5, 1, 4]$. In the neighborhood system $\mathcal{N}_2$, we consider the insertion by shift to the left of the task T_5 from position p_1 in position p_2, the new sequence is $Y_i'' = [7, 6, 9, 3, 8, 2, 5, 1, 4]$. Finally in the last neighborhood system $\mathcal{N}_3$, the task in the position p_2 i.e. T_2 is inserted in position p_1 with shift to the right, the new sequence is $Y_i''' = [7, 6, 2, 5, 9, 3, 8, 1, 4]$.

3.3 The GA Algorithm

The GA technique is an optimization method inspired by the establishment of the genetic code in humans. This approach is based on obtaining the genetic code of a living being from two parents, this code will be obtained by taking from each parent a set of genes to form its own genes. Three phases are to be considered: the selection phase of two parents, the crossing phase and the mutation phase. This is an individual population algorithm where the procedure is repeated until a stop criterion is met to obtain the correct gene code. By simulating this phoneme, the algorithm can be adopted to find a good solution in an optimization problem. The GA algorithm is applied in [19, 20] in solving FSSP. We present a description of our approach in the Algorithm 3 to solve the FND modeled FSSP case.

Algorithm 3: The GA for FND problem.

input : Generate the population $\mathcal{Z} = \{Z_1, Z_2, ..., Z_U\}$ the of sequences.
Evaluate each sequence Z_u *in the* $\mathcal{Z}$
$Z_{GA} = argmin_{1 \leqslant u \leqslant U}\{C_{\max}(Z_u), Z_u \in \mathcal{Z}\}$
while *Stopping criteria not satisfied* **do**
  for $u = 1$ *to* U **do**
    Selection phase: select randomly the sequence Z_l.
    Crossing phase: generate two new sequence $Z_{u'}$ *and* $Z_{l'}$
    Mutation phase: generate two new sequence $Z_{u''}$ *and* $Z_{l''}$
    Update the population $\mathcal{Z}$ *and* Z_{GA}
  end
end
output: Z_{GA}

In Fig. 4, we represent an illustrative example of GA algorithm using parents made up of the two sequences Z_u=[4, 8, 5, 9, 6, 2, 3, 1, 7] and Z_l=[9, 3, 8, 5, 1, 2, 7, 6, 4] from the selection phase . In the crossing phase, we will fix a part of each sequence, then we will complete the rest of the other sequence, we thus obtain the sequences $Z_{u'} = [3, 8, 5, 9, 6, 2, 1, 7, 4]$ and $Z_{l'} = [4, 8, 9, 5, 1, 2, 6, 3, 7]$ respectively. In the mutation phase, we apply the position inversion procedure from the section phase while keeping a fixed part of each sequence, we obtain the sequences $Z_{u''} = [7, 1, 3, 9, 6, 2, 5, 8, 4]$ and $Z_{l''} = [4, 6, 7, 5, 1, 2, 8, 3, 9]$.

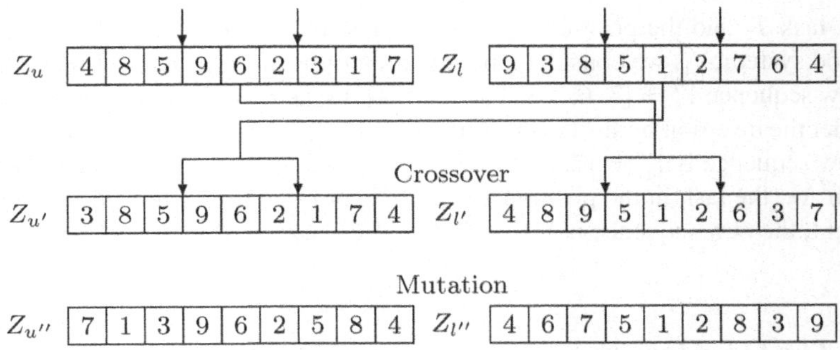

Fig. 4 An example of crossover and mutation procedure in GA

4 Experimental Result

In order to verify the efficiency and robustness of our algorithms, we have conducted a set of comparative studies between our methods. Numerical simulation is performed on a set of instances varying the number of tasks n and the number of processors m. We have dealt with two problem cases: the small and medium size instances and the large size instances. For the first type of instance, we consider that $n \in \{20, 30, \ldots, 90\}$ and for the second type of instance, we consider that $n \in \{100, 150, 200, \ldots, 450\}$. For both types of problems we have taken the number of processors $m \in \{5, 8, 10\}$. This choice allows us to extend the study to more than five processors for a possible extension of the number of steps necessary for data processing. The processing times t_{ji} are generated according to a uniform law in the interval $[1, 99]$ units of time. We consider the average relative percentage of deviation (ARPD) between our algorithms for an average of five replications for each instance.

$$\text{ARPD} = \frac{1}{5} \times \sum_{r=1}^{5} \left(\frac{C^r_{\max} - C^*_{\max}}{C^*_{\max}} \right) \times 100\% \tag{6}$$

The expression of ARPD is given by the expression of the equation (6), where $C^r_{\max}$ represents the value of the algorithm for a replication r and $C^*_{\max}$ designates the best value obtained in five replication of all algorithm. In order to limit the calculation time during the comparative study between the algorithms, we choose the time limit defined by: $T_{\max} = \rho \times n \times m$ milliseconds.

In Table 2, we give the value of ARPD for three case studies $\rho = 20$ and $\rho = 40$ and $\rho = 60$ for the problems considered of small and of medium size. This analysis is made while respecting the time limit T_max of computation imposed. We notice that the IG algorithm realizes the smallest value of ARPD is 1.28% and that the ABC algorithm records the largest value of ARPD 6.45 %. Also we find that out of the seventy-two cases tested IG is the best performing with a success rate of 70.83%.

The same analysis is performed for instances considered relatively large size. In Table 3, we represent the value of ARPD for three case studies $\rho = 20$ and $\rho =$

Table 2 ARPD value for small and medium size instances, the best values are in bold

$n \times m$	$\rho = 20$			$\rho = 40$			$\rho = 60$		
	IG	ABC	GA	IG	ABC	GA	IG	ABC	GA
20 × 5	**4.56**	5.26	4.87	**4.28**	5.77	4.98	**3.56**	4.56	4.33
20 × 8	4.22	**4.18**	4.33	4.32	**4.23**	4.45	4.31	**4.12**	4.15
20 × 10	**3.12**	3.45	3.22	**3.08**	3.15	3.22	**3.02**	3.23	3.12
30 × 5	**5.23**	6.35	5.56	**5.12**	6.23	5.44	**5.08**	6.09	5.23
30 × 8	5.66	6.45	**5.34**	5.34	6.32	**5.23**	5.12	6.28	**5.23**
30 × 10	**5.12**	6.16	5.08	**5.07**	6.09	5.06	**4.98**	5.03	5.02
40 × 5	4.33	4.66	**4.22**	4.25	4.55	**4.12**	4.13	4.45	**4.09**
40 × 8	**4.51**	4.78	4.68	**4.31**	4.67	4.53	**4.25**	4.54	4.37
40 × 10	**4.88**	4.98	4.89	**4.73**	4.79	4.67	**4.66**	4.74	4.65
50 × 5	3.45	3.89	**3.28**	3.33	3.78	**3.25**	3.25	3.51	**3.18**
50 × 8	**3.45**	3.67	3.88	**3.32**	3.56	3.75	**3.27**	3.45	3.66
50 × 10	**3.29**	3.54	3.65	**3.12**	3.32	3.18	**3.08**	3.23	3.11
60 × 5	**3.52**	3.69	3.49	**3.42**	3.65	3.35	**3.16**	3.37	3.22
60 × 8	**3.17**	3.27	3.56	**3.13**	3.22	3.45	**3.11**	3.19	3.23
60 × 10	**2.63**	2.88	2.66	**2.51**	2.78	2.52	**2.43**	2.69	2.45
70 × 5	2.67	2.87	**2.55**	2.53	2.65	**2.44**	2.35	2.45	**2.28**
70 × 8	**2.81**	3.03	2.88	**2.78**	2.96	2.78	**2.25**	2.45	2.31
70 × 10	**2.32**	2.47	2.34	**2.28**	2.34	2.37	**2.13**	2.23	2.19
80 × 5	2.26	2.32	**2.25**	2.21	2.28	**2.13**	2.15	2.28	**2.11**
80 × 8	**2.47**	2.69	2.55	**2.45**	2.55	2.53	**2.33**	2.44	2.33
80 × 10	**2.54**	2.58	2.67	**2.32**	2.45	2.63	**2.28**	2.34	2.51
90 × 5	**1.89**	1.93	1.92	**1.78**	1.88	1.89	**1.67**	1.78	1.83
90 × 8	1.67	1.78	**1.65**	1.56	1.76	**1.46**	1.45	1.62	**1.32**
90 × 10	**1.38**	1.67	1.64	**1.33**	1.54	1.55	**1.28**	1.52	1.43

40 and $\rho = 60$. For this category of problem, we notice that IG gives from the minimum value of APRD of 0.03% in the case of $\rho = 60$. On the other hand, the ABC algorithm realizes the maximum value of ARPD 4.16%. Based on the simulation study performed for this category of problems, the IG algorithm recorded the highest percentage of success who is from 70.83%.

In Fig. 5, we show the efficiency of our algorithms by plotting the convergence curves. We give the result for an instance assumed to be relatively small. Indeed, the simulation concerns the instance 30 × 5 for a limit computation time of 160 seconds. We see that the IG algorithm is the better algorithm compared to the other two algorithms.

Similarly in Fig. 6, we represent the convergence curves of the algorithms for an instance 100 × 10. The calculation time limit is 100 s which we consider largely sufficient for the stabilization of our algorithms towards their good solutions. In this simulation, we can see the good performance of the IG algorithm.

Table 3 ARPD value for large size instances, the best values are in bold

$n \times m$	$\rho = 10$			$\rho = 20$			$\rho = 30$		
	IG	ABC	GA	IG	ABC	GA	IG	ABC	GA
100 × 5	**3.56**	4.16	3.82	**3.38**	4.07	3.88	**3.26**	3.66	3.43
100 × 8	3.35	3.44	**3.28**	3.32	3.45	**3.22**	3.18	3.29	**3.11**
100 × 10	**2.82**	2.89	2.85	**2.65**	2.78	2.74	**2.46**	2.57	2.63
150 × 5	**2.33**	2.45	2.64	**2.23**	2.34	2.35	**2.17**	2.36	2.21
150 × 8	**2.45**	2.68	2.53	**2.32**	2.53	2.35	**2.21**	2.53	2.33
150 × 10	2.61	2.78	**2.43**	2.56	2.65	**2.34**	2.45	2.59	**2.28**
200 × 5	**1.79**	1.92	1.88	**1.67**	1.73	1.78	**1.55**	1.66	1.58
200 × 8	1.56	1.69	**1.42**	1.43	1.54	**1.31**	1.33	1.48	**1.24**
200 × 10	**1.45**	1.57	1.53	**1.34**	1.48	1.52	**1.33**	1.42	1.39
250 × 5	1.68	1.65	**1.64**	1.45	1.55	**1.34**	1.44	1.53	**1.24**
250 × 8	**1.27**	1.52	1.31	**1.21**	1.44	1.27	**1.13**	1.39	1.27
250 × 10	**1.17**	1.26	1.22	**1.14**	1.18	1.16	**1.11**	1.14	1.15
300 × 5	0.79	**0.68**	0.78	0.65	**0.58**	0.65	0.53	**0.47**	0.47
300 × 8	**0.77**	0.93	0.87	**0.64**	0.82	0.75	**0.48**	0.62	0.57
300 × 10	**0.45**	0.63	0.54	**0.32**	0.53	0.41	**0.29**	0.42	0.37
350 × 5	0.86	0.76	**0.45**	0.78	0.66	**0.38**	0.77	0.63	**0.33**
350 × 8	**0.44**	0.68	0.56	**0.33**	0.53	0.43	**0.22**	0.45	0.27
350 × 10	**0.19**	0.32	0.22	**0.17**	0.31	0.18	**0.13**	0.26	0.15
400 × 5	0.31	0.35	**0.23**	0.28	0.29	**0.25**	0.22	0.24	**0.21**
400 × 8	**0.19**	0.32	0.22	**0.16**	0.31	0.18	**0.14**	0.24	0.17
400 × 10	**0.22**	0.43	0.28	**0.17**	0.31	0.18	**0.11**	0.24	0.13
450 × 5	**0.14**	0.22	0.18	**0.11**	0.19	0.16	**0.09**	0.13	0.11
450 × 8	**0.09**	0.17	0.13	**0.05**	0.15	0.12	**0.04**	0.11	0.08
450 × 10	**0.07**	0.11	0.12	**0.05**	0.09	0.08	**0.03**	0.08	0.06

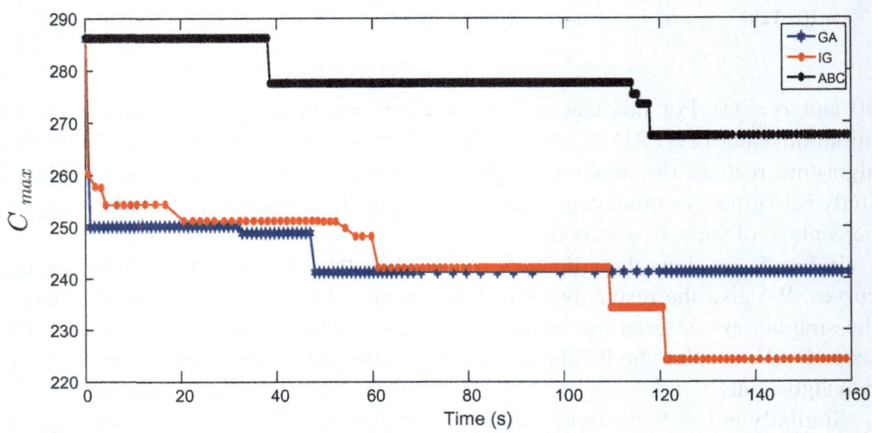

Fig. 5 Convergence curve $C_{\max}$ versus time for 30 × 5 instance

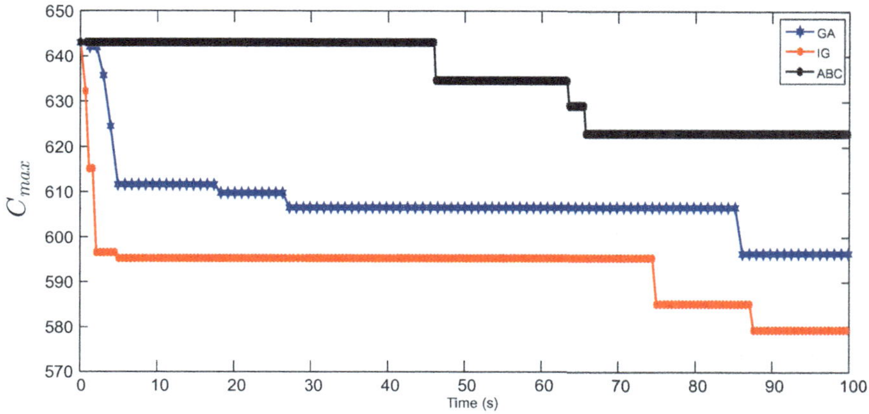

Fig. 6 Convergence curve $C_{\max}$ versus time for 100×10 instance

5 Conclusion

Recently, the fake news detection problem has caught the attention of university researchers. In this article, we take a new approach to modeling fake news detection by a flow shop scheduling problem optimization. The problem is modeled one by one set of processing phases arranged in series through which all the documents to be analyzed. We propose three metaheuristics: IG, ABC and GA algorithms in order to find the best document sequence to schedule. The fairly large number of documents to process requires that the entire document be completed as soon as possible. The objective is to minimize the maximum completion time called the makespan. An experimental study is carried out on a set of instances by varying the size of the number of documents to be analyzed. The simulation shows that the IG algorithm performs well compared to the other two metaheuristics.

References

1. Tacchini, E., Ballarin, G., Della Vedova, M. L., Moret, S., & de Alfaro, L. (2017). Some like it hoax: Automated fake news detection in social networks. arXiv preprint arXiv:1704.07506.
2. Granik, M., & Mesyura, V. (2017, May). Fake news detection using naive Bayes classifier. In *2017 IEEE First Ukraine Conference on Electrical and Computer Engineering (UKRCON)* (pp. 900–903). IEEE.
3. Ozbay, F. A., & Alatas, B. (2020). Fake news detection within online social media using supervised artificial intelligence algorithms. *Physica A: Statistical Mechanics and its Applications, 540*, 123174.
4. Zhang, J., Dong, B., & Philip, S. Y. (2020, April). Fakedetector: Effective fake news detection with deep diffusive neural network. In *2020 IEEE 36th International Conference on Data Engineering (ICDE)* (pp. 1826–1829). IEEE.

5. Kaliyar, R. K., Goswami, A., Narang, P., & Sinha, S. (2020). FNDNet?a deep convolutional neural network for fake news detection. *Cognitive Systems Research, 61*, 32–44.
6. Gilda, S. (2017, December). Notice of violation of IEEE publication principles: Evaluating machine learning algorithms for fake news detection. In *2017 IEEE 15th Student Conference on Research and Development (SCOReD)* (pp. 110–115). IEEE.
7. Al-Ahmad, B., Al-Zoubi, A. M., Abu Khurma, R., & Aljarah, I. (2021). An evolutionary fake news detection method for COVID-19 pandemic information. *Symmetry, 13*(6), 1091.
8. Sheikhi, S. (2021). An effective fake news detection method using WOA-xgbTree algorithm and content-based features. *Applied Soft Computing*, 107559.
9. Wang, Y., Wang, L., Yang, Y., & Lian, T. (2021). SemSeq4FD: Integrating global semantic relationship and local sequential order to enhance text representation for fake news detection. *Expert Systems with Applications, 166*, 114090.
10. Yuliani, S. Y., Abdollah, M. F. B., Sahib, S., & Wijaya, Y. S. (2019). A framework for hoax news detection and analyzer used rule-based methods. *International Journal of Advanced Computer Science and Applications, 10*(10), 402–408.
11. Pandey, A. C., & Tikkiwal, V. A. (2021). Stance detection using improved whale optimization algorithm. *Complex & Intelligent Systems, 7*(3), 1649–1672.
12. Thaher, T., Saheb, M., Turabieh, H., & Chantar, H. (2021). Intelligent detection of false information in arabic tweets utilizing hybrid harris hawks based feature selection and machine learning models. *Symmetry, 13*(4), 556.
13. Zhao, F., Zhang, L., Zhang, Y., Ma, W., Zhang, C., & Song, H. (2020). A hybrid discrete water wave optimization algorithm for the no-idle flowshop scheduling problem with total tardiness criterion. *Expert Systems with Applications, 146*, 113166.
14. Gmys, J., Mezmaz, M., Melab, N., & Tuyttens, D. (2020). A computationally efficient Branch-and-Bound algorithm for the permutation flow-shop scheduling problem. *European Journal of Operational Research, 284*(3), 814–833.
15. Ruiz, R., & Stützle, T. (2007). A simple and effective iterated greedy algorithm for the permutation flowshop scheduling problem. *European Journal of Operational Research, 177*(3), 2033–2049.
16. Ribas, I., Companys, R., & Tort-Martorell, X. (2011). An iterated greedy algorithm for the flowshop scheduling problem with blocking. *Omega, 39*(3), 293–301.
17. Nagano, M. S., de Almeida, F. S., & Miyata, H. H. (2021). An iterated greedy algorithm for the no-wait flowshop scheduling problem to minimize makespan subject to total completion time. *Engineering Optimization, 53*(8), 1431–1449.
18. Ruiz, R., & Maroto, C. (2005). A comprehensive review and evaluation of permutation flowshop heuristics. *European Journal of Operational Research, 165*(2), 479–494.
19. Aqil, S., & Allali, K. (2021). On a bi-criteria flow shop scheduling problem under constraints of blocking and sequence dependent setup time. *Annals of Operations Research, 296*(1), 615–637.
20. Andrade, C. E., Silva, T., & Pessoa, L. S. (2019). Minimizing flowtime in a flowshop scheduling problem with a biased random-key genetic algorithm. *Expert Systems with Applications, 128*, 67–80.

Case Studies and Frameworks

The Multiplier Effect on the Dissemination of False Speeches on Social Networks: Experiment during the Silly Season in Spain

Cristóbal Fernández-Muñoz, Ángel Luis Rubio-Moraga, and David Álvarez-Rivas

Abstract Fake News in the media and on social networks have grown exponentially in recent years, forming part of the new strategies of political communication and the dissemination of hate speech. The present work demonstrates in an experimental way the ease of creating and spreading fake news on social networks by putting into circulation, with the collaboration of the TYC communication agency, four false news on different topics without connection to the current news of the moment, following the typology of the so-called summer snakes, during the silly season. Existing profiles on social networks were used, especially on Twitter, as well as the collaboration of influencers for their dissemination. Likewise, complementary online actions were used to analyze the multiplication of reach and interaction, both organically and promoted. The findings confirm the ease of generating and disseminating disinformation on these channels, the high organic interaction obtained, but even more the formidable multiplier effect of up to 10 times its reach thanks to the false content promotion options offered by digital platforms based on a moderate investment in advertising, which is an additional element of controversy on the issue of disinformation in the digital field.

Keywords Fake news · Disinformation · Social media · Twitter · Online advertising · Digital contents

C. Fernández-Muñoz (✉) · Á. L. Rubio-Moraga · D. Álvarez-Rivas
Universidad Complutense de Madrid, Madrid, Spain
e-mail: cristfer@ucm.es

Á. L. Rubio-Moraga
e-mail: alrubio@ucm.es

D. Álvarez-Rivas
e-mail: dalvarez@ucm.es

M. Lahby et al. (eds.), *Combating Fake News with Computational Intelligence Techniques*, Studies in Computational Intelligence 1001,
https://doi.org/10.1007/978-3-030-90087-8_12

1 Introduction and State of the Question

The media lies are not an invention of our time but have existed since we exist. Not even in the journalistic field it is a novel phenomenon. Already in "the gazettes of the eighteenth century hoaxes and libels were a tool of power well known by kings and aristocrats" [18], although their use on a large scale and their scientific study did not take place until the twentieth century, when the so-called mass society was constituted. Already in the first third of the century, totalitarian regimes of one sign or another made use of falsehoods as a tool for propagandistic disinformation and the first historical confirmation of the use of the term disinformation dates from then [41] when the Russians who had emigrated to France after the First World War, reported how the Bolshevik political police made reference to disinformation to define the set of actions that sought to prevent the consolidation of the communist regime in Russia [8, 28, 47].

A few years later, the Minister for Public Enlightenment and Propaganda of Nazi Germany, Joseph Goebbels, would make disinformation a war strategy to disorient the enemy, deceive the populations that were going to invade and create false expectations of triumph in his own troops [4]. Meanwhile, in the United States a formidable scientific machine was put into operation with the creation of the Office of War Information (1942–1945), an agency created by the United States Government to consolidate government information services in the one in which researchers from different areas worked to control the flow of hoaxes and define persuasive communication strategies with the aim of convincing the troops and citizens [3, 9, 44].

Thus, the first systematic studies of the phenomenon of misinformation and rumor were carried out in the United States by a team of scientists from various fields who approached the phenomenon from a multidisciplinary perspective, considering the psychological, cognitive, and linguistic processes of communication involved in these phenomena. From Social Psychology, Allport and Postman [1], defined rumors as propositions related to everyday events, transmitted for everyone to believe in them, without there being any concrete data to verify their accuracy. Robert Knapp defined the rumor as a statement destined to be believed, which is linked to the present, a proposition to believe a topic spread without official verification, being seen as a special case of informal social communication, together with myth, legend, or the current mood [31].

Even though studies on disinformation were lavished throughout the twentieth century, researchers of this phenomenon carried the weight of information manipulation exclusively in the media and the attempts by the issuer to use them and to manipulate them to meet their objectives [19]. This position is consistent with the current critical approach towards the media and towards an increasingly questioned independence of the same, due to the progressive political and economic control over the press. As Rodríguez Andrés [41] states, this has led the media, especially large corporations, to become powerful means of social influence based on these economic or ideological interests, almost always for the benefit of the ruling classes [37, 38, 43]. Nor has the journalistic profession always contributed to dignify its practice.

Der Spiegel recognized that one of its journalists, Claas Relotius, invented reports [7]. Fourteen of the 60 publications of 2011 were lies.

And it was not the first case, unfortunately. Janet Cooke had already published falsehoods in the Washington Post [21] or Jayson Blair had admitted to having falsified 36 reports in the New York Times [5].

However, lies and false news are not the exclusive monopoly of the media as Jacques Derrida pointed out in his famous conference entitled "History of Lies", correcting the spotlight and pointing out the construction of intentional lies from political power. and other areas [15].

Thus, the spread of it is distributed among different actors (politicians, companies, media, users of social networks, etc.) that combine to serve the very act of lying, which is always intentional. To all this we now add the usual trend of civil society on social networks.

This, guided by emotional aspects or ideological reasons, fans the flame of lies through the individual or collective propagation of false content, bringing into play new factors such as speed, breadth, or universality, "vectors that enhance the boom in its dissemination protected in participation, anonymity, the hidden source and the difficulty to erase its trace" [2].

Social networks as communication tools have modified the traditional paradigm of information management [27] giving the possibility to any citizen to become an information source with a multiplied scope of their messages with hardly any geographical limitations and instantly. The rumor arises from people who do not have a voice, from groups interested in generating information in their favor or to detract from opponents [14]. It was therefore born as a communication strategy. It grows and develops in an insufficient information environment. Its main breeding ground is an uninformed society; takes hold with scenarios in which information is withheld, hidden, or manipulated. And despite being informal or not having a clear origin, the rumor seduces because it provides us with a better way of understanding the world [29]. It constitutes an escape from people with which they intend to build social reality as they would like it to be.

In this sense, there are very diverse psychosocial studies that have revealed the limitations of reason and the fact that individuals sometimes ignore the facts because they do not adapt to what they themselves think. Reason has always been considered the most valuable and effective brain function, that which, together with the cerebral frontal cortex and analytical thinking, differentiates our brain from that of other animals, and which allows us to decide through a precise analysis of the information and its variables. However, the truth does not always matter. The social psychologist Kunda [34], when developing the theory of motivated thinking, concluded that "it is most likely that people will reach the conclusions they want to reach". To defend our vision of the world we are discarding some data and collecting others in the direction that suits us until we reach the conclusion that interests us and that reinforces our worldview [42]. Motivational theories applied to persuasive communication decades ago concluded that subjects are more likely to reinforce our attitudes, ideas, and behavioral intentions than to modify them [12]. What can also be interpreted as a protective shield so that things fit in with what we already know about the world, avoiding cognitive disruptions. However, these cognitive biases leave individuals at the mercy of lies. Emotions sometimes play a more decisive role than reason itself and, in many cases, not only do they exert their influence in deciding but are also capable of penetrating the heart of a phenomenon since, when it comes to complex decisions, emotions know reasons that exceed the powers of reason [39].

Hoaxes take advantage of these neural nooks and crannies and are expressed as a phenomenon defined by its source (unofficial), its process (broadcast on chain) and its content (a news item referring to a current event) and, nevertheless, the veracity, by on the contrary, it is not part of their scientific definition, although they generally require a dose of truth [32, 40]. In fact, fake news spreads 70% faster than truthful information according to research from the Massachusetts Institute of Technology (MIT), published in the journal Science [48]. The digital setting is perfect for rumors. The hoaxes spread, thanks to the viral effect of social networks, in a more powerful way than, until now, we knew [26, 46].

The ways of spreading this type of communication, clearly harmful and unethical, but highly effective, are varied and depend on the resources available, with digital being the ones that are gaining more and more prominence [45] and in principle they require less investment [16, 30]. The very structure on which the web is based reinforces the tribal grouping of the individual, generating, as stated by García-Marín and Aparici [22], "new dangers for democracy by accentuating what separates the members of society and reinforcing the encapsulation of the ego".

It is not surprising then that "fake news" was the term of the year 2017 according to the Collins [11], and it is that, although the term had been gaining relevance for some time, it was the President of the United States Donald Trump who put it back fashionable, thus referring to the news that was not favorable to him [10]. Since then, it has become an issue that has political consequences for global geostrategy. Thus, the National Security Strategy of Spain has for the first time included "disinformation campaigns" within the planned destabilization strategies and political interference, both due to content and technical cyberattacks [25].

Given the relevance of the question raised in this section, it was necessary to investigate and verify the behavior of fake news through an experimental project that would allow us to contribute to analyzing their evolution by monitoring them on social networks. In this way, and through both a qualitative and quantitative methodology based on the design and implementation of an experimental work, it will seek to verify the functioning of the mechanisms that favor the development of false news in the digital environment and establish new study parameters on the phenomenon of misinformation in social networks.

2 Material and Methods

In order to investigate the behavior of hoaxes on social networks, an experimental project was designed in collaboration with the firm "Torres y Carrera Consultores de Comunicación" that would allow analyzing its evolution and dissemination results through the creation and monitoring of four false news which were released during the month of August 2020. The social networks chosen were Twitter, Facebook, and LinkedIn, depending on the theme of each news item, although finally only the results of Twitter were used for the analysis work. The most open social network focused on news, characterized by the management of information in real time and which serves precisely as a reference to the press itself.

The comparison of the four news items provides, from a qualitative perspective, limited knowledge [17], although it provides sufficient data to allow evaluating the possible future expansion of the object of study [24]. It is an exploratory work of the immediate behavior of the four stories that in turn allows to lay the foundations for a comparative study of the behavior of the news and the influence of the promotional support variables through advertising or the use of influencers on the results of reach and interaction with network users.

The selection of the experimentation period is due to the summer period, traditionally associated with the decrease in news and the appearance of the so-called "summer snakes", an expression referring to the irrelevant or surprising news published by some media during the holidays. social responsibility to be followed by the research team, as well as in order to isolate the variables of informative interest that could influence the experiment, the stories chosen did not present current thematic connections, especially avoiding political controversy, but they did consider variables that correlate favorably with the probability of becoming news in networks [23, 33]. Thus, the starting idea consisted of taking advantage of light-hearted and neutral themes, linked to entertainment or curiosities that were attractive to the user. Active public in social networks, prevailing in all case a credible content that ruled out a possible involution, by incredible, from the beginning of the diffusion. The preparation phase included the investigation of information sources around the topic to nurture the context of each story. Likewise, the profiles chosen were not false but rather existed previously, with an average of 981 followers, and their credibility was reinforced by prior publication of related content to give them credibility before publishing the false news.

In two of the stories, they had the support of influencers and, in turn, in three of the news stories, advertising support was used with investments of between 50 and 100 euros and a total investment of 250 euros. The sample with the combination of these variables allowed the analysis of news behavior, without promotional or organic support, with influencers or without them, and with advertising support or without promotion.

The four news items were: N.1 "A Spanish actress, the protagonist of Spider-Man 3", taking advantage of the informational hitch of the largest comic book festival in the world, San Diego Comic-Con, the false news was released emphasizing the fact that until then no Spanish actress had participated before in a Marvel super production. N.2 "The chimpanzee that plays Fortnite", spreading the hoax that Gogo, an eleven-year-old chimpanzee, could win over young people of the same age by playing the video game that was one of the main hobbies of the youngest during the world confinement of the population by COVID19. N.3 "Reggaeton elevates Spanish as a musical language" based on real news about the growth of Spanish in the world of

Table 1 Scope of fake news

	N.1	N.2	N.3	N.4	∑	AV
Twitter Profile Followers	1046	924	991	965	3.926	981,5
Total Posts	26	13	15	76	130	32,5
Organic publications	24	12	15	72	123	30,75
Paid publications	2	1	0	4	7	1,75
Influencers	Yes	No	Yes	No		
Investment	100 €	50 €	0	100 €	250 €	62,5 €
Total reach	108.033	41.800	3.818	125.600	279.251	69.813
Organic reach	36.131	1.224	3.818	40.006	81.179	20.295
Paid reach	71.902	40.576		80.564	193.042	64.347
Reach per € (Impression)	719.02	811.52		805.64	2.336	778.73

pop, but leading it to distortion and exaggeration, ensuring that it surpassed English and N.4. "A project to read the minds of workers" in the context of teleworking, technology, and workers' rights, it was sought to create a fictional story that would lead to think that companies were close to achieving control over the transmission of information between their workers through telepathy.

The field work was carried out with the team from the Torres y Carrera Consultores de Comunicación agency. In all cases, after three weeks of the hoax, it was revealed in the fourth week that it was false news, and the nature of the investigation was explained.

The statistical analysis of the results was carried out using the analysis of the behavior of the publications and the monitoring of search tags with the Twitter Analytics tool and the data crossing was executed with the SPSS (version 23.0 for windows 10) and Excel 14.0 programs. Microsoft. The universe offers a total of 130 publications generated for the experiment from 4 profiles on Twitter (Table 1) with an average publication frequency of three publications per week, and with a total reach of 279,251 impressions.

To determine the characteristics of the results achieved by the sample on Twitter, a descriptive analysis (summations, percentages and means) and crossing variables has been carried out. As an analysis factor of the reception of each story, the average interaction rate or engagement rate calculated on impressions has been used, that is, on the total number of people who have seen the content. This formula measures the amount of interaction (likes, comments, or shares) obtained by social content in relation to the reach obtained, that is, the impressions between users achieved by each publication.

The comparison between the results of the news has been made using the Chi-square test in the case of categorical variables. The Pearson correlation calculation has allowed us to know the continuous probability distribution with the different parameters representing the degrees of freedom of the random variables.

3 Analysis and Results

The 130 publications made from the Twitter profiles used for the experiment as well as from the influencers who collaborated in the experiment were distributed through the RT or share functions, identifying and tagging potential prescribers to favor their interaction with the content, for a total of 81,179 impressions. As a complement to organic dissemination, three sponsored campaigns were established on content to measure the increase in reach and interaction of the most relevant publications, adding 193,042 additional impressions. In no case was a content veracity control carried out by the platform before contracting the online advertising action. Table 1 describes the reach obtained on Twitter with the four fake news, the number of publications and the support or not of complementary advertising investment.

About the first story N. 1, a total of 26 tweets were disseminated (24 organic and 2 promoted). In organic activity, storytelling posts registered a good level of reach and engagement: an average of 1196 impressions, 105 interactions and an average interaction rate of 5.5%. The content that worked best was accompanied by audiovisual material: photomontages, GIFs and links to videos and news of interest, which correlates with other research in this regard [20, 36]. The results of promoted content showed a significant increase in the number of interactions. In the first publication, it went from 692 organic interactions to 16,435 with the promotion and its interaction rate from 16 to 51%.

In the false news N.2, the profile used @GabrielSegura78, was characterized to give credibility to the topic to be treated as a researcher and trainer of the supposed gamer chimpanzee. The publication strategy with a total of 13 publications focused on organic content in the first weeks of August 2020, reinforced with a strategy of a sponsored tweet with 50 euros the last week. Interaction with eSports or video game journalists and regular Fortnite players was encouraged. Monitoring hashtags such as #Gogo or #WildMonkeyFortnite were used. The organic tweet that worked the best was the one that included a YouTube demo video, obtaining 535 impressions and 30 interactions with an engagement rate of 5.6%.

In the hoax No. 3, organic behavior was analyzed without the support of promotional actions. The project was enriched with the participation of the selected profile in conversations related to music and the reggaeton genre, and around related or created content and hashtags ad hoc such as #GraciasReggaeton or #ElEspañolTriunfa. A total of 15 posts were disseminated on Twitter, reaching an engagement rate of 9.2%.

In the case of N.4, a profile characterized as a specialist in personnel selection in a temporary work agency was used to provide it with the greatest possible realism. After several organic publications, a promotional campaign was carried out for the published content, with an advertising investment of 100 euros. This budget allowed the audience to increase their interaction with the content with more than 125,000 impressions in total. The average interaction rate was 1.1%.

The activity generated in the four cases developed has shown that it is feasible to generate a conversation around false news on social networks with both organic and

Table 2 Fake news interaction

	N.1	N.2	N.3	N.4	Σ	AV
Total interactions	19.013	5.145	352	1.381	25.891	6.472
Organic interactions	1.428	107	352	313	2.200	550
Paid interactions	17.585	5.038	0	1.068	23.691	5.923
Mentions	7		4	19	30	10
Profile visits	2.965	1.214	602	3.088	7.869	1.967
ER on followers	1817.7%	556.8%	35.5%	143.1%		638.3%
Organic ER on followers	136.5%	11.6%	35.5%	32.4%		54%
Organic ER	4.0%	8.7%	9.2%	0.8%		5.7%
ER on total reach	17.6%	12,3%	9.2%	1.1%		10.1%
Investment/Reach	0.0014 €	0.0012 €	0	0.0012 €		0.0010 €
CPM	1.3908 €	1.2323 €	–€	1.2412 €		0.9661 €
Investment/Interaction	0.0057 €	0.0099 €	0	0.0936 €		0.0273 €

paid content, with a direct correlation between investment and scope of the hoax. With modest profiles in terms of the number of followers (931 on average) and only an average of 32 publications, a multiplication of impressions is observed (the average reach was 69,813 impressions) with interaction rates higher than 0.8%. As can be seen in Table 2, the interactions varied between 0.8 and 17.5%, which is very prominent given that it is generally estimated that the participation rate or percentage of engagement from 0.5% is a good RE rate on Twitter [13]. The advertising investment made correlates positively with the reach achieved multiplying by 778 impacts on average each euro invested. The cost per thousand or average CPM was € 0.96.

In general terms, it has been possible to verify that most of the publications (especially the sponsored ones) registered a good level of reach, oscillating between the 3,818 impressions of the hoax No. 3 with a 9.2% organic interaction rate, without advertising support, and the 125,600 impressions of package No. 4, with a 1.1% ER. Regarding this last variable, the hoax that registered the best results was undoubtedly No. 1 with an engagement rate of 17.6% and a total of 19,013 interactions compared to 352 for No. 3, despite not Having the support of advertising promotion actions, the interaction index on the reach achieved was notably higher than N.4, which, having a good reach, registered the lowest interaction rates both organic and on the total reach. Figure 1 shows the different levels of interaction achieved, both organic and considering that derived from advertising support.

The commitment to an exclusively organic strategy in case N.3 is also reflected in the results related to visits to the profiles created on Twitter. Thus, compared to the 602 visits in the case of the aforementioned campaign, the 2965 and 3088 visits to the profiles of hoaxes N.1 and N.4, respectively, stand out, in which a combined strategy of posts was chosen organic and promoted. In Fig. 2 the behavior in terms of scope of the four hoaxes can be observed, highlighting the case of N.4 and N.3, as those with the highest and lowest number of impressions, respectively.

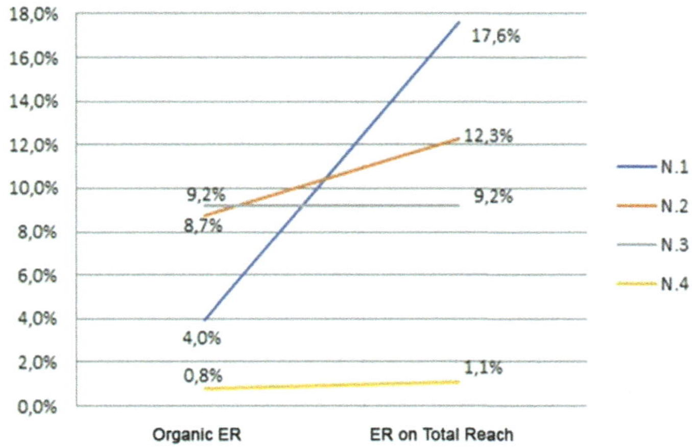

Fig. 1 Engagement rates (*Source* Own creation from analysis data)

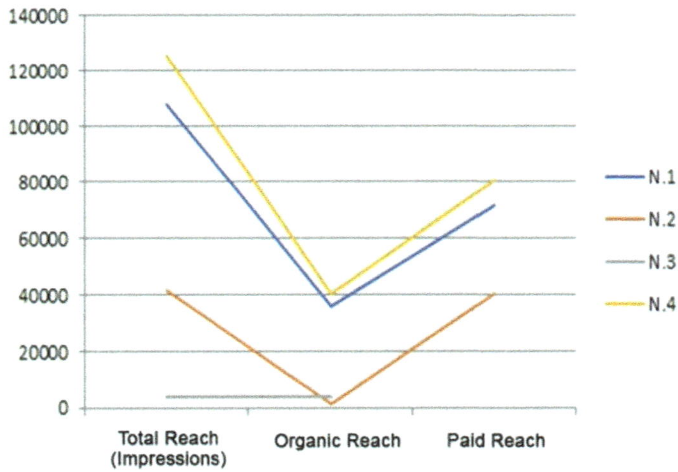

Fig. 2 Scope of fake news (*Source* Own creation from analysis data)

The Pearson coefficient in the correlation of the investment and scope variables rises to 0.9790, which shows the enormous interrelation between the two. However, in the case of investment and interaction, although the interaction rate on the reach increases in all cases with the advertising support, this dependence cannot really be sustained with the Pearson correlation, which remains at 0.5487, for what we can affirm that the advertising support multiplies proportionally the scope of the hoaxes, although it does not necessarily imply a greater interaction.

From the point of view of content, the publications that obtained the greatest impact were those that were accompanied by audiovisual material (photomontages, gifs, videos or links to videos and news of interest) or calls to action. Thus, in case N.1, this type of content provoked the reaction of a good number of users who even contributed the name of some Spanish actresses who were candidates for the Marvel production, while in the case of hoax N.2 it was achieved Including the firm recruitment of seven opponents to face the chimpanzee, 67 other people were interested in taking part in the challenge.

4 Discussion and Conclusions

The falsehood and dissemination of lies do not seem to be governed in our days by the classic rules of manipulation since, although the resources and mechanisms of said manipulation persist, new practices typical of the digital world have been generated that amplify the scope and, with it, the potential dangers of these digital hoaxes, [45].

The participation of the media is no longer essential for the creation and consolidation of conspiracies or theories that challenge previously contrasted and empirically proven data. Any politician, company or pressure group can be configured as a means of communication in itself and spread their messages through social networks without any prior filter on the data they handle and disseminate. In addition, with a minimal investment (just a total of 250 euros was invested in the experiment) it is possible to multiply the impact and reach a not inconsiderable number of impressions (the accumulated in the project was 250,000 impacts).

The rubbish has reached the pinnacle taking advantage of the rise of interactive digital media, which, in turn, coincides with the decline of traditional media. This phenomenon has shown how far the traditional Internet media live and how the strategies carried out by them have only increased their loss of credibility. Society needs the truth of knowledge to survive, and the phenomenon of fake news has returned to focus on what it means to do Journalism, that is, to control the powerful in an ethical and rigorous way. To act as a counter-power and not to become a Fourth Estate. Being connive with the factual powers, and the dominant elites, has a high cost in disaffection towards information professionals [7].

The cases studied in the Culebras Project show that co-responsibility in the dissemination of a hoax fundamentally concerns 4 actors, who should address the phenomenon with the importance it deserves: citizens as individuals/users/people (prosumers) who consume, generate and distribute information; to large technology companies (parent companies and subordinates), which have become true amplifiers and channels such as some of those analyzed here: Facebook, Twitter, LinkedIn, Google, YouTube…; the traditional media (increasingly transmedia); to the governments, whose priority has come to be managed by their intelligence services; and international multilateral organizations that affect transparency and the quality of democratic systems. And all those parties involved must provide answers to contain and prevent the spread of lies that undermine the pillars of democracy itself.

International organizations are marking these issues as priorities on their agendas, with a maximum priority, as in the case of UNESCO with the celebration of the International Day of Universal Access to Information (right to know day), where greater transparency is demanded with civil society and the prevailing need for public access to information to formulate sound public policies and save lives in health and social terms.

States, administrations, and public powers must implement tools of greater depth, solidity and credibility that reinforce democracy, turning disinformation into a category of State challenge, with the same commitment as cybersecurity. Disinformation in society is a very powerful weapon and as an example, the National Intelligence Center (CNI), through its National Cryptological Center (CCN), the entity in charge of cybersecurity, has designed a guide against hoaxes. Also fighting the profit of the "click farms" that are created and spread the lie for spurious purposes. Care must be taken in trying to combat lies with tools such as the Penal Code, in crimes that combat disaffection and criticism of the public representatives they manage. The temptation of power to eliminate dissent or stark criticism and jeopardize the very right to freedom of expression is historical. We are in the right direction when the 2014 Law on Transparency and Access to Information, and regional and local regulations, is applied. The infodemic virus only serves to fuel mistrust in institutions and in the social, democratic, and legal state. One responsibility of the public powers is to literate and educate people in information and communication technologies. It is surprising to see how, alongside apparently inconsequential falsehoods, opinions are constantly published that demonstrate the bad digital education of citizens.

It is vital to reinforce and anchor the value of educating in communication, which represents more self-criticism in people, promoting empathy and critical autonomy, teaching listening as an inexcusable step [6].

Technology companies must delve into self-regulation and self-control of false profiles, verification of information, limit the possibility of reproduction of mass messages and heavy investments in personnel to filter and contrast. COVID-19 has helped internet companies join their efforts to create information contrast networks or WhatsApp donate a million dollars to the International Content Verification Network (IFCN) for the Coronavirus Fact Alliance, of which more than 100 organizations from 45 different countries are already part.

And the people, the citizens, the users and consumers of the networks and the information, are the best dam against lies. The answers to this infodemic have in each responsible user the best vaccine. As conscientious citizens we should avoid forwarding or generating information that raises doubts, we can bet on activism in the co-responsibility of the communicative act, which involves at least two times: listening/reading and response.

References

1. Allport, G., & Postman, L. (1947). The psychology of rumor. Henry Holt

2. Alonso, M., & García Orta, M. (2015). Noticias falsas en Internet: Difusión viral a través de las redes sociales. COBCIBER. Oporto: Observatório de Ciberjornalismo. Retrieved December 12, 2020
3. Álvarez Fernández, J., & Secanella, P. (1991). Desinformación. In Á. Benito (Ed.), *Diccionario de Ciencias y Técnicas de la Comunicación* (pp. 365–375). Ediciones Paulinas.
4. Balfour, M. (1979). *Propaganda in war 1939–1945: Organization, policies and publics in Britain and Germany*. Routledge & Kegan Paul.
5. Blanks Hindman, E. (2005). Jayson Blair, The New York Times, and Paradigm Repair. *Journal of Communication*, 225–241. https://doi.org/10.1111/j.1460-2466.2005.tb02669.x
6. Buitrago A., García Matilla A., & Gutiérrez Martín A. (2017). Perspectiva histórica y claves actuales de la diversidad terminológica aplicada a la educación mediática,EDMETIC, ISSN-e 2254–0059, Vol. 6, No 2, (Issue dedicated to: Media education and digital competence), págs. 81–104
7. Carbajosa, A. (2019). Class Relotius: El escándalo 'Der Spiegel': paren la rotativa, todo es mentira. El País. Retrieved December 12, 2020, from https://elpais.com/elpais/2019/02/12/eps/1549973689_120344.html
8. Cathala, H. (1986). Le temps de la désinformation. París: Stock
9. Chomsky, N., & Herman, E. (1995). *Los guardianes de la libertad: Propaganda, desinformación y consenso en los medios de comunicación de masas*. Grijalbo Mondadori.
10. CNN (2017). President-elect Donald Trump refused to take a question from CNN reporter Jim Acosta, calling him fake news. Retrieved July 20, 2020, from CNN: https://youtu.be/VqpzkqGxMU
11. Collins Dictionary (2017). Why Fake News? Retrieved from Collins. Language Lovers Blog: https://blog.collinsdictionary.com/language-lovers/etymology-corner-collins-word-of-the-year-2017/
12. Conejera Idígoras, M., Donoso Christie, D., Moyano Díaz, E., Peña Herborn, J., & Saavedra Ponce de León, F. (2003). Comunicación persuasiva y cambio de actitudes hacia la seguridad de tránsito en peatones. Revista Latinoamericana de Psicología, 35(1), 77–90. Retrieved November 30, 2020, from https://www.redalyc.org/articulo.oa?id=80535107
13. Contentcal (2020). What is a Good Social Media Engagement Rate? Retrieved 30 December 2020, from https://www.contentcal.io/blog/what-is-a-good-social-media-engagement-rate/
14. Contreras Orozco, J. (2001). Rumores: voces que serpentean. Revista Latina de Comunicación Social, 40(4). Retrieved December 12, 2020, from https://www.redalyc.org/articulo.oa?id=819/81944009
15. Derrida, J. (1997). *Historia de la mentira: Prolegómenos*. Universidad de Buenos Aires.
16. Di Domenico, G., Sit, J., Ishizaka, A., & Nunan, D. (2021). Fake news, social media, and marketing: A systematic review. *Journal of Business Research, 124*, 329–341. https://doi.org/10.1016/j.jbusres.2020.11.037
17. Elman, C. (2008). Symposium on qualitative research methods in political science. *The Journal of Politics, 70*(1), 272–274. https://doi.org/10.1017/s0022381607080206
18. Fernández, M. (2014). La expansión del rumor en los medios digitales. In F. Sabés, & J. Verón, Universidad, Investigación y Periodismo Digital (pp. 19–36). Zaragoza: Asociación de Periodistas de Aragón. Retrieved December 6, 2020
19. García Avilés, J. (2009). La desinformación. In J. Herrero (Ed.), *Manual de teoría de la información de la comunicación* (pp. 327–346). Universitas.
20. García-Avilés, J. A., & Arias Robles, F. (2016). Géneros periodísticos en los formatos visuales de Twitter: una propuesta de tipología. *Textual & Visual Media*, 101–132
21. García Márquez, G. (1981, Abril 28). ¿Quién cree a Janet Cooke? El País. Retrieved November 30, 2020, from https://elpais.com/diario/1981/04/29/opinion/357343203_850215.html
22. García-Marín, D., & Aparici, R. (2019). La posverdad: el software de nuestra era. In R. Aparici, & D. García-Marín, La posverdad. Una cartografía de los medios, las redes y la política (pp. 25–44). Barcelona: Gedisa
23. García-Perdomo, V., Salaverría, R., Kilgo, D., & Harlow, S. (2018). To share or not to share: The influence of news values and topics on popular social media content in the United

States, Brazil, and Argentina. *Journalism studies, 19*(8), 1180–1201. https://doi.org/10.1080/1461670X.2016.1265896
24. Gerring, J. (2004). What is a case study and what is it good for? *American Political Science Review, 98*(2), 341–354. https://doi.org/10.1017/s0003055404001182
25. González, M. (2017, Diciembre 1). Las campañas de desinformación, nueva amenaza para la seguridad nacional. El País. Retrieved August 2, 2020, from https://elpais.com/politica/2017/11/30/actualidad/1512066298_815549.html
26. Gravanis, G., Vakali, A., Diamantaras, K., & Karadais, P. (2019). Behind the cues: A benchmarking study for fake news detection. *Expert Systems with Applications, 128*, 201–213. https://doi.org/10.1016/j.eswa.2019.03.036
27. Herrero-Curiel, E. (2011, December). El periodismo en el siglo de las Redes Sociales. Vivat Academia (117), 1113–1128. https://doi.org/10.15178/va.2011.117E.1113-1128
28. Jacquard, R. (1988). *La desinformación: Una manipulación del poder*. Espasa.
29. Kapferer, J. (1987). *Rumores. El medio de difusión más antiguo del mundo.* Emecé.
30. Kapusta, J., & Obonya, J. (2020). Improvement of misleading and fake news classification for flective languages by morphological group analysis. *Informatics*. https://doi.org/10.3390/informatics7010004
31. Knapp, R. (1944). A Psychology of Rumor. *Public Opinion Quarterly, 8*(1), 23–37. Retrieved December 12, 2020
32. Kumar Vishwakarma, D., Varshney, D., & Yadav, A. (2019). Detection and Veracity analysis of Fake News via Scrapping and Authenticating the Web Search. *Cognitive Systems Research* (58). https://doi.org/10.1016/j.cogsys.2019.07.004
33. Kümpel, A-S., Karnowski, V., Keyling, T. (2015). News sharing in social media: A review of current research on news sharing users, content, and networks. *Social media+society, 1*(2), 1–14. https://doi.org/10.1177/2056305115610141
34. Kunda, Z. (1990). The case for motivated reasoning. *Psychological Bulletin, 108*(3), 480–498. https://doi.org/10.1037/0033-2909.108.3.480
35. Lehrer, J. (2007). *Proust was a Neuroscientist*. Canongate Books.
36. Oluwaseun A., Deepayan B., & Shahrzad Z. (2018). Fake news identification on twitter with hybrid CNN and RNN models. In Proceedings of the 9th International Conference on Social Media and Society (SMSociety '18). Association for Computing Machinery, New York, NY, USA, pp. 226–230. https://doi.org/10.1145/3217804.3217917
37. Ortega, F. (2006). *Periodismo sin información*. Tecnos.
38. Otte, M. (2010). *El crash de la información: Los mecanismos de la desinformación cotidiana.* Ariel.
39. Pascal, B. (1940). *Pensamientos*. Espasa Calpe.
40. Reis, J., Correia, A., Murai, F., Veloso, A., & Benevenuto, F. (2019). Supervised learning for fake news detection. *IEEE Intelligent Systems, 34*(2), 76–81. https://doi.org/10.1109/MIS.2019.2899143
41. Rodríguez Andrés, R. (2018). Fundamentos del concepto de desinformación como práctica manipuladora en la comunicación política y las relaciones internacionales. Historia y Comunicación Social, 231–244. https://doi.org/10.5209/HICS.59843
42. Salas, J. (2018, Enero 28). ¿Por qué no cambiamos de opinión aunque nos demuestren que estamos equivocados? El País. Retrieved December 12, 2020, from https://elpais.com/elpais/2018/01/26/ciencia/1516965692_948158.html
43. Serrano, P. (2009). *Desinformación*. Península.
44. Solbés, J. (1988). Media business: argent, idéologie, désinformation. París: Messidor
45. Talwar, S., Dhir, A., Singh, D., Singh Virk, G., & Salo, J. (2020). Sharing of fake news on social media: Application of the honeycomb framework and the third-person effect hypothesis. *Journal of Retailing and Consumer Services, 57*. https://doi.org/10.1016/j.jretconser.2020.102197
46. Vatsalan, D., & Arachchilage, N. (2020). Understanding the strategies of creating fake news in social media. *Preprints*. https://doi.org/10.20944/preprints202011.0369.v2

47. Volkoff, V. (1986). La désinformation, arme de guerre. París: Julliard
48. Vossoghi, S., Roy, D., & Aral, S. (2018). The spread of true and false news online. *Science, 359*(6380), 1146–1151. https://doi.org/10.1126/science.aap9559

Detecting News Influence in a Country: One Step Forward Towards Understanding Fake News

Cristian Pop and Alexandru Popa

Abstract The concept of "fake news" has been referenced and thrown around in news reports so much in recent years that it has become a news topic in its own right. At its core, it poses a chilling question—what do we do if our worldview is fundamentally wrong? Even if internally consistent, what if it does not match the real world? Are our beliefs justified, or could we become indoctrinated from living in a "bubble"? If the latter is true, how could we even test the limits of said bubble from within its confines? We propose a new method to augment this process, by speeding up and automating the more cumbersome and time-consuming tasks involved. Our application, NewsCompare takes any list of target websites as input (news-related in our use case, but otherwise not restricted), visits them in parallel and retrieves any text content found within. Web pages are subsequently compared to each other, and similarities are tentatively pointed out. These results can be manually verified in order to determine which websites tend to draw inspiration from one another. The data gathered on every intermediate step can be queried and analyzed separately, and most notably we already use the set of hyperlinks to and from the various websites we encounter to paint a sort of "map" of that particular slice of the web. This map can then be cross-referenced and further strengthen the conclusion that a particular grouping of sites with strong links to each other, and posting similar content, are likely to share the same allegiance.

Keywords Web crawler · Web scraper · Text similarity · Social networks

C. Pop (✉) · A. Popa
University of Bucharest, Bucharest, Romania
e-mail: alexandru.popa@fmi.unibuc.ro

A. Popa
National Institute of Research and Development in Informatics, Bucharest, Romania

M. Lahby et al. (eds.), *Combating Fake News with Computational Intelligence Techniques*, Studies in Computational Intelligence 1001,
https://doi.org/10.1007/978-3-030-90087-8_13

1 Introduction

The topic of fake news is in the collective consciousness for some time now, due to its alleged impact on swaying public opinion on important issues, going so far as to potentially influence election results [2] in some cases. We find entire articles devoted to studying their impact [27], and methods of detection [9]. While some of the conclusions in these articles may be merely tentative, there are still some hard-to-dispute facts we can start using as a basis. For instance, we know that more than two thirds of Americans report getting at least some of their news on social media according to a Pew Research study [12] from 2017. Worldwide, 48% of people surveyed reported believing a fake news story was real before finding out it was fake, according to an Ipsos report [24]. Interestingly, the same report finds that 63% of people are confident in their own ability to identify fake news, while only 41% are confident that the average person can do the same. Are people in general overly confident about themselves, or too cynical about others? Hard to say, but nevertheless an interesting idea to explore.

Enterprising research out there has already found insightful characteristics of fake news, with one paper going so far as to draw parallels between fake news and satire [22]. This could not be easily done without the appropriate technology to gather large quantities of data, and analyzing it in new and creative ways. Taken to its logical conclusion, such research could eventually lead to heuristic algorithms able to detect and filter out fake news, a monumental breakthrough in and of itself. While not being naive enough to ignore all the challenges (one could easily imagine the rise of an "arms race" between fake news manufacturers and detectors, akin to the current system of viruses and antiviruses), this is one idea that we have found immensely motivating in our quest to push the boundaries of what can currently be done.

Relatedly, examining how news sources disseminate their content, how this fits in to their respective ecosystem, and how they continuously adapt in order to keep up a working business model, are all intriguing subjects in their own right. We know from existing research that newspaper publishers are aggressively trying to expand into the digital realm, going as far as adopting a "digital first" approach, but the data shows they are still heavily reliant on print in terms of revenue [30]. Exclusively online news outlets on the other hand do not have the luxury of print to fall back on, so we expect them to make that much more of an effort in establishing a foothold in the online market to draw revenue from. This is actually supported by some of our findings, see [37] for a specific example.

Since the topic of fake news is a complex one, it can hardly be expected to be tackled end-to-end over the course of a single article. More research is always welcome, and our understanding of it can only deepen in proportion with the number of researchers shining a spotlight towards it. Of course, any new research should ideally be done in a non-partisan fashion so that new studies can present objective conclusions, which are less likely to be dismissed offhand (especially by laypeople) in an increasingly polarised world [16]. That being said, it may be hard to even know how to begin tackling this issue, considering the sheer amount of data out there that needs to be collected, stored and whittled down into manageable chunks, to fit the scope of various investigations. As such, we want to do our part in reducing this barrier to entry, to build upon the works of others and at the same time provide a stepping stone for other people coming up with innovative research ideas that would otherwise be difficult to implement on account of technical challenges.

2 Related Work

We find similar work already out there, albeit with slightly a different application and purpose. Of course, we are not the first to consider the potential of data analysis, and the usefulness of providing enthusiastic people with investigative acumen with tools they could put to good use. Gray et al. [18] offer a particularly accessible guide aimed at journalists wishing to take charge and initiate their own data-heavy investigations. There are also repositories [11] dedicated to collecting large troves of documents and other data sets, opening them up to be analyzed by interested parties. What we try to offer is a slightly "meta" spin, by enabling investigations into the supposed investigative outlets themselves. Keeping tabs on the behaviour of entities tasked with shaping public opinion, either deliberately or unwittingly, should arguably rank fairly high as far as research topics go.

The issue of scraping social media data is explored in some detail by Marres and Weltevrede [29], who note that scraping is currently a prominent technique for the automated collection of online data, promising to offer new opportunities for digital social research. There is a fair amount of hype surrounding scraping as a herald of the coveted "revolution" in social research brought on by the advent of the Internet. What makes the technique special is allowing research to be done as an ongoing process, rather than a finished process. Of course, their application involved scraping just a handful of pages and charting very specific changes on said pages over time. Our application's current focus is a lot more generic, aiming to target a large number of distinct websites, and tries to avoid any kind of specialization that could prove restrictive for a general use case. Of course, future development can still be done to address various special cases with some minor tweaks.

The same article by Marres and Weltevrede [29] mentions a service used at the time, ScraperWiki [8], aiming to serve as a platform for developing and sharing scrapers. It has since been renamed to QuickCode, as it "isn't a wiki or just for scraping any more". ScraperWiki is mentioned a handful of times among the various works we have looked at in preparation for this article, but not so much since its rebranding as QuickCode. It is not entirely clear if the platform remains as accessible as it once was for the casual researcher at the time of writing. We could not find other similar platforms worth noting, therefore if web crawling/scraping research is indeed an underserved niche, our proposed solution should help plug that gap.

Other interesting research seeks to employ scraping to analysis with a more predictive application in mind. Lerman and Hogg [26] have tried come up with a model that is able to predict future news popularity starting from a data set acquired from scraping entries on a popular social media platform. Their work is greatly helped by the particular structure of their chosen platform (i.e. digg.com), where it is to pick up on early user voting results on new entries, extrapolating from there and estimating future popularity based. This should be easy to replicate on other sites with similar voting systems (e.g. reddit.com), but a great deal more creativity is required to do something similar on a more generic set of websites. That is, unless we can distill our set of target websites to include only ones with a very well defined set of characteristics, or choose some other metric to apply statistical modeling on and derive predictive benefit out of.

Yet another direction of research is sentiment analysis, as explored by Balahur and Steinberger [3] specifically for the use case of news articles. They employ the freely accessible Europe Media Monitor (EMM) family of applications [23], which at the time was retrieving between 80,000 and 100,000 articles per day in about 50 languages, scraping about 2200 hand-selected online news sources and a few specialist websites (these numbers have increased in recent years). A fairly impressive data set, unless it happens that our target websites fall outside of these news sources, which is where our application fills in the gap by allowing any number of custom entries to scrape on a regular basis. We estimate that some fairly involved tweaks would be required to add a similar sort of functionality to the processing side of our application, but the website content as currently gathered by our scraper should already lend itself well to the task.

A more niche approach, coming from what looks like fledgling research from Vargiu and Urru [42], involves figuring out how to pick out the most relevant contextual ads, based on insight gleaned from from scraping existing web pages. This does not necessarily apply solely to news sites, but it does give us an idea of at least one of the lucrative directions this kind of research can develop into. The amount of automation already out there in the advertising world should give us pause for thought, however. A solid business model right now could prove to be overinflated and unsustainable in the long term. According to a 2014 study by Association of National Advertisers [31], bots now comprise an estimated 23% of all online video ad viewers, and 10% of all static display ads. Rushkoff presents an eloquent, yet grim (and possibly somewhat alarmist) view in his book [40] on the topic:

> *Consider the irony: malware robots watch ads, monitored by automated tracking software that tailors each advertising message to suit the malbots' automated habits, in a human-free feedback loop of ever-narrowing "personalization". Nothing of value is created, but billions of dollars are made.*

With that in mind, we should be far more interested in creating something of value, rather than chasing ephemeral gains.

3 Our Results

What we try to add to the existing body of work is effectively a new solution in the form of a fast, efficient, mostly automated application able to gather vast amounts of information about websites, in as generic a form as possible. Our aim is to have an information dump that is easily to compile, and greatly simplifies the work of future researchers who need large sample sizes to interpret and derive conclusions from, according to various specific use cases. Some of these use case ideas have already been at least tentatively explored in articles mentioned in this introduction. We are confident that a good deal of research endeavors would have benefited from the kind of data dump we can now provide, and yet more research can benefit from it going forward.

We also put the application to the test on an individual use case to start with (i.e. Romanian news websites), to at least overcome the most glaringly obvious issues and challenges before releasing to the general public. A good deal of effort has been made in ensuring the application has more than just a niche appeal about it, and that it can be run reliably for long stretches without much manual interference. However, we also expect (and welcome) any constructive criticism and bug reports that get us closer to a flawless product. Despite not coming from a sociology background, we try our hand at interpreting the results we get from our use case, at least to the extent that we are aware of what characteristics to look for (see [37] for more details).

In the process of developing the app, some of the biggest hurdles that had to be handled were caused by the flaky and unpredictable nature of web content in general. By far, the element of human error involved in setting up websites seems to be the biggest source of issues with setting this sort of automated solution. Simple typos can lead to cascading failures (sometimes in spectacular fashion) when improperly interpreted by our heuristic algorithms. These failures are typically only obvious when they get to the point where they manifest among a noticeable segment of our result set. As such, there is a wide range of special case handling baked into our application code. While probably not fully exhaustive, we can reasonably expect that scenarios that are yet to be discovered should not have a statistically significant impact on results.

4 Preliminaries

According to the comprehensive primer by Olston and Najork [32], a web crawler (also known as a robot or a spider) is a system for the bulk downloading of web pages. Web crawlers are used for a variety of purposes, and the one we are most interested in here is an application of data mining, where we analyze web pages for statistical properties, and try to perform various data analytics. The web crawler starts off with a list of URLs to visit, otherwise known as *seeds*. This list can be quite small to begin with, as we expect it to grow exponentially. As the crawler visits these URLs, it identifies all the hyperlinks in the page and adds them to the list of URLs to visit, known as the *URL frontier* in some publications [28]. URLs from the frontier are recursively visited according to a set of policies. If the crawler is performing archiving of websites, as we do, it copies and saves the information contained within as it goes.

Once we get the process of acquiring a large data set of website content out of the way, the next step is to do more in-depth processing and extract more valuable information out of it. What we want to do is implement a kind of plagiarism detector to point out the more glaring similarities between articles on different websites. From the outset, it is clear this can turn into a time and resource-intensive task, and we need to be somewhat clever in order to avoid exponential complexity spiraling out of control and rendering the whole thing unfeasible.

Without even delving into algorithms, it should be make intuitive sense that most naive implementations would require too much processing to allow it to scale well, and there is at least some minimal research required to avoid wasting much time reinventing the wheel. For instance, a brute-force method of comparing 100 pieces of text one by one would require 4950 separate comparisons after a quick calculation ($n(n-1)/2$ where n = 100). Any optimization we implement along the way to reduce the number of comparisons performed can have a significant impact on the overall time. The particular algorithm we implement for performing the comparison is also crucial, to the extent that we can find one to process chunks of website text at least as fast as they are coming in from the web crawler side.

5 Overview of the Application

The back-end runs as an executable JAR file, so the machine running it needs to have Java runtime installed (version 8 or later). We also need to set up a PostgreSQL database for it to use, which can be done by following the steps listed on the GitHub project page readme file [35]. Once started, it will start automatically start crawling any sites listed in its database (this will be empty to start with). While crawling, it scans for new links to visit, and download the text content from every website visited to a local folder, where it will be indexed and processed to find similarities. More technically-inclined users will be able to connect to the back-end database directly to view real-time changes and make any low-level tweaks where it makes sense to do so.

The front-end is a Javascript-based single-page app (SPA) serving a number of key functionalities:

- Listing all websites discovered by the web crawler
- Allowing specific websites to be toggled as special interest, Romanian news websites for our use case, causing them to be queried more often for snapshots (at minimum every 1 h)
- Drawing a graph to visualize links to and from our special interest websites, allowing nodes to be added or removed in order to minimize clutter
- Listing instances where similar text content was detected on different websites
- Displaying various statistics about the web crawler's activity.

Note that the front-end can only be accessed while the back-end is running.

6 Technical Details

All the code written for the application is freely available on GitHub [35, 36] for anyone to examine or make use of, either as-is or by building upon it to suit some other purpose. In its current form, it should be accessible to most developers (particularly coming from a Java background), by following the instructions listed in the "Readme" files. For reference, the desktop machine used for all development and testing work is running a 64-bit octa-core (16-thread) CPU with 64 GB of RAM, and an NVMe solid-state drive for storage. The application is not particularly memory-intensive, but due to its multithreaded nature it does benefit significantly from multiple cores and high speed CPUs. Depending on the number of websites targeted and the frequency of snapshots taken, available storage could start to become a concern. For just over 100 websites each snapshot folder seems to add up around 1 gigabyte, including website text content and generated indexes.

Note that the back-end project is the most important component, and will be referred to interchangeably as "NewsCompare" or "the application" throughout this article. It can be picked up from scratch, and used on its own just by tweaking configuration values and keeping an eye on logging output. The front-end is effectively just a convenient way of interacting with the data set and provides some visualization of the application results. A prebuilt version of the back-end component is available on GitHub [34], with all the default settings we used throughout our testing configured at compile time.

In Sect. 6.1 we try to give a comprehensive run through the complexities of developing a web crawling solution nearly from scratch, which may prove helpful to anyone interested in rolling their own implementation. Section 6.2 similarly deals with how we set up an inverted index [6] for our set of text documents in order to perform fast searches and comparisons between them. This whole section should be a good starting point for anyone trying to better understand our publicly available code, to either modify or improve it. We try to note various issues and improvements already tackled, and also lay out some potential quality-of-life improvements for the future.

6.1 Web Crawling

Recall from Sect. 4 that a web crawler is a system for the bulk downloading of web pages. In our particular case, we try to visit every hyperlink at least once, but place a much higher emphasis on a manually curated list of websites where we want frequent snapshots saved. How often we are able to take these snapshots largely depends on how quickly we can run through this list on every iteration. As long as we keep it relatively short, and only visit a small set of websites on every iteration, we can afford to schedule our crawler to run fairly often. This in turn allows taking frequent snapshots, which are useful for record-keeping or auditing purposes. The main graphs and reports that we generate should typically be based on the most recent snapshot, unless a specific comparison between snapshots is otherwise required.

One of the immediately useful side effects of web crawling is that we automatically get to compile a list of directional links between the websites we start off with, and the ones discovered along the way. This allows us to effectively map a limited section of the visible web, and visualize it as a directed graph, with the websites serving as nodes and links as directed edges. This can serve as a basic sanity check on whether our results look valid and useful, but can also lead to basic conclusions in their own right (provided we have some interest in web architecture to begin with). Having readily available access to basic graph details, like node degree and connectivity allows us to see how our results line up with existing research, and potentially put it to the test. See [37] for some actual examples of insight derived from our particular set of target websites.

6.1.1 Challenges

Successive requests to the same server can lead to getting blacklisted or banned if the time between requests is too short. Should this occur on some websites (and slip by unnoticed), it could potentially skew our result set. A naive implementation of a web crawler might overlook this (or a malicious actor could ignore it entirely), but it stands to reason that most servers will rightfully seek to defend themselves against perceived acts of aggression, at least to the extent of limiting damage and maintaining high uptime. Any behavior that would not realistically be carried out by a human could raise red flags, causing web servers to start dropping requests. Based on our own empirical observation, imposing a mandatory delay of 100–200 between successive requests seems to yield good results.

URL normalization [5] is an important requirement, at least to the degree where we are satisfied that it covers our target websites properly. Any shortcomings in this area could lead to visiting semantically equivalent URLs, resulting in wasted effort and potential over-representation in our result set.

From the existing research, we adapt a few interesting ideas from a proposed algorithm [4] for a systematic and robust method of URL normalization, however in the final implementation we rely largely on simple string manipulation using regular expressions. Some of the more notable steps we employ are:

- Converting the scheme and host to lower case
- Removing the default port (e.g. 80 for `http`)
- Removing the fragment (`#`) and query (`?`) components of the URL
- Removing the protocol component (`http://` or `https://`)
- Removing www as the first domain label (where it exists).

Filtering out non-web content helps keep our data set smaller and more focused. We employ several basic methods here, based on string pattern matching in website

URLs. The links we filter out here are not web pages, so we know from the start they are unlikely to provide useful information in keeping the crawling process going:

- Filtering by file extension, e.g. links ending in ".jpg", ".doc", ".avi", ".mp3" etc.
- Filtering non-HTTP(S) protocols, e.g. links starting with "mailto://", "skype://", "whatsapp://" etc.
- Filtering specific string formats, e.g. links formatted in a way consistent with phone/fax numbers, or email addresses.

"Crawler traps" can be a significant time drain if unnoticed for extended periods. As Olsten and Najork mention [32], there exist "websites that populate a large, possibly infinite URL space on that site with mechanically generated content". The example they give is that of a web-based calendaring tool, where each month has its own page and a hyperlink to the next and previous months.

For our given set of websites, the biggest danger we notice is that of websites linking to various external indexing services. For instance, a website could link to its own entry on archive.org, which sends our crawler down an unfeasibly long chain of links that do not really improve our result set if followed. We are not directly interested in travelling the entire breadth of other existing indexing services or aggregators, we have started to maintain a list of exclusions for the crawler to avoid. See Sect. 6.1.3 for an idea on improving this process.

Thread-safe methods for reading and writing to data storage, in the context of using a single, traditional SQL database for data storage. In this case, using a transaction isolation level of "repeatable read" in PostgreSQL [38] appears to be enough to ensure the data integrity with only a moderate slowdown.

6.1.2 Optimization

Parallelization is something generally well-suited to web crawling activities. Intuitively we should be able to fetch content for most websites independent of each other, so the work can be done on separate threads. Some experimentation may be required with the number of threads assigned to each individual task in order to achieve this result.

Java's parallel streams [10] are a handy tool for quickly implementing parallel processing. Where other, more low-level, solutions would require us to handle the dividing of a problem into subproblems, then combining the results of the solutions ourselves, the Java runtime does this largely automatically. While it does not automatically guarantee operations perform faster (in some cases, quite the opposite, due to overhead), it makes it quite easy to make small code tweaks and find an optimal solution through trial and error. If we already follow the functional programming paradigm, it could mean something as simple as replacing calls to `stream()` with `parallelStream()`, configuring the thread pool size, and running performance tests. See Table 1 for the results of a test run comparing the effect of various configuration settings on the same sample set of 334 web pages.

Table 1 Effects of multithreading on web crawling speed (334 pages sample size)

No. of threads	Run time (s)	Average pages/s
1	665.3	0.50
5	239.4	1.39
10	130.2	2.56
20	74.5	4.48
50	27.2	12.27
100	25.6	13.04
200	33.4	10.00

Table 2 Effects of fetching webpages slices sequentially versus overlapping

Mode	Pages	Successes	Success rate	Time (s)	Avg. pages/s
Sequential	10434	9533	91.36%	788.95	13.22
Overlapping	9163	8296	90.53%	212.49	43.12

When crawling deeper than surface-level (i.e. more than just the website home page), we need to be cautious about our parallel tasks from inadvertently making too many simultaneous requests to the same server, as we mention in Subsection 6.1.1. Our approach here is divide our entire set of target web pages we wish to crawl during a regular run (typically around 10–20,000 for our 334 target websites) into "slices", with each slice containing at most one page from the same parent domain. These slices can be visited entirely sequentially, which is nice and safe (but also slow), or we can try to find a way to have them run in a partially overlapping fashion, which would be more optimal but also place us at a slightly higher risk of getting blacklisted for excessive requests. In our particular case, we settle on configuring each thread to run at a slightly different delay (in 100 ms increments), which does not seem to impact the rate of successful requests, and the speed improvement is more than threefold for our use case (Table 2).

Separating tasks, i.e. having the raw data retrieval separate from the actual processing, should increase overall throughput. Since network speed and latency can vary wildly by server, hitting a particularly slow website might otherwise bottleneck the entire process.

Depending on the particular use case, we can wait until our targeted websites are fully retrieved before starting the processing, or we can run the two tasks roughly in parallel, with the processing lagging slightly behind. Some factors influencing this decision include whether we want to extract useful information from partial results, or if we think we could squeeze some extra performance and have the CPU cycles to spare for it (i.e. if fetching websites is not already keeping us at 100% load, or close to it).

Table 3 Effects of timeout value on crawler speed and success rate

Timeout (s)	Run time (s)	Requests	Successes	Percentage (%)
0.001	504.0	11420	9766	85.51
1	530.4	11420	9732	85.21
5	622.9	11420	9760	85.46
10	740.4	11420	9773	85.57

Batch processing is one of the less obvious points, since in typical work loads with small sample sizes the performance impact is negligible. But something like building a large array of objects that are saved in a single call to the database, instead of saving each one individually gives us a massive speed advantage, particularly in the context of multiple threads seeking concurrent database access.

Timeout periods should be configured to a sensible value, to prevent having threads locked up in useless waiting periods for more than is absolutely necessary. This value can be arrived at through trial and error, by keeping track of the number of successful server responses over multiple trial runs. We expect this number to increase along with the timeout period, but we should see the improvement rate drop off sharply after a certain point. It is precisely this point of diminishing returns that gives us the best trade-off between results and performance.

We present the results of several web crawler test runs, each with different timeout values, where we target 334 websites and request 20 distinct pages from each of them (Table 3).

As we expect, the largest timeout values correlate with the largest number successful requests, but the improvement rate is marginal at best. The difference in percentage points is so insignificant enough that it may be explained away by random chance (or possibly, random background CPU usage on the test machine at the time). As such, we are comfortable in reducing the timeout to the bare minimum for our purposes.

6.1.3 Future Improvements

Automating "crawler trap" detection is a good starting step for improving the robustness of the application, allowing it to run independently with greater confidence. Since there is no real way to predict how often this kind of issue can surface, the way we mitigate it currently is by keeping an eye on the application's log output on a regular basis. We need to manually add any newly discovered "trap" to our list of excluded domains, so this is somewhat time-consuming. As soon as we find the trade-off acceptable, we can look at implementing a heuristic algorithm limiting the crawler's traversal of a particular domain past a specific threshold, making the entire process more automated.

Smarter timeouts would improve general crawling speed, especially over long stretches of time, but the impact can vary between marginal and significant depending on the set of websited targeted. By keeping track of each website's recently failed requests, we can place servers that seem to be unreachable (temporarily or otherwise) on cooldown, querying them less often and reducing the amount of time wasted waiting on timeouts overall.

The cooldown value can be set to an arbitrary value to start with, but ideally should be arrived at after some experimentation. We do not want to unwittingly restrict certain websites from our data set too harshly and risk skewing our conclusions. However, since any websites affected by this optimization are unresponsive to begin with, the risk of this should be fairly low.

Database replication adds a fair degree of complexity to the entire architecture, but at the same time provides a small-to-moderate boost in performance, by separating the application's responsibilities among multiple databases hosted on potentially multiple servers (or virtual machines). We expose a good number of API endpoints, some for displaying various statistics on the crawler's progress and results, others to provide a better visualization on our data set (or sections thereof). The SQL queries involved in retrieving this data take up to several seconds to run in some cases, largely due to the volume of data involved, and this constitutes extra load on our current "single point of failure" database.

PostgreSQL provides a very powerful solution in this regard [39], allowing us to do near real-time streaming replication of data from our master database to a standby one. The former can keep handling all the "heavy lifting" required by the web crawler, while the latter is used as a read-only source for reporting purposes. We also get the added bonus that the standby database can be automatically promoted at any time to master status, should the original master suffer an unrecoverable failure, significantly improving uptime and reliability. We do not include an implementation of database replication in the current version of the app, as it would further complicate the setup process for anyone seeking to reproduce (or build upon) our findings. It is however worth mentioning, in the interest of laying out the various pros and cons for interested parties.

6.2 Text Comparison

As mentioned in Sect. 4 most naive approaches for text comparisons on a large scale require too much processing time. Computing a kind of string similarity coefficient based on Levenshtein distance [7] (i.e. finding the smallest number of insertions, deletions, and substitutions required to change one string or tree into another) potentially gets us the results we are interested in, but is still very much a brute force approach. The most obvious shortcoming is that we are effectively doing the same work over and over by processing each string from scratch on every comparison. The first big improvement would be to introduce an initial, preparatory step of distilling strings into their base components for easier comparison later on.

Donald Knuth gives a very well-written primer in his famous book *The Art of Computer Programming* [25] on how inverted indexes are used to set up fast searching through text strings. To put it succinctly, we set up our index by making a list of unique terms in each individual block of text, and keep track of where the term is located within the text. From here, we can boil down every word to its most basic form (e.g. plural to singular, conjugated verb to infinitive form etc.) to reduce the size of our list of terms while improving representation. Additionally, we can filter out so-called "stop words", which are the most frequent and almost useless words (e.g. "a", "I", "the" for English), further lowering the noise in our search results.

Luckily, we are able to avoid much of the complexity of implementing our own inverted index solution by co-opting the open-source project Apache Lucene [13] into the application. It comes with a wide array of language analyzers (including Romanian), making it suitable both for our particular use case and improving the odds of our application becoming useful as a generic tool for future researchers. By making good use of Lucene's "more like this" functionality [15], we can avoid making an inordinate amount of one-to-one comparisons between items in our data set. This largely mitigates one of the concerns stated earlier, and means the number of comparisons we do (as well as the time taken for each comparison) should scale linearly rather than exponentially.

At this point we are able to perform the indexing and comparison steps at a manageable pace, something in the order of minutes instead of days for around ten thousand text files. However, we still have significant noise in our result set, so we need to further refine our algorithms. To this end, Abid et al. [1] suggest n-grams, i.e. sequences of words of length n, are a much better choice than single words for indexing and searching. Indeed, we observe a much tighter result set after switching to tri-grams, and the set itself is small enough to be discernible by a quick skim through (no longer requiring us to scour through millions of resulting combinations).

6.2.1 Challenges

Most challenges in this area stem from the fact that we are attempting to adapt a number of rough, heuristic algorithms to make sense of fairly nuanced text generated by humans (i.e. an extremely limited application of natural language processing). We want it to be useful, so the signal to noise ratio needs to be high, without excluding any useful results and reduce our overall accuracy. For instance, some of the conclusions coming from the app may be accurate (two pieces are text are very similar), but effectively useless at the same time (e.g. copied and pasted cookie policies, privacy policies, GDPR statements etc.). Conversely, two sources may be very similar content-wise, but the individual website's HTML structure could make it difficult to pick out particularly relevant blocks of text, causing it to slip under our radar.

Website architectures can be quite varied, and we want to keep any assumptions about particular approaches in this field at a minimum, so that the application ca be as generic as possible. In particular, subdomains can be somewhat tricky to deal with, we need to remember at all times to consolidate results belonging to the same top domain as a single source. After all, our stated purpose is to find similarities between wholly distinct websites, to point out the spread of content, and we do not concern ourselves with reused content between different sections of the same website. This consolidation step goes a long way towards improving our signal-to-noise ratio and making the more interesting results shine through.

Relevant content is sometimes hard to discern from surrounding context. Looking at any given news website, there is a lot of content displayed on page, but there is often surprisingly little space alotted to the actual content, i.e. the news article itself. The sidebars are typically reserved for internal/external links, advertising, and various widgets seemingly designed to provide some kind of use to the reader. We can discern that many design patterns favor drawing the user's attention, keeping them engaged and encouraging repeated visits, even when it might come into conflict with the main stated purpose of the site. While humans can quickly learn to intuitively pick up on useful content, automating this kind of processing into our algorithms can be quite tricky and time-consuming. In particular, the rise of interstitial advertising, and a general tendency to break up news into fragments and sprinkle vaguely related content between them needs to be accounted for. We will not go into whether or not an entire article is effectively an advertisement, as that falls somewhere outside the scope of the current research.

6.2.2 Optimization

Parallelization is already mentioned in 6.1.2 with regard to how it dramatically improves web crawling performance. The same rules apply here, even though we may not be able to find the same number of truly independent tasks that can be run in parallel. We hit a plateau of diminishing returns fairly quickly, but the performance gains are still worth pursuing as long as they are not too time-draining or significantly impact the readability or maintainability of the resulting code. We sit at a comfortable level of throughput right from the start, in no small part thanks to inherent optimizations present in the software library we employ [13].

6.3 Technologies Used

We aim to avoid using any proprietary or license-based software, so that all of our code can remain public. We are grateful to the open source community for the multitude of varied and powerful tools at our disposal, and we can at least state that we do not feel hamstrung by our decision. An honorable mention should be made to **Apache Nutch** [14], a fully featured web crawling solution that could help future tech-minded people to co-opt web crawling into their projects. We do not make use of Nutch in our case, mainly because we wanted to have tighter control over the crawler's behavior, and were comfortable enough in rolling our own lower-level implementation.

6.3.1 Back-End

We use **Spring Boot** [41] to quickly and easily get a RESTful web service [33] up and running using Java, but with minimum boilerplate and configuration out-of-the-box. It ties in well with **PostgreSQL** [19], which is used for mostly for persistence, but also storage to some degree. We need to save a limited amount of data from the websites explored by the web crawler to our database, some of which is used to inform future crawling iterations. **Hibernate** [20] makes it easier to perform the mapping between our Java classes and database tables, while **Flyway** [17] allows us to create our database structure in incremental migrations that can be easily replayed on a new machine when setting it up from scratch.

The crawler component uses **jsoup** [21] to create all of its network connections and also parse the resulting HTML pages using methods that allow for familiar CSS-like selectors. We also make local text dumps of the bulk of website contents, which are afterwards picked up by our implementation of **Apache Lucene** [13], creating indexes for quick text searches and comparisons.

6.3.2 Front-End

We use **Knockout** to build a simple yet dynamic JavaScript interface that pulls data from our application's endpoints and displays them in a more user-friendly fashion. The graph page uses an implementation of **vis.js** to help visualise website data as an interactive graph, again using data pulled from the back-end. **Webpack** is used to create a browser-friendly bundle of our own JavaScript source files, together with any node packages we use, as well as any other assets (e.g. CSS files).

7 Conclusions and Future Work

In this chapter, we described an application that aims to make a first step towards the fake news detection. Our application scraps the pages on several websites specified as input by the users and detects the similarity between these webpages.

There is a lot of future work since the structure of the news sources is unpredictable. For example, since the introduction of GDPR policy, many websites include pop-ups that can disturb the behaviour of our crawler. Thus, the crawler has to be redesigned to cope with these "obstacles". In the future, we aim to use the 'robots.txt' feature that many news websites have in order to obtain a more structured picture of the information that we retrieve.

Another idea that we plan to implement in the future is to use the timestamps provided by our crawler in order to determine how the information is propagated through the Internet.

References

1. Abid, M., Usman, M., & Waleed Ashraf, M. (2017). Plagiarism detection process using data mining techniques. *International Journal of Recent Contributions from Engineering, Science & IT (iJES), 5*, 68.
2. Allcott, H., & Gentzkow, M. (2017). Social media and fake news in the 2016 election. *Journal of Economic Perspectives, 31*, 211–236.
3. Balahur, A., & Steinberger, R. (2009). Rethinking sentiment analysis in the news : from theory to practice and back. In *Proceedings of WOMSA*.
4. Bar-Yossef, Z., Keidar, I., & Schonfeld, U. (2009). Do not crawl in the dust: Different urls with similar text. *ACM Trans. Web, 3*(1), 3:1–3:31.
5. Berners-Lee, E. A. (2005). *URI Generic Syntax. RFC 3986, The Internet Engineering Task Force*.
6. Black, P. E. (2008). *Dictionary of algorithms and data structures*. https://xlinux.nist.gov/dads/HTML/invertedIndex.html. [Online; Accessed 23-February-2019].
7. Black, P. E. (2008). *Dictionary of algorithms and data structures*. https://xlinux.nist.gov/dads/HTML/Levenshtein.html. [Online; Accessed 23-February-2019].
8. T. S. C. Company. Scraperwiki. https://scraperwiki.com/, 2019. [Online; Accessed 25-February-2019].
9. Conroy, N. J., Rubin, V. L., & Chen, Y. (2015). Automatic deception detection: Methods for finding fake news. In *Proceedings of the 78th ASIS&T Annual Meeting: Information Science with Impact: Research in and for the Community*, ASIST '15 (pp. 82:1–82:4). American Society for Information Science.
10. O. Corporation. Parallelism. https://docs.oracle.com/javase/tutorial/collections/streams/parallelism.html, 2019. [Online; Accessed 25-February-2019].
11. O. Crime and C. R. Project. Investigative dashboard. https://investigativedashboard.org/, 2019. [Online; Accessed 25-February-2019].
12. Shearer, E., & Gottfried, J. (2017). *News use across social media platforms 2017*. http://www.journalism.org/2017/09/07/news-use-across-social-media-platforms-2017/. [Online; Accessed 16-February-2019].
13. A. S. Foundation. Apache lucene. http://lucene.apache.org/, 2019. [Online; Accessed 23-February-2019].
14. A. S. Foundation. Apache nutch. https://nutch.apache.org/, 2019. [Online; Accessed 25-February-2019].
15. A. S. Foundation. Morelikethis (lucene 7.7.0 api). https://lucene.apache.org/core/7_7_0/queries/org/apache/lucene/queries/mlt/MoreLikeThis.html, 2019. [Online; Accessed 23-February-2019].
16. Funk, C. (2017). Mixed messages about public trust in science. *Issues in Science and Technology, 34*(1).
17. B. GmbH. Flyway db. https://flywaydb.org, 2019. [Online; Accessed 25-February-2019].
18. Gray, J., Chambers, L., & Bounegru, L. (2012). *The data journalism handbook: How journalists can use data to improve the news*. O'Reilly Media.
19. P. G. D. Group. Postgresql. https://www.postgresql.org, 2019. [Online; Accessed 25-February-2019].
20. Hat, R. (2019). Hibernate orm. http://hibernate.org/orm. [Online; Accessed 25-February-2019].

21. Hedley, J. (2019). jsoup java html parser. https://jsoup.org. [Online; Accessed 25-February-2019].
22. Horne, B .D., & Adali, S. (2017). This just in: Fake news packs a lot in title, uses simpler, repetitive content in text body, more similar to satire than real news. *CoRR*, abs/1703.09398.
23. E. S. HUB. Europe media monitor - newsbrief. http://emm.newsbrief.eu/NewsBrief/clusteredition/en/latest.html, 2019. [Online; Accessed 25-February-2019].
24. Ipsos. Fake news – ipsos perils of perception report. https://www.ipsos.com/en-au/fake-news-ipsos-perils-perception-report, 2018. [Online; Accessed 16-February-2019].
25. Knuth, D. E. (1997). *The Art of Computer Programming, Volume 1 (3rd Ed.): Fundamental Algorithms*. Redwood City, CA, USA: Addison Wesley Longman Publishing Co., Inc.
26. Lerman, K., & Hogg, T. (2010). Using a model of social dynamics to predict popularity of news. In *Proceedings of the 19th International Conference on World Wide Web*, WWW '10 (pp. 621–630). New York: ACM.
27. Lazer, D. M. J., Baum, M., Benkler, Y., Berinsky, A. J., Greenhill, K. M., Menczer, F., Metzger, M. J., et al. (2018). The science of fake news. *Science, 359*, 1094–1096.
28. Manning, C. D., Raghavan, P., & Schütze, H. (2008). *Introduction to Information Retrieval*. New York: Cambridge University Press.
29. Marres, N., & Weltevrede, E. (2012). Scraping the social: Issues in live research. *Journal of Cultural Economy*.
30. Myllylahti, M. (2017). Does digital bring home the bacon? In *International Communication Association, San Diego*, 05 2017.
31. A. of National Advertisers. The bot baseline: Fraud in digital advertising. https://www.whiteops.com/hubfs/ANA_WO_Bot_Baseline_2014-1.pdf, 2014. [Online; Accessed 25-February-2019].
32. Olston, C., & Najork, M. (2010). Web crawling. *Found. Trends Inf. Retr., 4*(3), 175–246.
33. Pautasso, C., Zimmermann, O., & Leymann, F. (2008). Restful web services vs. big web services: Making the right architectural decision. In *17th International World Wide Web Conference (WWW2008)* (pp. 805–814), Beijing, China, April 2008.
34. Pop, C. (2019). Backend jar file. https://github.com/buxomant/NewsCompareBackend/blob/master/out/artifacts/NewsCompareBackend_jar/NewsCompareBackend.jar. [Online; Accessed 25-February-2019].
35. Pop, C. (2019). Newscompare backend. https://github.com/buxomant/NewsCompareBackend. [Online; Accessed 25-February-2019].
36. Pop, C. (2019). Newscompare frontend. https://github.com/buxomant/NewsCompareFrontend. [Online; Accessed 25-February-2019].
37. Pop, C., & Popa, A. (2019). Newscompare—A novel application for detecting news influence in a country. *SoftwareX, 10*, 100305.
38. PostgreSQL Global Development Group. Postgresql documentation: 10: 13.2. transaction isolation. https://www.postgresql.org/docs/10/transaction-iso.html, 2019. [Online; Accessed 16-February-2019].
39. PostgreSQL Global Development Group. Postgresql documentation: 10: 26.2. log-shipping standby servers. https://www.postgresql.org/docs/10/warm-standby.html, 2019. [Online; Accessed 21-February-2019].
40. Rushkoff, D. (2016). *Throwing rocks at the Google bus: How growth became the enemy of prosperity/Douglas Rushkoff*. New York: Portfolio/Penguin New York.
41. P. Software. Spring boot. https://spring.io/projects/spring-boot, 2019. [Online; Accessed 25-February-2019].
42. Vargiu, E., & Urru, M. (2013). Exploiting web scraping in a collaborative filtering-based approach to web advertising. *Artificial Intelligence Research, 2*, 01.

Factors Affecting the Intention of Using Fintech Services in the Context of Combating of Fake News

Lam Oanh Ha, Van Chien Nguyen, Do Dinh Thuy Tien, and Bui Thi Bich Ngoc

Abstract In the context of technological revolution 4.0, the development of financial technology has made important contributions in reducing transaction costs, saving transaction time and promoting economic growth. Financial technology has been applied in many fields in payment, savings management, investment, asset management, peer-to-peer lending and data analysis, while limiting the spread of fake news where social networks with fast speed and great impact. The purpose of the study is to analyze the factors affecting the intention to use Fintech services in an emerging country in Asia. Research shows that minimizing the impact of fake news through upgrading service quality, security, and trustworthiness of Fintech will have a positive impact on the intention to use Fintech services in Vietnam. In addition, four factors including: usefulness (SHI), ease of use (DSD), social influence (XH), and communication about Fintech services also have a positive impact on intention to use. Fintech services.

Keywords Fintech · Fake news · Intention · Trust

1 Introduction

Fintech, a compound word of 'Financial' and 'Technology', is a process of applying technological solutions in everyday financial services [1]. In today's period, India, UK and China are the countries that have seen major changes in the digital economy from Fintech [2–4]. In the study of the Vietnam microfinance working group (2018), Fintech is used to talk about computer techniques or technologies that financial institutions apply in the process of providing products and services to customers customer. At the same time, many studies have shown that financial technology has

L. O. Ha · D. D. T. Tien · B. T. B. Ngoc
Department of Finance and Banking, Thu Dau Mot University, Thu Dau Mot City, Binh Duong, Vietnam

V. C. Nguyen (✉)
Institute of Graduate Studies, Thu Dau Mot University, Thu Dau Mot City, Binh Duong, Vietnam
e-mail: chiennv@tdmu.edu.vn

M. Lahby et al. (eds.), *Combating Fake News with Computational Intelligence Techniques*, Studies in Computational Intelligence 1001,
https://doi.org/10.1007/978-3-030-90087-8_14

been applied in many fields such as: payment (cryptocurrency, mobile payment, QR code payment), saving, investment and management assets, blockchain and related applications, credit, peer-to-peer lending, insurance and data analytics [5–8]. Fintech service use is the process by which individuals conduct transactions using mobile phones or digital devices connected to the Internet [9, 10].

Fintech is not only a trend of the world but also a direction of the banking and financial industry in Vietnam. Developing financial services on the basis of applying information technology is one of the important measures for Vietnamese financial institutions to improve their competitiveness in the face of international integration. Besides, Fintech also gives technology companies a big piece of cake when they search and develop software solutions to support or directly provide alternative Fintech services with new utilities, fast processing speed and low cost [11, 12].

According to the Vietnam Statistical Yearbook, Vietnam's population is more than 96 million people, of which the proportion of the population aged 15–64 accounts for 69.3% [13]. At the same time, the Digital Marketing 2019 report of WeareSocial and Hootsuite also shows that 66% of the Vietnamese population uses the internet, equivalent to 64 million people [14]. Moreover, Vietnam is one of the top three growing smartphone markets in Southeast Asia with the percentage of mobile subscribers using smartphones reaching 60% in 2017 and by 2021, the number of smartphone subscribers in Vietnam will increase 3 times higher than 2017. At the same time, Neilsen [15] states that the percentage of smartphone users in rural areas is 68%, while this figure is 84% for big cities. The above figures are proof that Vietnam is a potential market for Fintech. Solidiance [16] has shown that the transaction value of the Vietnamese Fintech market reached $4.4 billion in 2017 and forecast this figure to be $7.8 billion in 2020. According to Fintech Singapore [17], by the end of 2020, Vietnam is home to more than 120 Fintech startups, some prominent Fintech service providers in Vietnam can be mentioned as: Payoo, Momo, Moca, ViettelPay, Onepay, VinID, Zalopay, Baokim.vn, napas, Airpay.

However, in the context of the 4.0 technological revolution, fake news is spread through social networks due to internet connection, globalization makes the amount of information spread at a fast speed, the amount of information that each individual can accept more. According to the evaluation of [18], in life, each individual's decisions are always influenced by emotional factors (sentiments), which can create too excited or too pessimistic psychology in the future financial transactions. In that context, fintech has the ability to remove fake news through information processing, screening and building reliability of the system. As a result, the financial technology provider continuously improves the information quality of the system, in order to create benefits for users with high reliability, meeting transaction needs, and saving costs and transaction time. Using financial technology, organizations provide payment services in the tourism industry, shopping and paying online, updating stock prices through online applications.

In the context of Vietnam, an emerging country in Asia, the proportion of Internet and smartphone users is high, the habit of using online services is increasing, along with the increasing number of Fintech service providers. The format and quality are continuously improved, and fake news is also spreading more and more complex, affecting the quality of services in fintech. Carrying out this study, we will evaluate the factors affecting the intention to use fintech services in the impact of fake news in Vietnam, as a case study in developing countries, and at the same time, appropriate policy suggestions for these countries in the era of rapid change of the industrial revolution 4.0.

In addition to the introduction, the remainder of this study consists of: part 2 and part 3, which discusses the literature review and research model, and part 4, which discusses research methods and data. Meanwhile, research results will be discussed in Sect. 5, conclusions and some recommendations are discussed in Sect. 6.

2 Literature Review

The Innovation Diffusion Theory (IDT) model was developed by [19]. The author argues that, IDT explains innovation and customers who realize the benefits of that innovation will accept the new product. Meanwhile, Engel et al. [20] proposed a research theory of consumer behavior EKB -ENGEL-KOLLAT-BLACKWELL, emphasizing the social normative value factor affecting buyer behavior.

In 1960, Fishbein with the TRA—Theory of Reasoned Action, then developed by the research of [21] is considered to be the pioneer research on consumer behavior. This theory suggests that "attitudes and subjective standards" are the two basic factors that influence an individual's intentions. In 1991, Ajzen [22] developed TRA into TPB-Theory of Planned Behavior, arguing that the factor "perception controls behavior" also affects people's intentions.

According to the theory of Technology Acceptance Model-TAM was developed by [23] and revised and supplemented in 1989 and 1993. TAM shows that the attitude of technology users is influenced by two factors. The factors are perceived usefulness and perceived ease of use. In 2003, Venkatesh et al. [24] built and developed a unified theory of acceptance and use of UTAUT technology, four factors affecting behavioral intention are: performance expectations, expectations about effort, social influence and material conditions. In the marketing mix model 4P in research by [25] shows that consumer's consciousness is influenced by marketing factors such as: product or service, price, distribution, and promotion.

3 Research Models

In this study, the author synthesizes previous research based on 7 scales, including a scale of trust that represents an individual's trust when using fintech services, a scale of trust that represents the role of a person using fintech services before fake news, reflected in the level of safety when using the service.

3.1 Trust Scale (Symbol TN)

The scale shows the confidence of individuals when using services and trust in Fintech. Specifically, when they feel safe in using Fintech services, it is evident that the higher level of trust, the more individuals avoid risks according to fake news. This scale is built on the work of [24], the questions are also tested by the work of [26, 27].

H1: There is a positive relationship between perceived trust and intention to accept Fintech services.

3.2 Perceived Usefulness Scale (Symbol HI)

The scale represents an individual's ability to subjectively evaluate when using Fintech's services that will increase their work efficiency. The scale was developed based on the studies [24, 28–32], demonstrated that usefulness affects the intention to accept using a product or service.

H2: There is a positive relationship between perceived usefulness and intention to accept Fintech services.

3.3 Easy-To-Use Cognitive Scale (Symbol DSD)

The scale indicates the perceived ease of using Fintech services (is the degree to which individuals perceive the difficulty or ease of using Fintech products/services). The scale due to ease of use is built on the research of [28], the questions have been tested by many other studies such as [24, 31–33] achieved good research results.

H3: There is a positive relationship between perceived ease of use and the intention to accept Fintech services.

3.4 Scale of Social Influence (Symbol XH)

This scale shows the extent to which an individual perceives important people and other people around them that they should use Fintech's new products/services. The scale of social influence is indicated by the student's assessment of 5 questions. This scale is built based on 4 questions in the study of [24]. The questions are also tested by the work of [25, 27].

H4: There is a positive relationship between social influence and the intention to accept Fintech services.

3.5 Innovation Scale (Symbol TM)

The scale shows the confidence of individuals when using services and trust in Fintech. Innovation represents a preference for novelty and challenges in life. The innovation of individuals when using Fintech services is the tendency to use new products and prefer product differences. The innovation scale was built based on the research of [34, 35].

Traditional Fintech services with the development of effective support of information technology and modern technical means have formed diverse and modern products such as payment services, services. Credit. Many modern Fintech services contain a high content of information technology, the intellectual content in the service accounts for a large proportion, so to use modern Fintech services requires individual skills with the support of tools and equipment. The scale also adds two more scales, to better assess the tendency to like to use new products and to like to learn about new technologies, learn new products and services.

Hypothesis H5: There is a positive relationship between perceived innovation and intention to accept Fintech services.

3.6 Fintech Service Communication Scale (Symbol DV)

Communication is the communication process to share information, knowledge and experience about Fintech products and services, communication is done at any opportunity so that individuals can see or see the information. messages about Fintech products and services in the media. The service communication scale is built based on the research of [36–38].

Hypothesis H6: There is a positive relationship between perceived innovation and intention to accept Fintech services.

3.7 Scale of Intention to Accept and Use Services (Symbol YD)

The scale represents the intention to accept and use Fintech services as a motive for taking action, making decisions about whether to accept and use or not to accept Fintech services in the future. The scale of intention to accept and use the service is built based on the study of [24, 27, 28, 39–42].

On the basis of the theory of intention to use and the results of the review of relevant domestic and foreign studies, the author proposes a model of this study including 6 factors affecting the intention to use Fintech services in specific research in Vietnam. The observed factors and variables are shown in the descriptive statistics and the research model is summarized in Fig. 1

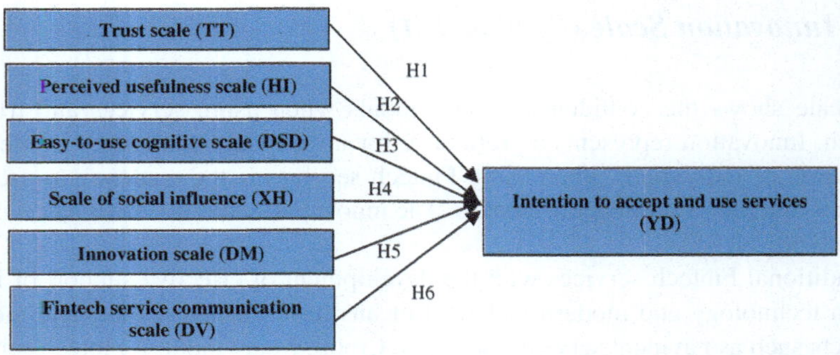

Fig. 1 Research model. *Source* Author's compilation

4 Research Methodology and Data

Qualitative research is the process of reviewing relevant documents in order to find out the factors affecting students' intention to use Fintech services. The authors conducted individual interviews and group discussions of 10 experts who are finance professors, leaders of Viettel pay, and individuals using Fintech services. Qualitative research has achieved the purpose of checking the suitability of the scale and screening the independent variables, preliminarily determining the relationship between the dependent variable and the independent variables. At the same time, the authors received expert suggestions on using words and completing the content of the quantitative questionnaire.

Quantitative research is the process of collecting primary data through a survey in Vietnam that has never used, has been or is using Fintech in payment. The official survey was conducted from April to December 2020, through online interviews or sending online surveys through applications such as Microsoft team, Zoom, Zalo, Facebook. The total number of usable survey responses is 281 votes, the satisfactory sample size according to the standards of [43] is 5 times larger than the number of observed variables (minimum 215 votes).

The collected information is processed by SPSS 25.0 software through the following steps:

Step 1: Descriptive statistics to make statistics about demographic information of research subjects.

Step 2: Test the reliability of Cronbach's Alpha to eliminate inappropriate variables in the research model. According to Hair et al. (2010), Cronbach's Alpha coefficient from 0.6 to 0.8 is a good measurement scale.

Step 3: Exploratory factor analysis—EFA aims to gather many interdependent observations into a more meaningful set of variables. EFA analysis needs to meet Factor loading > 0.5; 0.5 < KMO < 1; Bartlett's test has Sig < 0.05; variance extracted Total Variance Explained > 50% and Eigenvalue > 1.

Step 4: Regression analysis to determine the impact level of factors.

T-test, ANOVA test were also performed to ensure the appropriateness of the research hypothesis.

5 Research Results

5.1 Descriptive Statistics

Regarding the gender structure, according to the survey results of 281 respondents, 130 are male, accounting for 46%, and 151 are female, accounting for 53%, as shown in Fig. 2

An evaluation study based on a survey of student groups in many provinces and cities in Vietnam, Fig. 3 shows that first-year students accounted for 2.85%, second-year students accounted for 9.96%, and third-year students accounted for 27.05%. In which, the group of 4th-year students (final year) accounted for a fairly high proportion of 60.14%, this is the main age group that is usually responsible for their own economic activities, so they also often use Fintech services to support its operations.

Surveying a group of students in many provinces and cities in Vietnam, Fig. 4 shows that 46.98 percent of students are in the economic sector (business administration, banking and finance, and accounting) (132 people), 12.81% students majoring in engineering and technology (36 people), 6.76% students majoring in

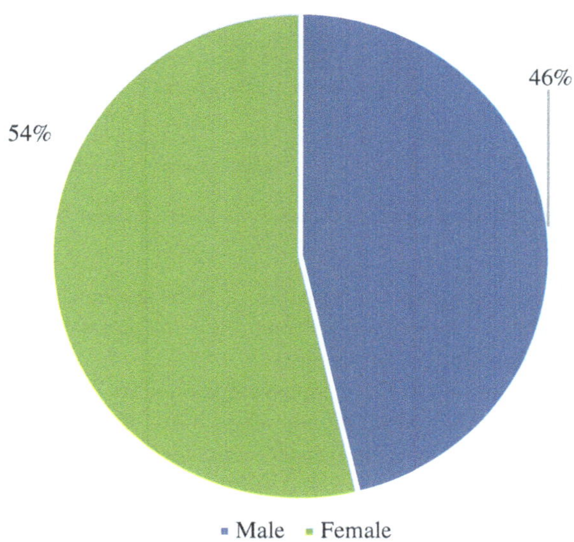

Fig. 2 Gender of survey subjects. *Source* Author's analysis

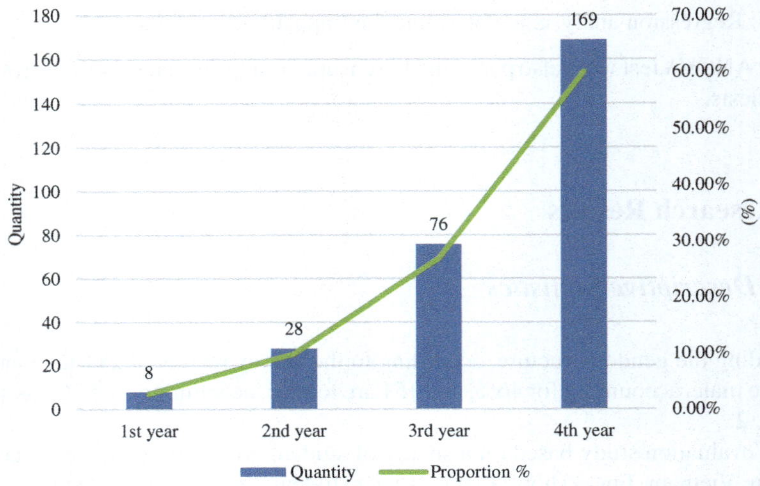

Fig. 3 Evaluation of survey subjects. *Source* Author's analysis

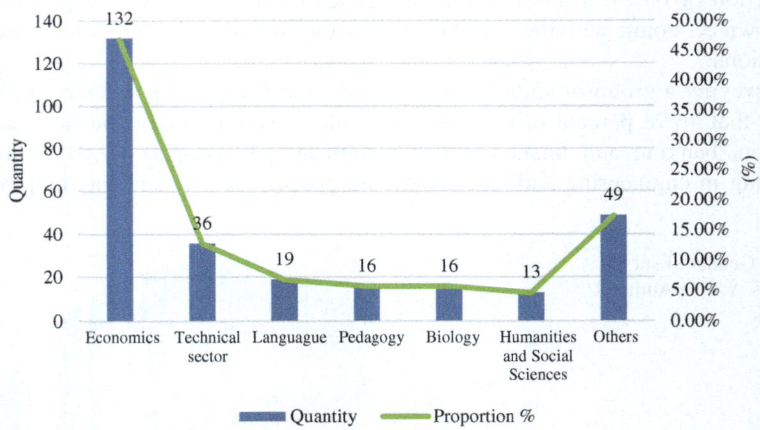

Fig. 4 Subject's training majors. *Source* Author's analysis

foreign languages (19 people), 5.69% students majoring in pedagogy and 5.69% students majoring in biotechnology (16 people), 4.6% students majoring in social and humanities (13 people).

According to Fig. 5, the average monthly income of 45.9% of the sample is from 1 to 3 million VND/month, about 6.4% of the sample has an income of 9,000,000 VND/month or more. It can be concluded that the majority of students have a monthly income ranging from 1 to 3 million VND.

Figure 6 shows that the number of surveyed students studying in Ho Chi Minh City. Hanoi, City. Da Nang, City. Ho Chi Minh City, Binh Duong Province, City. Can

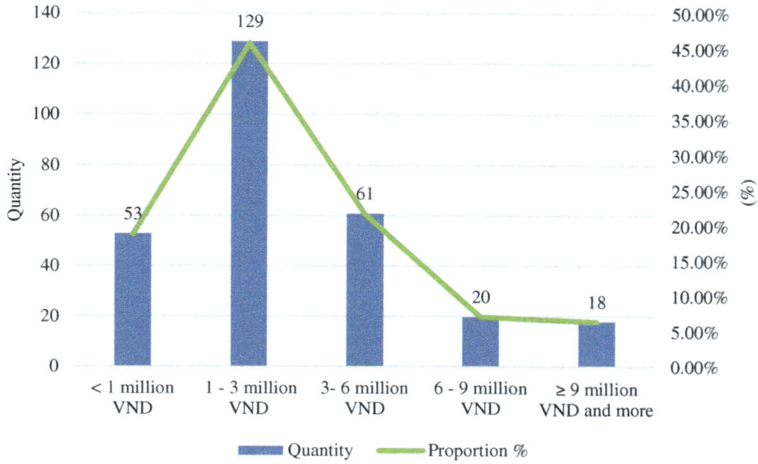

Fig. 5 Survey subject's monthly income. *Source* Author's analysis

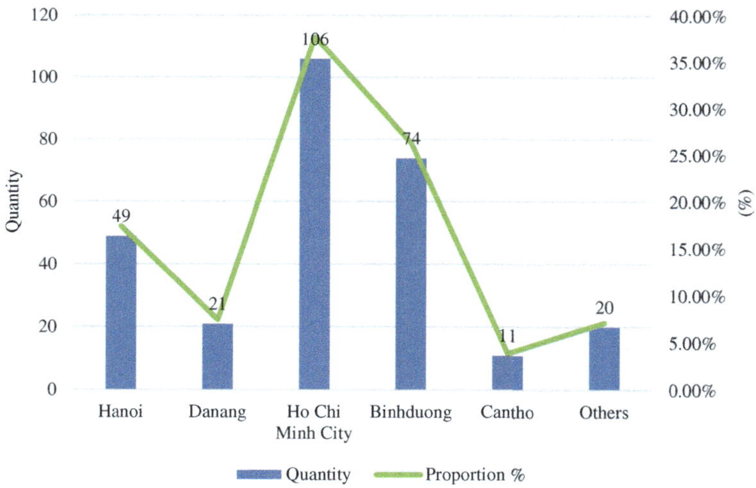

Fig. 6 Study/work place of the surveyed subjects. *Source* Author's analysis

Tho and other provinces are 49, 21, 106, 74, 11 and 20 respectively; respectively: 17.4%, 7.5%, 37.7%, 26.3%, 3.9% and 7.1%.

In the survey, Fig. 7 shows that 60.85% of students participating in the survey use momo wallet, followed by zalopay, Viettel pay and Airpay with 34.16%, 16.37% and 11.39% respectively. In which, Payoo is the oldest Fintech application in Vietnam but no students participate, which can be explained by Payoo's aim to provide financial technology services for businesses [17].

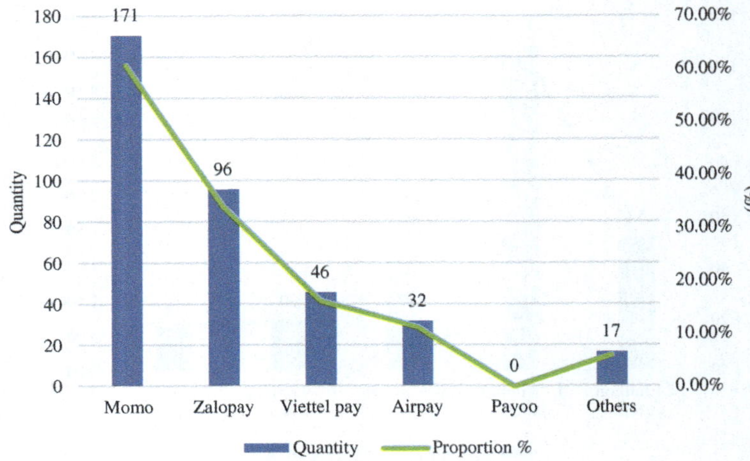

Fig. 7 Fintech application the respondents by using. *Source* Author's analysis

Regarding the purpose of using Fintech, Fig. 8 shows that 178 (63.35%) students participated in the survey using Fintech for the purpose of money transfer. 100% of students use it to pay bills, shop online. Only about 17.79% (50 students) save on Fintech applications, 2.85% (8 students) invest in securities and no students borrow money through Fintech.

In addition, the study also assessed that over 60% of survey participants confirmed that the reason they choose to use Fintech is also the benefits that Fintech brings to users such as: ease of use, convenience during the epidemic season and the trend

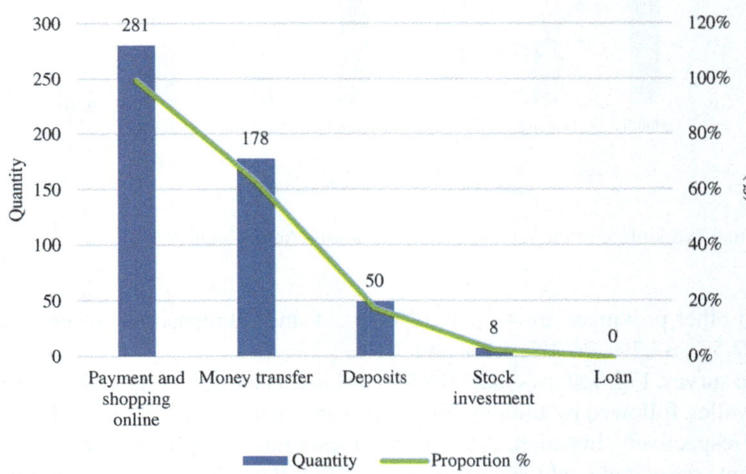

Fig. 8 Purpose of using Fintech by respondents. *Source* Author's analysis

of technological revolution 4.0, have many incentives, no money transfer fee, have many promotions, get real refund, fast, no system error. In addition, over 67% of respondents pointed out that the risks and disadvantages of using Fintech are: unregulated services, low ability to ensure customer information, and poor customer care support. results, App errors, 24-h money detention (sometimes even losing money but not being able to make transactions), poor security, service staff using customer information for personal gain, easy to cheat in online shopping. The quality of the products customers who buy through the service may be of lower quality than the advertised ones, and are limited in linking with shopping websites. In particular, 221 respondents (78.6%) said that when using Fintech applications, they face the risk of information loss. Only about 22.8% of people believe that Fintech is still a legal risk due to the lack of strict management of the law.

The descriptive statistical results of each scale are shown in Table 1

5.2 EFA—Exploratory Factor Analysis

Testing the scale, the results show that the scales are reliable, the correlation coefficients with the total variable are higher than 0.640 (variable DSD3). The lowest Cronbach's Alpha is the perceived scale of ease of learning and using Fintech services (0.763). The type of observed variable DSD4 (feeling that Fintech services are flexible and easy to apply) due to the Cronbach's Alpha coefficient of the scale will be higher if this variable is removed (Table 2).

Exploratory factor analysis EFA for factors affecting the intention to use Fintech services, 6 factors are extracted at eigenvalues 2134 and the extracted variance is 78.858%. Thus, the extracted variance is satisfactory because it is greater than 50%, the EFA for the remaining factors is suitable because KMO = 0.905 (>0.5), Sig. of Bartlett test = 0.000 (<0.05). Thus, after assessing the scale affecting the intention to use Fintech services, there are 6 factors and 38 observations (Tables 3 and 4).

5.3 Regression Results

Regression results show that the innovation factor (TBDM) has no statistical significance with 90% confidence (P-value > 0.1). Therefore, this factor will be removed from the model to build the best model. The research model, after being included in the regression analysis, proved the following hypotheses:

H1: There is a positive relationship between perceived trust and the intention to accept Fintech services.

H2: There is a positive relationship between perceived usefulness and the intention to accept Fintech services.

H3: There is a positive relationship between perceived ease of use and the intention to accept Fintech services.

Table 1 Descriptive statistics of variables

Code	Scale	Min	Max	Mean	Std. Dev
Trust scale (TT)					
TT1	When using Fintech services, I believe that my personal information is kept confidential	1.00	4.00	3.0356	0.74557
TT2	When using Fintech services, I believe that my transaction is guaranteed	1.00	4.00	3.0000	0.79732
TT3	When using Fintech services, I believe that my privacy will not be disclosed	1.00	4.00	2.9537	0.77551
TT4	When using Fintech services, I believe that the Fintech environment is safe	1.00	4.00	3.4306	0.65710
Perceived usefulness scale (HI)					
HI1	Using Fintech services increases business productivity and efficiency	1.00	5.00	3.4448	0.99534
HI2	Using Fintech services saves more time	1.00	5.00	3.3808	0.96039
HI3	Using Fintech services makes it quick and convenient for currency and credit transactions	1.00	5.00	3.8861	0.90308
HI4	Use Fintech services in line with business needs	1.00	5.00	3.4413	0.98067
HI5	Using Fintech services is useful and convenient for business activities	1.00	5.00	3.4342	0.99827
Easy-to-use cognitive scale (DSD)					
DSD1	Easy to learn and use Fintech services	1.00	4.00	2.9359	0.84694
DSD2	Making transactions with Fintech services is clear and easy to understand	1.00	4.00	2.6228	0.87426
DSD3	Fintech services can be used expertly	1.00	4.00	2.6299	0.81834
DSD4	Feel that Fintech services are flexible and easy to apply	1.00	4.00	2.4413	0.80463
DSD5	Feel that every service Fintech provides meets the needs of students	1.00	4.00	2.6548	0.90142
Scale of social influence (XH)					
XH1	People who are important to me think that I should use a new and more modern Fintech service	1.00	4.00	2.8612	0.74063
XH2	People who are familiar with me think that I should use a new and more modern Fintech service	1.00	4.00	3.3665	0.66881
XH3	People who influence my behavior think that I should use a new and more modern Fintech service	1.00	4.00	2.8577	0.77068
XH4	Most of the people around me think that I should use the new and more modern Fintech service	1.00	4.00	2.9146	0.76052
XH5	I see a lot of people using banking services. I think I should use new and more modern banking service for my operation	1.00	4.00	2.9110	0.77177
Innovation scale (DM)					
DM1	I often search for information about Fintech services	1.00	4.00	3.5125	0.59223

(continued)

Table 1 (continued)

Code	Scale	Min	Max	Mean	Std. Dev
DM2	I like to go to places where I get a lot of information about new Fintech services	1.00	4.00	3.0036	0.76297
DM3	I like magazines that introduce and advertise new Fintech services	1.00	4.00	3.0071	0.77456
DM4	I took the first opportunity to learn about Fintech's new service	1.00	4.00	3.0000	0.76064
DM5	I always love to learn new services and new benefits of Fintech services	1.00	4.00	2.9893	0.77221
DM6	I often learn new technologies to use modern Fintech services	1.00	4.00	3.0214	0.73648
DM7	I am often interested in learning new information and using Fintech's services	1.00	4.00	3.0285	0.73137
Fintech service communication scale (DV)					
DV1	I was carefully introduced by Fintech staff, detailing how to use banking products and services	1.00	5.00	3.6584	0.80889
DV2	I was carefully introduced by Fintech staff, detailing the benefits of banking products and services	1.00	5.00	3.5516	0.84832
DV3	At transaction offices, there is usually a direct student care department to answer and guide me on utilities and procedures for implementing Fintech services	1.00	5.00	3.5552	0.84388
DV4	Before using Fintech services, I often find out information about products and services through leaflets, radio and television…	1.00	5.00	3.6406	0.87157
DV5	Before using Fintech services, I often find out information about products and services through social media, wards…	1.00	5.00	3.6299	0.82702
DV6	Fintech is always ready to provide information about Fintech services to students through a communication system of communes, wards, etc., a system of local service providers such as Women's Union, Farmers' Union…	1.00	5.00	3.5943	0.80123
DV7	Fintech always has staff on hand to answer my questions and guide me in making documents and procedures to use the right products	1.00	5.00	3.6655	0.82927
DV8	I was carefully introduced by Fintech staff, detailing the benefits of banking products and services	1.00	5.00	3.6406	0.79439
DV9	Information about products and services that Fintech provides helps me realize the usefulness of Fintech services	1.00	5.00	3.6014	0.83101
DV10	Information about products and services that Fintech provides makes me aware of the ease of using Fintech services	1.00	5.00	3.5907	0.87001

(continued)

Table 1 (continued)

Code	Scale	Min	Max	Mean	Std. Dev
DV11	Information about products and services that Fintech provides helps me to be more aware and interested in new Fintech services	1.00	5.00	3.6157	0.84619
DV12	Information about products and services that Fintech provides helps me feel more confident when using banking services	1.00	5.00	3.5979	0.81407
DV13	Information about products and services that Fintech provides affects social relations	1.00	5.00	3.6050	0.81318
Intention to accept and use services (YD)					
YD1	I intend to use more new services that Fintech provides in the next 3 months	1.00	5.00	3.6406	0.82958
YD2	I intend to use Fintech services often in the future	1.00	5.00	3.6121	0.85088
YD3	I plan to use more Fintech services provided in the near future	1.00	5.00	3.6406	0.82958
YD4	I intend to increase my understanding and use of new and modern Fintech services in the future	1.00	5.00	3.6121	0.85088

Source Author's analysis

H4: There is a positive relationship between social influence and the intention to accept Fintech services.

H5: There is a positive relationship between perceived innovation and intention to accept Fintech services (Tables 5, 6 and 7).

With the above results, five factors that have a direct and positive influence on the intention to use Fintech services in the case of Vietnam are: perceived usefulness (HI), perceived ease of use (DSD), and trust (TT), social influence (XH), communication about Fintech services (DV). In which, the communication factor about Fintech services is the factor that has the strongest impact on the intention to use Fintech services in Vietnam with a regression weight of 0.848. Next is perceived usefulness (regression weight 0.128). Then there are the factors: social influence, trust and perceived ease of use. In particular, the trust factor of individuals when using Fintech services when they feel safe in the context of the risks of fake news that can be encountered in Fintech transactions, research shows that the Trust has a positive impact on the intention to use Fintech in the context of Vietnam, which means that the evidence, the reduction of fake news through the improvement of Fintech service quality will have a positive impact on the intention to use Fintech.

The research results show that the intention to use Fintech services in Vietnam is most strongly influenced by the communication factor about Fintech services, this result is suitable because students are the young generation aged 18–23 years old, they are often monitored and easily oriented by propaganda activities about products and services. The awareness factor of innovation is not statistically significant, which can be explained by Fintech, although applying technology platforms and constantly innovating, but there are not many novelties.

Table 2 Cronbach's alpha of research variables

Variables	Scale mean if item deleted	Scale variance if item deleted	Correcred item—total correlation	Cronbach's alpha if item deleted
Anpha = 0.859 Trust				
TT1	9.3843	3.787	0.660	0.839
TT2	9.4199	3.552	0.689	0.829
TT3	9.4662	3.693	0.659	0.841
TT4	8.9893	3.711	0.836	0.775
Alpha = 0.943 Perceived usefulness				
HI1	14.1423	12.294	0.830	0.932
HI2	14.2064	12.614	0.812	0.935
HI3	13.7011	12.282	0.945	0.913
HI4	14.1459	12.454	0.818	0.935
HI5	14.1530	12.280	0.829	0.933
Alpha = 0.839 Easy-to-use cognitive				
DSD1	10.3488	6.735	0.802	0.760
DSD2	10.6619	6.953	0.707	0.787
DSD3	10.6548	7.605	0.598	0.818
DSD4	10.8434	8.218	0.457	0.852
DSD5	10.6299	7.027	0.656	0.802
Alpha = 0.884 Social influence				
XH1	12.0498	6.369	0.672	0.870
XH2	11.5445	6.149	0.855	0.831
XH3	12.0534	6.094	0.721	0.859
XH4	11.9964	6.268	0.679	0.868
XH5	12.0000	6.179	0.693	0.865
Alpha = 0.915 Innovation				
DM1	18.0498	13.212	0.921	0.888
DM2	18.5587	12.933	0.731	0.903
DM3	18.5552	12.912	0.721	0.904
DM4	18.5623	13.047	0.710	0.905
DM5	18.5730	12.846	0.738	0.902
DM6	18.5409	13.178	0.712	0.905
DM7	18.5338	13.285	0.695	0.906

(continued)

Table 2 (continued)

Variables	Scale mean if item deleted	Scale variance if item deleted	Correcred item—total correlation	Cronbach's alpha if item deleted
Alpha = 0.969 Fintech service communication				
DV1	43.2883	72.727	0.828	0.966
DV2	43.3950	72.604	0.794	0.967
DV3	43.3915	72.232	0.827	0.966
DV4	43.3060	72.206	0.799	0.967
DV5	43.3167	72.660	0.813	0.966
DV6	43.3523	72.686	0.840	0.966
DV7	43.2811	72.560	0.818	0.966
DV8	43.3060	73.185	0.809	0.966
DV9	43.3452	72.255	0.840	0.966
DV10	43.3559	72.209	0.801	0.967
DV11	43.3310	71.829	0.855	0.965
DV12	43.3488	72.321	0.854	0.965
DV13	43.3416	72.390	0.850	0.966
Alpha = 0.954 Intent to use Fintech services				
YD1	10.8648	5.725	0.886	0.940
YD2	10.8932	5.617	0.890	0.939
YD3	10.8648	5.725	0.886	0.940
YD4	10.8932	5.617	0.890	0.939

Source Author's analysis

Table 3 KMO and Bartlett's Test

Kaiser–Meyer–Olkin measure of sampling adequacy		0.905
Bartlett's Test of Sphericity	Approx. Chi-Square	8873.754
	Df	703
	Sig	0.000

Source Author's analysis

The majority of Vietnamese students have low income, so the usefulness of products and services with low cost is one of the factors affecting the intention to use. The survey results show that 77.2% (217 students) think that they may lose information when using Fintech services, but they still choose to use it because of benefits such as fast, safe payment. during the COVID-19 season, free money transfer, many promotions, easy to use, can connect with friends and relatives.

Table 4 Rotated component matrix

Variables	Component					
	1	2	3	4	5	6
DV12	0.861					
DV9	0.856					
DV13	0.855					
DV1	0.853					
DV11	0.850					
DV3	0.849					
DV6	0.844					
DV7	0.837					
DV2	0.831					
DV5	0.830					
DV8	0.829					
DV4	0.818					
DV10	0.800					
DM1		0.921				
DM5		0.815				
DM3		0.792				
DM2		0.791				
DM6		0.786				
DM4		0.785				
DM7		0.756				
HI3			0.923			
HI5			0.879			
HI1			0.876			
HI4			0.871			
HI2			0.844			
XH2				0.864		
XH3				0.807		
XH1				0.806		
XH5				0.801		
XH4				0.766		
TT4					0.848	
TT2					0.827	
TT1					0.799	
TT3					0.789	
DSD1						0.899

(continued)

Table 4 (continued)

Variables	Component					
	1	2	3	4	5	6
DSD5						0.829
DSD2						0.798
DSD3						0.793

Extraction Method Principal Component Analysis
Rotation Method Varimax with Kaiser Normalization
Source Author's analysis

Table 5 Model summary

Model	R	R Square	Adjusted R Square	Std. Error of the Estimate	Durbin-Watson
1	0.864[a]	0.746	0.740	0.40136	2.018

a. *Predictors* (Constant), DV, DSD, TT, DM, XH, SH
b. *Dependent Variable* YD
Source Author's analysis

Table 6 ANOVA analysis

Model	Sum of Squares	df	Mean Square	F	Sig
Regression	129.627	6	21.605	134.117	0.000[b]
Residual	44.138	274	0.161		
Total	173.765	280			

a. *Dependent Variable* YD
b. *Predictors* (Constant), DV, DSD, TT, DM, XH, SH
Source Author's analysis

Table 7 Coefficients

Model	Unstandardized Coefficients		Standardized Coefficients	t	Sig	Collinearity Statistics	
	B	Std. Error	Beta			Tolerance	VIF
(Constant)	–0.648	0.206		–3.147	0.002		
TT	0.072	0.041	0.057	1.741	0.083	0.860	1.163
HI	0.128	0.029	0.142	4.356	0.000	0.867	1.154
DSD	0.061	0.034	0.055	1.798	0.073	0.987	1.013
XH	0.089	0.042	0.069	2.127	0.034	0.873	1.145
DM	0.035	0.042	0.027	0.834	0.405	0.901	1.110
DV	0.848	0.037	0.762	22.959	0.000	0.841	1.189

Dependent Variable YD
Source Author's analysis

6 Conclusions

In the context of technological revolution 4.0, financial technology has been applied in many fields in payment, savings management, investment, asset management, peer-to-peer lending and data analysis. At the same time, fake news is spread through social networks at a faster rate. The purpose of the study is to analyze the factors affecting the intention to use Fintech services in an emerging country in Asia. Research shows that mitigating the impact of fake news through improving service quality and trust of Fintech will have a positive impact on the intention to use Fintech services. In addition, the usefulness (HI), ease of use (DSD), social influence (XH), communication about Fintech services (DV) also have a positive impact on the intention to use Fintech services.

Through the research, we have some suggestions to Fintech service providers as follows: it is necessary to promote communication activities about the service to attract the attention and use of individuals in society, and increasing the utilities and incentives for users of Fintech services, enhancing security to improve trust for service users. In addition, Fintech is not only a trend of the financial industry but also promotes non-cash payment activities—in line with the theme of developing countries in general and Vietnam in particular, therefore, the State agencies need to build a legal corridor to ensure the operation of suppliers as well as the safety of Fintech service users.

References

1. Arner, D. W., Barberis, J., & Buckley, R. P (2015). The Evolution of FinTech: A New Post-Crisis Paradigm? University of Hong Kong Faculty of Law, Research Paper No. 2015/047
2. Chua, C. J., Lim, C. S., & Aye, A. K. (2019). Factors affecting the consumer acceptance towards Fintech products and services in Malaysia. *International Journal of Asian Social Science, 9*(1), 59–65.
3. Cham, T. H., Low, S. C., Lim, C. S., Aye, A. K., & Raymond, L. L. B. (2018). Preliminary study on consumer attitude towards Fintech products and services in Malaysia. *International Journal of Engineering and Technology, 7*(2.29), 166–169
4. Kim, Y., Park, Y. J., Choi, J., & Yeon, J. (2015). An empirical study on the adoption of Fintech service: Focused on mobile payment services. *Advanced Science and Technology Letters, 114*(26), 136–140.
5. Stewart, H., & Jürjens, J. (2018). Data security and consumer trust in FinTech innovation in Germany. *Information and Computer Security, 26*(1), 109–128.
6. Wonglimpiyarat, J. (2017). FinTech banking industry: A systemic approach. *Foresight, 19*(6), 590–603.
7. Datta, P. A. (2011). Preliminary study of ecommerce adoption in developing countries. *Information Systems Journal, 21*, 3–3.
8. Donner, J., & Tellez, C. A. (2008). Mobile banking and economic development: Linking adoption, impact, and use. *Asian Journal of Communication, 18*, 318–322.
9. Barnes, S. J., & Corbitt, B. (2003). Mobile banking: Concept and potential. *International journal of mobile communications, 1*(3), 273–288.
10. Barnes, S.J., & Scomavacca, E. (2004). Mobile marketing: The role of permission and acceptance. *International Journal of Mobile Communications, 2*(2), 128–139

11. Romānova, I., & Kudinska, M. (2016). Banking and Fintech: A challenge or opportunity? *Contemporary Studies in Economic and Financial Analysis, 98*, 21–35.
12. Vietnam Microfinance Working Group (2018), Report Application of Financial Technology (Fintech) in Microfinance activities towards Universal Finance in Vietnam, Vietnam. Venkatesh, V., Morris, M. G., Davis, G. B., Davis, F. D. (2003), User acceptance of information technology: Toward a unified view. *MIS Quarterly, 27*(3), 425–478
13. General Statistics Office of Vietnam (2020). Vietnam Statistical Yearbook 2019, Statistical Publishing House.
14. Kemp, S. (2020). Digital 2020 October Global Statshot Report—produced in partnership with Hootsuite and We Are Social. Vietnam digital 2020. Retrieved from https://wearesocial.com/blog/2020/10/social-media-users-pass-the-4-billion-mark-as-global-adoption-soars
15. Nielsen (2017). Vietnam Smartphone insights Report. Avaiable at https://www.nielsen.com/wp-content/uploads/sites/3/2019/04/Web_Nielsen_Smartphones20Insights_EN.pdf
16. Solidiance (2018). Report Unlocking Vietnam's Fintech Growth Potential, Vietnam
17. Fintech Singapore (2020), Vietnam Fintech Report 2020
18. FintechFutures (2017). Fintech: beware the fake news. Available at https://www.fintechfutures.com/2017/08/fintech-beware-the-fake-news/
19. Rogers, E. M. (1962). *Diffusion of innovations* (1st ed.). Free Press.
20. Engel, J., Kollatt, D., & Blackewll, R. (1978). *Consumer Behavior*. Dryden Press.
21. Ajzen, I., & Fishbein, M. (1970). The prediction of behavior from attitudinal and normative variables. *Journal of experimental social Psychology, 6*(4), 466–487.
22. Ajzen, I. (1991). The theory of planned behavior. *Organizational Behavior and Human Decision Processes, 50*(2), 179–211.
23. Davis, F. D. (1986). A technology acceptance model for empirically testing new end-user information systems. Cambridge, MA
24. Venkatesh, V., Morris, M. G., Davis, G. B., & Davis, F. D. (2003). User acceptance of information technology: Toward a unified view.*MIS quarterly*, 425–478
25. Kotler, P., & Armstrong, G. (2004). Principles of marketing (Vol. 10th)
26. Amin, H., Hamid, M. R. A., Lada, S., & Anis, Z. (2008). The adoption of mobile banking in Malaysia: The case of Bank Islam Malaysia Berhad (BIMB). *International Journal of Business and Society, 9*(2), 43.
27. Foon, Y. S., & Fah, B. C. Y. (2011). Internet banking adoption in Kuala Lumpur: An application of UTAUT model. *International Journal of Business and Management, 6*(4), 161.
28. Davis, F. D. (1989). Perceived usefulness, perceived ease of use, and user acceptance of information technology. *MIS quarterly*, 319–340
29. Taylor, S., & Todd, P. (1995). Decomposition and crossover effects in the theory of planned behavior: A study of consumer adoption intentions. *International journal of research in marketing, 12*(2), 137–155.
30. Chitungo, S. K., & Munongo, S. (2013). Extending the technology acceptance model to mobile banking adoption in rural Zimbabwe. *Journal of business administration and education, 3*(1), 51–79.
31. Hang, D. M., Thao, N. T., Hoai, D. T., & Thu, N. T. L. (2018). Các nhân tố tác động đến quyết định sử dụng dịch vụ fintech trong hoạt động thanh toán của khách hàng cá nhân tại Việt Nam. *Tạp chí khoa học và đào tạo ngân hàng, 194*, 11–19.
32. Linh, N. H. N., & Tuyen, D. Q. (2020), Factors affecting the intention to use Fintech services in Vietnam. *Economics, Management and Business, 275*
33. Lee, C. C., Cheng, H. K., & Cheng, H. H. (2007). An empirical study of mobile commerce in insurance industry: Task–technology fit and individual differences. *Decision support systems, 43*(1), 95–110.
34. Agarwal, R., & Prasad, J. (1998). A conceptual and operational definition of personal innovativeness in the domain of information technology. *Information systems research, 9*(2), 204–215.
35. Manning, K. C., Bearden, W. O., & Madden, T. J. (1995). Consumer innovativeness and the adoption process. *Journal of consumer psychology, 4*(4), 329–345.

36. Pikkarainen, T., Pikkarainen, K., Karjaluoto, H., & Pahnila, S. (2004). Consumer acceptance of online banking: An extension of the technology acceptance model. *Internet Research*
37. Rogers, E. M. (1995). *Diffusion of Innovations* (4th ed.). The Fess Press.
38. Rogers, E. M. (2003). *Diffusion of Innovations* (5th ed.). The Fess Press.
39. Davis, F. D. (1993). User acceptance of information technology: System characteristics, user perceptions and behavioral impacts. *International journal of man-machine studies, 38*(3), 475–487.
40. Davis, F. D., Bagozzi, R. P., & Warshaw, P. R. (1989). User acceptance of computer technology: A comparison of two theoretical models. *Management Science, 35*(8), 982–1003.
41. Chen, C. D., Fan, Y. W., & Farn, C. K. (2007). Predicting electronic toll collection service adoption: An integration of the technology acceptance model and the theory of planned behavior. *Transportation Research Part C: Emerging Technologies, 15*(5), 300–311.
42. Fishbein, M., & Ajzen, I. (1975). *Belief, attitude, intention and behavior: An introduction to theory of research.* Addison-Wesley.
43. Hair, J. F., Black, W. C., Babin, B. J., Anderson, R. E., & Tatham, R. L. (2010). *Multivariate data analysis.* Prentice-Hall.

36. [illegible] Buildings: [illegible]
37. [illegible]
38. [illegible]
39. [illegible]
40. [illegible]
41. Chen, C. [illegible]
42. [illegible] Addison-Wesley.
43. [illegible] Prentice Hall.

Crowd Sourcing and Blockchain-Based Incentive Mechanism to Combat Fake News

Munaza Farooq, Aqsa Ashraf Makhdomi, and Iqra Altaf Gillani

Abstract With advancing technology, social media sites have become narratives for news surfacing the digital world. Social media users generally have no qualm for forwarding whatever comes their way. The flexibility of social media platforms offers no way to prove the credibility of a shared piece of information. The authenticity of the news and maintaining netiquette over social media sites have become precedence areas. This has challenged the technology in maintaining the ethics and objectivity of journalism. This chapter presents a shared, decentralized framework that implements an alternate vision to the customary way people share information. In particular, we provide an innovative blockchain and crowd sourcing-based framework for demonstrating the provenance of news surfacing through digital media. Under this platform, the truth is sourced from the crowd, stored, and consequently transformed into a real-time interface. Moreover, using the immutable feature of blockchain, all users are made accountable and valued based on their reputation. Further, nodes are provided with incentives for their rightful behavior and are penalized for their malicious actions. Thus, we aim to offer a much-needed layer for secure, reliable information exchange in our present-day social media infrastructure that does not rely on a single source of truth.

1 Introduction

Reading and watching the news has been an important way for individuals to keep themselves informed about the world and current affairs. Traditionally newspapers, radio, and television were the medium through which people consumed information.

Munaza Farooq and Aqsa Ashraf Makhdomi both authors have equal contribution.

Note: A short version of this draft has been accepted in Intl. Conf. on Advances in Cyber Security (ACeS) 2021.

M. Farooq (✉) · A. Ashraf Makhdomi · I. Altaf Gillani
National Institute of Technology Srinagar, Jammu and Kashmir, India
e-mail: iqraaltaf@nitsri.ac.in

M. Lahby et al. (eds.), *Combating Fake News with Computational Intelligence Techniques*, Studies in Computational Intelligence 1001,
https://doi.org/10.1007/978-3-030-90087-8_15

Significantly few people controlled the editing, publication, and amplification stages with the traditional news process, so news's credibility was preserved. Nowadays, the mainstream media is mostly being ruled by massive corporations leading towards the centralization of news and its presentation of facts to benefit their financial or political agenda. With the considerable level of corporate media concentration, people have lost their trust in the mainstream media as these networks demonstrate their political preferences leading towards group thinking.

With trust in traditional news networks at a low point, people have started moving from mainstream media towards social media for accessing news. The evolution of social media has revolutionized the way we exchange information. It has provided us with an open and distributed platform where users can freely share their content and express themselves. However, with diverse sources and online intermediaries present, infiltration in news-related information management and communication has a considerable impact on shaping individuals' opinions. This has unfolded yet another issue of *reliable information* and *accountability* over the social media networks.

Fleeting news has become part of the static searchable realm[16]. The boundaries between news production and information creation and sharing are gradually blurring in the current online news environments [22]. Although sharing is a natural human experience but when somebody shares an article, there is no way to prove its credibility.

With an increase in online social media usage to network and communicate with people, it has become a conspicuous source for the propagation of disinformation. The low cost, easy access and rapid diffusion of information on social media have enabled the wide propagation of fake news on social media [19]. Because of that, malicious actors of society are using it as a tool to create a narrative that benefits them and suits their agenda. Modern social media platforms are broken and fail to focus on the quality of the user-generated content due to centralized architecture, the situation being akin to a country with no law enforcement agencies. Social media platforms miss adequate regulation, and their roles and responsibilities are still not clearly defined [6]. Although certain social media platforms have added the feature of content moderation, such as flagging the content as instigative or otherwise, but the individual propagating such news can still get away easily.

Lately, the issue of fake news and the role of social media sites in regulating such menace have impelled several research projects. Most of the research has focused on using various artificial intelligence [13] or machine learning [1, 3, 7] algorithms for curbing the issue. Distributed ledger technologies and specifically blockchain, also provides opportunities that can help in combating the fake news [11, 12, 15]. These technologies enable privacy, security, and trust in a peer-to-peer network in a decentralized fashion without any central managing authority [6]. The main characteristics such as immutability, provenance and distributed consensus of blockchain help in recording the series of events from their inception to conclusion on a distributed ledger through the consensus of majority of participants. While the contents of the distributed ledger are immutable, one can always prove the provenance of the content stored in it.

We aim to offer a much-needed layer for secure, reliable information exchange in our present-day social media infrastructure, which does not rely on a single source of truth. This can be achieved by harnessing the three distinctive attributes of blockchain, i.e., *immutability, data provenance,* and *distributed consensus*. The core idea is to provide a decentralized platform with reliable and accountable information exchange, where every activity carried out over the network is stored immutably, and data provenance of the activity can always be proved. We have proposed a platform where users can freely express themselves and be accountable for their actions. Unlike present-day social media networks where due to their centralized architecture and lack of regulations, preserving the netiquette over the Internet seems an unfathomable task. Our model provide users freedom of expression as well as preserve the boundary between freedom and abuse of social media. It rewards users for being truthful and penalizes them for being misleading. In particular, our major contributions can be summarized as follows:

- We propose a decentralized crowd sourcing and blockchain-based framework for sharing news on social media platforms.
- We use an immutable database for recording events associated with the decentralized application, making sure that the provenance of data can always be tracked down in our proposed model.
- We incorporate the characteristic feature of human behavior associated with fake news and propose an innovative reward generation mechanism to combat the issue.
- We introduce a new consensus rule based on Byzantine Generals Problem [10] to improve the existing blockchain-based fake news solutions by using active users and validators to agree upon the validity of news.
- Finally, we also implement a prototype model on the Rinkeby test network of Ethereum. The experimental results indicate feasible performance by our proposed model especially under the popular Sybil and Byzantine Generals attack.

The rest of the chapter is organized as follows. Section 2 discusses the related work on fake news using blockchain and other applied methods. Section 3 discusses the preliminaries required for understanding our model. Section 4 presents the design of our proposed model and it details out the appropriate consensus mechanism and reward generation architecture. Section 5 introduces the prototype implementation. Section 6 discusses the experiments conducted on the test network and performance evaluation of the model. Section 7 discusses the limitation and application of our model. In the end, Sect. 8 concludes the chapter highlighting its key contributions.

2 Related Work

Fake news detection has attracted the interest of researchers in recent years and a number of different approaches have been proposed. In addition, conventional content moderation techniques have been incorporated by several social online platforms (e.g., Mozilla, Facebook, Twitter) and trade associations (e.g., EACA, IAB Europe and WFA). All these have made progress in their commitment to tackle fake news [6]. For instance, Google has announced the Google News Initiative to support the news industry in quality journalism. However, these techniques used by the social media platforms for flagging content as instigative or otherwise assumes a centralized regulator which removes the content immediately.

Machine learning based algorithms were the foremost approaches developed to check the fake news menace. In this approach, fake news detection is formulated as a classification or regression problem, with the former being used more frequently. However, the classification is usually restricted to binary classification where the news is checked for being either real or fake. The challenge here is related to the partly fake or partly real news. Similarly, regression problems face the challenge of converting the discrete labels of news datasets to numeric scores of truthfulness. Tacchini et al. [20] proposed an automated fake news detection system for classifying the facebook posts as hoaxes or non hoaxes based on the users who "liked" the post. The paper has used two classification techniques namely logistic regression and boolean crowd sourcing algorithms. The paper suggests that information diffusion pattern can be an effective tool in automatic hoax detection systems. Ozbay et al. [13] proposed a two step verification process for identifying fake news on social media networks. In the first step they converted the un-structured data sets into the structured data sets, and in the second step they applied supervised artificial intelligence algorithms to decide the status of news. Experimental results indicate feasible performance by their model.

Although, all these machine learning based approaches have been effective in fake news detection to a greater extent, however, most of these are centralized with a single source controlling identification, prevention and detection of fake news. This is troublesome as the decision made by a single central authority mostly tends to be skewed. To overcome this challenge, the decentralized architecture of blockchain is effective as there is no central authority to skew the results and the control is completely distributed.

In this distributed direction, Qayyum et al. [15] proposed an open protocol for tracking the credibility of news by introducing the concept of Proof of Truthfulness (PoT), where any node in the network can verify whether a content is or not part of a blockchain. In particular, they store the content in a Merkle tree i.e., a binary tree built using hash pointers. In another work, Murimi [12] proposed Blockchain Enhanced Framework for Social Networking (BEV-SNS), a framework for incentivizing user behavior on SNSs to achieve two objectives: control over data access, and creation of value through SNS transactions. To illustrate this framework, author provides examples with SNSs that lie along the spectrum of anonymity and showed that the framework could be scaled for future use in a variety of collocated spaces.

In terms of blockchain implementation, several works have been proposed to trace the origin of news article by using a decentralized blockchain framework [4, 18]. Moving the work ahead in this direction, blockchain has been used a tool along with other frameworks [2, 14, 17] to combat the spread of fake news. Huckle and White [8] proposed an Ethereum framework with standardized metadata for the verification of the authenticity and the provenance of digital media. Such a propotype uses

Interplanetary File System (IPFS), a P2P content addressed file system. However, their proposed system to find fake resources is limited to multimedia. On this platform, developers can get any real-world data in the form of an API while also being sure the data is verified, i.e., a result of consensus, for which rules are set by the developer himself. The data cannot be stored or controlled by a 3rd party, and all funds are securely stored on the blockchain. Our proposed approach for the fake news detection is also based on blockchain distributed architecture. We will discuss the approach in detail in Sect. 4.

3 Preliminaries

In this section, we first present a preliminary discussion about the basic features of blockchain systems which is then followed by some of its related important terminologies.

3.1 Why Blockchain?

Our proposed framework as discussed before, primarily relies on Blockchain, a peer to peer decentralized network. Under this network the data gets stored on a cryptographically secure, append only immutable ledger through the consensus of majority of participants. The key features of Blockchain technology which make it useful with respect to our targeted problem of fake news detection are as follows:

- *Decentralization*: The distributed ledger prevents the centralization of news by storing it securely across a number of interconnected systems. Unlike in centralized news systems where news production is controlled by a handful of people, each node in our model can participate in news production or verification process, eliminating concentration of power in the hands of few people.
- *Provenance*: Provenance helps us in recording the history of data, from its outset to completion on the distributed ledger. By storing all the details of the news, its publisher, the users and validators which voted for it, on the blockchain, our model can examine the people which are spreading fake news.
- *Immutability:* Each news stored in the block is connected to the previous blocks via a digital signature which means making any change to the current news without changing all the previous blocks is not possible, thus, making it immutable. This immutability feature of blockchain helps us in tracing users and validators and making them accountable for their actions.

3.2 How Does the Blockchain Make Its Decisions?

In centralized systems, all major decisions including the preservation of data against security threats are made by the central authority which regulates the transaction rules. However, in blockchain based systems as there is no central authority we need a network of random people to make decisions and guarantee safe information exchange and avoid fraud issues such as double spending attacks [9]. These bunch of random people need to collaborate and come to an agreement that results in a decision which is better for the network's interest. All such coordination and decision making in blockchain is done by the *consensus algorithms*. Some of the well-known consensus algorithms are as follows:

- Proof of Work
- Proof of Stake
- Delegated Proof of Stake
- Proof of Authority

Proof of Work: Proof of Work is a consensus algorithm in which the block producer called miner needs to solve a mathematical puzzle in order to add block to the blockchain. The mathematical puzzle requires intensive computational power, as it can only be solved by brute force. In Bitcoin, the mathematical puzzle directs miners to find a random number called nonce, such that the hash of the sum of data in the block and the random number is less than the difficulty determined by the system. The miner which first solves the puzzle, broadcasts the result to the entire network, which consecutively is validated by other miners. This mechanism makes sure that someone who has put in adequate computational power or has done a lot of "work" in speculating the nonce will earn the right to update the block of transactions.

Proof of Stake: The Proof of Stake was created as an alternative to the Proof of Work, to tackle with its intense computational power requirement. In Proof of Stake algorithm the block producer is selected in a pseudo-random way based on the amount he is willing to put at stake. If he validates the block correctly, then the amount he had put at stake is returned to him and he is rewarded with the transaction fees proportional to what he had put at stake, but if he validates the block incorrectly his stake is lost.

Delegated Proof of Stake: In Delegated Proof of Stake users of the network vote for a few delegates that will secure the network on their behalf. These delegates are responsible for achieving consensus during the generation and validation of new blocks. The delegates also called witnesses or block producers are fixed in number and are selected in round robin order, by the users of network, who uprightly get number of votes proportional to the number of tokens they own on the network.

Proof of Authority: Proof of Authority is a consensus algorithm in which block validators i.e. the people responsible for verifying transactions in the network are selected based on their reputation score. The algorithm relies on a limited number of block validators which are randomly selected if their reputation score is above the threshold value.

3.3 *Smart Contract*

Smart contracts incorporate a set of rules that automatically enforce agreement among the appropriate stakeholders without the involvement of interceder. When these predefined rules are agreed upon the code executes and produces the output. The Smart Contracts are rooted in the blockchain network which implies once they have been implemented they cannot be altered by any authority making them immutable, transparent and letting their provenance to be tracked down. In our proposed model, Smart Contracts define the consensus mechanism and outline appropriate reward generation mechanism to the desired set of participants. The proposed set of rules cannot be amended by any authority as they have been permanently recorded on the blockchain.

4 Our Model

In this section, we present our proposed model in detail. In particular, we first discuss our proposed framework. After that, we define our proposed consensus rule, and finally we present the reward mechanism that is used by our model.

4.1 *Proposed Framework*

We have used blockchain to create a decentralized news network, where a distributed ledger records information about transactions in a block, and the blocks are interlinked through secure cryptographic functions. The distributed ledger based on blockchain acts as an immutable record of timestamped transactions, enabling our model to track the provenance of data so that users know that their data has not been altered. We define transactions as the set of actions performed by the user in our model. For example, transactions include account creation, uploading posts, access to posts, upvoting, downvoting, and network affordances to facilitate communication. Along these lines, the blockchain acts as an immutable database with an implicit trust mechanism, determining the integrity and transparency of news without third party involvement. Each action of user has a signature associated with it which ensures that the content published by the author or voting done by users will be stored permanently on the distributed ledger and will be made available to other users in the same form as it was published. It becomes impossible for a user to change its "*incorrect votes*" into "*correct votes*" or to alter the "*fake news*" and make it appear "*true* " after publishing it. Other users can always analyse the news source from the blockchain and be sure about its validity. This indelible recording of events in the system is essential for users being accountable for the content they share, i.e., a user can't simply delete an inappropriate post once he has posted it. Authorized users will always trace a post to a certain user adding the feature of accountability to the system. Further as the documents are distributed across various nodes, it ensures that news authenticity can be tracked down by the peer nodes, giving users a secure chain of transactions to check their sources. The distribution of transactions in blockchain ensures that the consensus process is not limited to a handful of authorities but each entity in our model can determine the status of news.

Apart from providing the features of "accountability" and "provenance" which help our proposed model to determine the origin of data and at the same time make users answerable for their actions, blockchain can also execute *Smart Contract*, which automates agreement between various nodes of the network. Smart Contracts play a significant role in combating fake news as they enforce agreement among all the network participants without intermediary's involvement. The agreements include set of rules that enforce the consensus mechanism and reward generation (see Sects. 4.3 and 4.4) required to authenticate the news article, and thereupon propose rewards and punishments to the desired participants.

Even though blockchain guarantees provenance and immutability of data, the nodes can still act in a malicious manner and detect the status of news incorrectly. These nodes can either act individually or they could collude with the peer nodes in order to advance in their direction of getting successful. To determine the status of news in the presence of these nodes, each node is associated with a reputation score, which helps our model to determine their integrity and thereupon resolve the authenticity of news. Further the nodes are administered with the amount of voting proportional to their reputation score. Thus by using reputation score along with blockchain, our model provides the means for storing and retrieving news articles along with their status (*Verified/Fake*) with absolute credibility. Blockchain makes sure that the provenance of news article could be followed by going through the blocks in a sequential manner, and the nodes in turn distinguish between fake and verified news. Since social media platforms are crowded with news articles it would not be possible to verify the authenticity of each news article. Our model verifies $(u + v) * n_{\mathrm{u+v}}$ number of news in parallel, where u is the number of users in our model, v is the number of validators and $n_{\mathrm{u+v}}$ is the number of news articles that have been assigned to each user /validator node. By allocating each user multiple news our model increases the number of news that can be verified simultaneously. Further in order to resolve conflict regarding the news that need to be verified first, our model gives priority to those news which have become viral.

4.2 System Architecture

Figure 1 illustrates the framework of our proposed model. It comprises of a Decentralised Application (dApp), which provides an interface to the user for retrieving and uploading the information on the blockchain, and Smart Contracts, which adds the functionality of news verification and reward generation mechanisms to it. Whenever a user accesses the blockchain via dApp, the data generated by him is stored in the

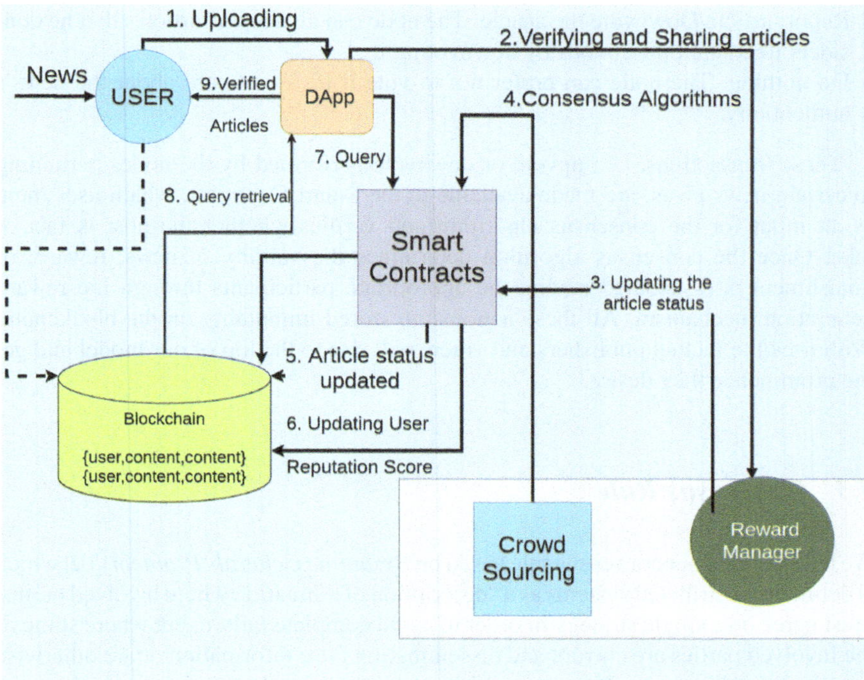

Fig. 1 Proposed framework

blockchain which brings in the feature of accountability in our proposed model. The users are termed as nodes and can belong to one of the following categories:

- *User Nodes*: They represent the core part of the decentralized network and are responsible for capturing the data, building the consensus and delivering verified data. Any user on joining the network can act as a user node and start uploading or verifying the news. These nodes earn incentives for their correct actions and are penalized for their mischievous behavior.
- *Validator nodes*: They are important for checking the authenticity of a post along the user nodes to avoid the 51% attack [5], to which blockchain consensus algorithms are susceptible. A user node can upgrade to a validation node after he has earned reputation score greater than the threshold defined for the network.

The user and validator nodes are responsible for deciding the authenticity or fakeness of a news article by means of a consensus mechanism defined in the next section. The news article gets verified/fake tag after an agreement is achieved among nodes in the stipulated time period which is derived through the sliding window (see 4.3). Whenever the sliding window is still, active users and validators need to derive the news's credibility and choose one of the following options:

- Upvote the article: The node can upvote the article, which he thinks contains true news

- Report article/Downvote the article: The node can also report articles that he considers to contain falsehoods by downvoting it.
- Do nothing: The node can prefer not to vote if he is not sure about the news's authenticity.

These transactions, i.e., upvote or downvote performed by the nodes pertaining to certain news posts, are made available to the Smart Contracts, which uses them as an input for the consensus algorithm and verifies whether the post is fact or fake. Once the consensus algorithm determines the validity of news, reward, or punishment is distributed among the appropriate participants through the reward generation mechanism. All these actions are stored immutably on the blockchain. With time the factual publishers and voters will rise to the top of our model and get the prominence they desire.

4.3 Consensus Rule

We have defined our consensus rule based on *Byzantine Generals Problem* [10] which is defined in distributed systems as a "description of a situation where involved parties must agree on a single strategy in order to avoid complete failure, but where some of the involved parties are corrupt and disseminating false information or are otherwise unreliable". This means for any m, Algorithm $A(m)$ reaches consensus if there are more than $3m$ generals and at most m traitors. This implies that the algorithm can reach consensus as long as $\frac{2}{3}$rd of the actors are honest [10]. Since Blockchain is also a distributed system, to reach a consensus on a news article and to avoid the 51% attack [5], wherein a user or a group of users maliciously control the network for a complete failure of the network our model proposes consensus rules which are modified with the number of users connected to the network (u) and number of validators (v), based on the $\frac{2}{3}$rd rule of Byzantine Generals Problem.

The consensus algorithm states that for associating verified tag to a news article $\frac{2}{3}rd$ of active users in the network and $\frac{1}{3}rd$ of the active validators should have upvoted the article and for associating fake tag to a news article $\frac{2}{3}rd$ of active users in the network or $\frac{1}{3}rd$ of the active validators should have downvoted the article within a random time slot T and if they do not agree during this time slot then a backoff procedure is implemented wherein another time slot is selected with wider duration and this process will continue until the active users and/or validators agree upon the authenticity of the news. For determining the time window T during which active users and validators will find out the authenticity or fakeness of the news we have used a *sliding window approach*, (see Fig. 2), wherein we define a contention window that will move forward until consensus is reached. Figure 3 illustrates the flow of actions in our proposed consensus mechanism. In this figure, we can see that after a news post is uploaded on the dApp, we note the current time unit t_i as the time for the start of consensus. We start a counter for users and validators to reach consensus, by generating a random number r_i. The random number will denote the

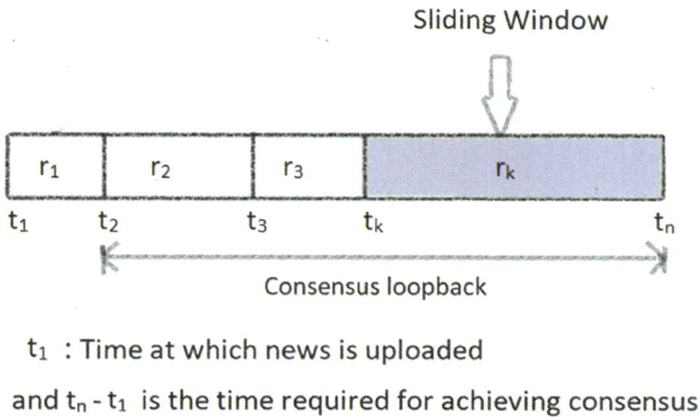

Fig. 2 Sliding window approach

time for which active users and validator nodes could decide the status of news. By selecting the time interval after which news status is determined at random the number of nodes that decide the authenticity of news will be dynamic, as each node who is active during the time period can join the network and vote for news article. This dynamic time interval will prevent collusion among nodes as they do not know the ratio of user and validator nodes they need to collude with in order to change the status of news. The random number $r_i \in (1, 2^i) * z$ where z is the lease time defined by the network. The lease time can be setup in minutes for the news articles which need to be verified quickly or it could be set to days for the news where reliability is more important. Initially $i = 1$ so the time for validators and users to agree upon the news's authenticity or fakeness lies in the range (1, 2) i.e., random number r_1 could be equal to 1 *unit* or 2 *unit* where *unit* can be in minutes, hours or days. During this time interval (t_1 *to* $t_2 = t_1 + r_1$), if $\frac{1}{3}rd$ of active users and $\frac{2}{3}rd$ validators agree that the news is authentic, then consensus is achieved, and news gets the verified tag and if the $\frac{1}{3}rd$ of active user or $\frac{2}{3}rd$ validator nodes say that the news is fake then the news gets fake tag. However if the active user and/or validator nodes could not reach to a decision during this time interval then a backoff procedure is implemented wherein i is incremented which in turn will increase the range of random number from 2 to 4 i.e., the random number (r_2) will now lie in between 1 and 2^2 i.e. $r_2 \in (1, 4)$. Again after 1, 2, 3 or 4 *units* of time from the current time slot i.e. after $t_3 = t_2 + r_2$ units , it will be checked if the user and validator nodes which were active from the beginning (t_1) could agree on a decision or not. If $\frac{1}{3}rd$ of user nodes and $\frac{2}{3}rd$ of validator nodes which were active during the time period t_1 to t_3 agree that the news is authentic then the news gets the authentic tag and if $\frac{1}{3}rd$ of user nodes or $\frac{2}{3}rd$ of validator nodes which were active during the time period agree that the news is fake then the news gets fake tag. Otherwise i will be incremented and the consensus will be checked after $t_3 + r_3$ where $r_3 \in (1, 8)$. The process will continue until consensus regarding news status is achieved among user and validation nodes i.e. until $\frac{2}{3}rd$ of validator

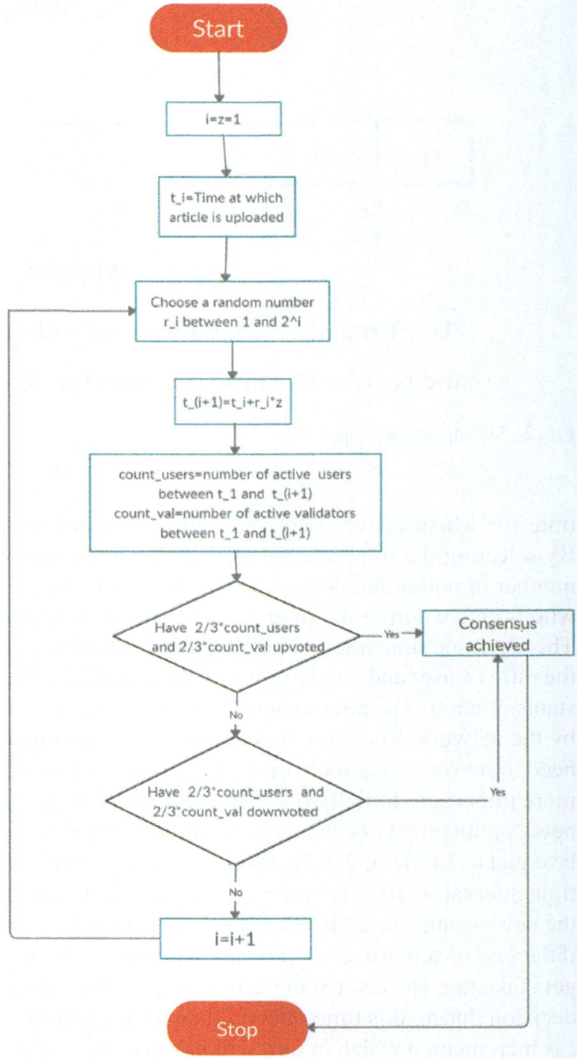

Fig. 3 Consensus mechanism

nodes and/or $\frac{1}{3}rd$ of user nodes agree regarding the status of news. This exponential increase in the range of random numbers gives users and validators more time to agree and thus decide upon the authenticity of the news. So each time consensus is not reached, the probability that the backoff time will become longer increases, and the time for collaboration among nodes in the network will increase. Since the time at which network will check the consensus among nodes and mine no one knows the results due to the random time slot based contention window method, our model does not allow nodes to collude with each other.

4.4 Reward Generation Mechanism

Previous researches have looked at bots to tackle fake news as they found bots broadcast high volume of redundant information, and make it appear credible to users. But an economic brief titled as "The spread of true or false news online" conducted by MIT in 2017 concluded that it is human behavior that contributes more to the differential spread of falsity and truth than automated robots do. This implies that the misinformation containment policies should emphasize on behavioral interventions, like labeling and incentives, rather than focusing exclusively on curtailing bots. Therefore, our model proposes using an individual's reputation score to emphasize regulating human behavior. In particular, *reputation score* measures the credibility of a user and warns the people of the network when it is falling off.

We propose a dynamic reputation system where an initial score of zero is assigned to each node, and the score evolves as the node verifies trustworthy news. The change in reputation score varies dynamically for each news post depending upon the number of user and validator nodes which took part in the consensus process. Our model gives a reputation score of $\{-2I, 0, I/0\}$ to correct, unknown and incorrect decisions made by the user/validator nodes respectively. The incentives (I) vary dynamically for each news post depending upon the number of upvotes or downvotes the news has got at the time consensus is achieved. The incentives for a particular news post are a logarithmic function of X, where X = number of upvotes for the news if news is authentic and number of downvotes if the news is fake i.e. $I = \log(X)$. We have used Incentive as log function of X to reduce its exponential growth in the network. Every user node who has participated in the news post's consensus will see an increase in the reputation score by I units, i.e., if a user has a reputation score of R_{t-1} units, after verifying a news correctly, his reputation score will be ($R_t = R_{t-1} + I$) units. The reputation score of validator node is not increased on their correct detection as it could advance their reputation score to such a degree that it becomes impossible for other nodes to surpass it. This could concentrate the power of voting among a handful of nodes within a few iterations. However, for incorrect detection, both user and validator nodes are penalized for their actions and their reputation score is deducted by $2I$ units, i.e., if their reputation score before the transaction was R_{t-1}, the reputation score after the transaction will be ($R_t = R_{t-1} - 2I$) units. Thus our incentive mechanism makes sure that the malicious behavior of node costs them twice as compared to what they had gained from honest behavior. If the nodes did not vote for the news post their reputation score remains unchanged. Thus by not providing reward to validator nodes on their correct news processing and further decreasing it on their incorrect detection our model does not allow domination and the control of overall network by a handful of nodes. Also our system sets different scaling factors for different behaviors to make the punishment for deceptive behaviors larger than the reward for honest ones. The scaling will prevent users from voting incorrectly and losing twice at what they gained through a reward mechanism.

Further our model categorizes nodes dynamically into user or validator nodes after each news post. It does so by comparing the reputation score of the nodes with the dynamic threshold of the network. The threshold (t_h) for our model is the running average of all the incentives I of the news post.

$$t_h = \frac{(n_t - 1) * I_{t-1} + \log(X_i)}{n_t}$$

where n_t is the number of news articles our model has verified till time instant t, I_{t-1} denotes the running average of all the incentives till time instant $t - 1$ and t_h denotes the new running average. As can be seen by the equation, threshold increases with an increase in the number of news posts, as the running average will increase, which will eventually make our network more reliable. After the status of news is determined the nodes are categorized into user or validator by comparing their reputation score with the threshold.

- If the reputation score of a user node surpasses the threshold ($R_t > I_t$) it is labelled as *validator* node.
- If reputation score of validator node does not exceed the threshold ($R_t \leq I_t$) it is downgraded to *user* node.

This dynamic categorization of nodes ensures the nodes cannot continue with their wrong over a period of time as they would be checked for their behavior after each status of news is decided. By defining the threshold as running average we are making sure that users which have actively taken part in the consensus process are adept to act as validator nodes, henceforth more value is assigned to their votes in the consensus process. Initially, the network will not have adequate number of validator nodes, so our model will rely on web scrapping from leading reputed dailies to verify the news. User votes will only increase or decrease their reputation score and will not be used for authentication purposes. If the user votes in accordance with the leading dailies, its reputation score will increase, otherwise its reputation score will be decreased. Those users who are consistent in voting and vote according to the leading dailies, their reputation score will surpass the threshold defined by the network, and be given a validator tag. By the time the network will have 33% of validator nodes, web scraping will be replaced by achieving consensus among validator nodes and user nodes. Thereupon credibility of news will be decided based upon the consensus among these nodes.

5 Prototype Implementation

We have developed our system as a decentralized application (dApp) which is an end to end to application on the blockchain that offers access to individuals, applications, and frameworks, not necessarily known to one another to execute peer to peer transactions. It has a user interface as its front-end and a back-end that incorporates blockchain and Smart Contract, with blockchain giving it decentralized architecture.

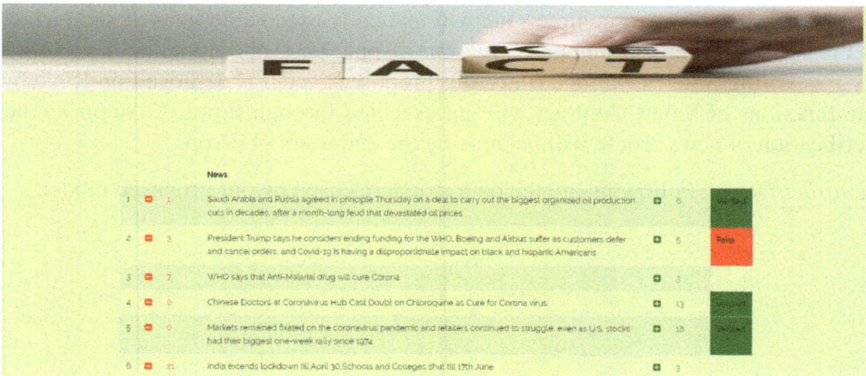

Fig. 4 Front end of our dApp

5.1 Implementataion and Basic Components of dApp

The decentralized web based application has been developed using *MetaMask, Truffle,* and *Web3.js.* While MetaMask facilitates access of blockchain to dApp with normal browser, Truffle provides set of tools for developing Ethereum Smart Contracts, and Web3.js is the official Ethereum JavaScript API which is used to interact with Ethereum Smart Contracts.

The proposed dApp has an interactive interface that allows for viewing on desktops or tablets. It consists of three main components:

- The front-end component, which interacts with the user.
- An Ethereum blockchain, that stores the user data and metadata about media resources.
- Ethereum Smart Contracts, written in Solidity, which read and write metadata about media objects on the blockchain

Front End Component: Front end component has been developed using HTML, CSS, jQuery and JavaScript. It consists of a user interface that helps the user keep itself updated with the authentic news. Apart from viewing, users can also post news that will be fact checked according to the consensus rules specified above. The user also has the option of upvoting or downvoting the article once, where upvoting the article increases the positive count of the article and downvoting increases its negative count. When the consensus on news post is reached, the front end will display the news's status, and thereupon user could not upvote or downvote it (Fig. 4).

Ethereum Blockchain: Our model's backend is Rinkeby, the world wide blockchain test network for debugging Ethereum Smart Contract. It stores the user data and metadata about media resources. Every activity carried out over the network is stored immutably on the Rinkeby test network, and the activity's provenance can always be proved.

Smart Contract: Smart contracts are automatically executable lines of code which get called whenever user interacts with our dApp. Our proposed model uses Solidity for writing Smart Contract. Since our proposed model is deployed on Rinkeby, the functions of Smart Contract will get executed through Proof of Authority, but verification of news article will be done by the end users of dApp.

Contract Design Following are the data structures used by our proposed model:

```
struct Article
{
  int art_id;
  address author ;
  bytes32 name;
  bytes32 title;
  bytes32 content;
  int verified;
  address[] pos_nodes;
  address[] neg_nodes ;
 }

struct nodes
 {
  int rep_score;
  int [] pos_art;
  int [] neg_art;
 }
  mapping (int256 => Article ) public Articles;
  mapping (address => nodes) public nodes;
```

Article data structure contains all the metadata related to news article like its id, author who has published it, the users which have voted for it and its status(Verified/Fake). Each article has an article id and all the metadata related to article can be accessed through this id via mapping Articles. The nodes data structures stores the reputation score of each user/validator node along with the articles he has upvoted or downvoted till now. The nodes mapping structure helps our model to connect MetaMask addresses to blockchain.

Figure 5 shows the details of transaction when a user upvotes a news article. The transaction gets appended to a 6296239 Rinkeby block with a nonce of 356. All the details regarding the transaction are displayed on Etherscan as is displayed by the figure.

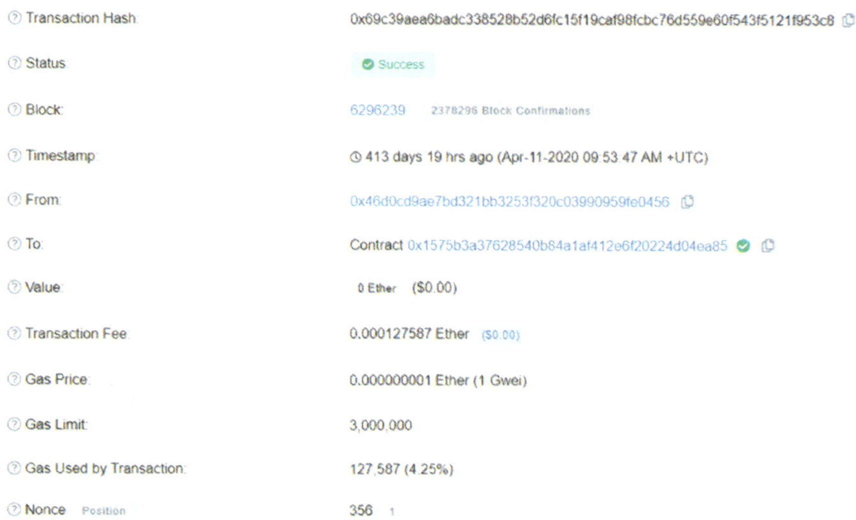

Fig. 5 Transaction hash

5.2 *Overall Working of dApp*

Figure 6 captures the overall working of our decentralized App. When a user posts an article via dApp, a block is created, which contains information about user's account ID, hash of the article, previous hash and the timestamp. Once the article is posted, active validators and users can take part in voting process. When users or validators vote for an article, another block is created that stores the hash of the events, i.e., votes that could be upvote or downvote and the account ID of users who voted, timestamp, and the hash of previous block. After a random time slot it is checked whether the consensus has been reached or not. If consensus is reached, another block is added to blockchain which updates the status of news i.e., news gets verified or fake tag. However, if the consensus is not reached during the time interval than another random time slot is selected wherein more users and validators are appended to vote for the news post and this process will continue until consensus is reached. Once the consensus is reached, the reputation score of the participants is updated.

5.3 *Test Implementation*

For testing purposes, we deployed our proposed model on the Rinkeby test network. In particular, we generated several accounts on Rinkeby to simulate the writers, users, and validators' role. We provided each user with the required ether (cryptocurrency of Ethereum) to pay for our model's different types of fees. We considered 100

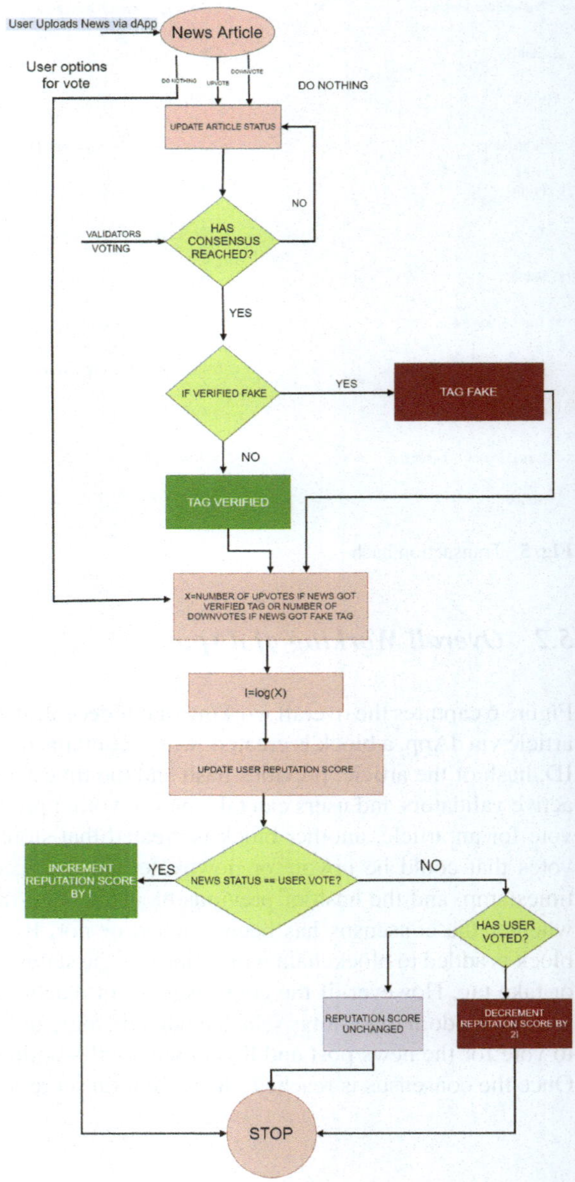

Fig. 6 Flow diagram of overall working of our proposed model

Table 1 Gas consumption for different functions

Function	Gas consumed
Post news	164281.6
Up vote	60382.73
Down vote	60415.266
Consensus	81467

users and a lease time of 5 minutes for our prototype implementation. Initially, all the users had the same reputation score, so validation was performed by scraping articles from leading, reputed dailies. The user vote only increased or decreased their reputation score and was not used for verifying news article. If the user voted in accordance with the reputed dailies its reputation score got increased otherwise, its reputation score decreased. After 28 news posts were validated, the initial phase got over, and our network had 33 users which were consistent in voting and had voted in accordance with the reputed dailies. Their reputation score increased beyond the threshold defined by the network, and they became validator nodes. Thereupon, verification of news is achieved by consensus among validator and user nodes.

We analyze the performance of our proposed model from an experimental study. Performance can be evaluated by considering the complexity of each function in the Smart Contract. Performance determines the transaction fee required to pay to the block producer for executing the functions defined in the Smart Contract. Since validating the functions consumes block producers' resources, more transaction fee will be required if functions are complex. This is measured in terms of 'Gas' in Ethereum network. The transaction fee that the user pays to the block producer for executing its transaction is the gas consumption product which we described above and the gas price set by the user according to its will. We record all the gas consumption for each function of the Smart Contract. Table 1 illustrates results of the experimental study. It can be followed that, compared to the user and validator, the writer tends to require more gas. This fits our model design because writing news to the blockchain will need updation of bytes32 to blockchain which consume more gas. The fair amount of gas consumption for users and validators is able to convince blockchain users to take part in the system to prove the authenticity of news. The consensus algorithm needs to update the reputation score of nodes dynamically and determine the network's running average. Since this function deals with integers and integer modification consumes much less gas as compared to the bytes32, thus gas consumption of consensus function is found to be moderate.

6 Results and Discussion

There have been numerous research works considering the use of blockchain for combating fake news using inbuilt consensus mechanisms such as Proof of Work, Proof of Authority etc. We have proposed a novel consensus mechanism along with an end to end blockchain network. In order to evaluate the proposed model, we take into account its performance based on time complexity, scalability and its fairness in order to avoid malicious behaviour. We also explore the efficiency of our model when subject to various distributed system attacks.

6.1 Performance Evaluation

We analyze the performance of our consensus protocol based on the following metrics:
Time Complexity: To evaluate the performance of our model we need to determine the time it takes to execute various functions. The process of upvoting, downvoting or reading a news article takes constant amount of time as these functions involve operations of reading or writing to the blockchain once. However, the time taken to decide upon the authenticity of news builds upon the consensus function which in turn depends upon the agreement among user and validator nodes. If our model includes m validator nodes and n user nodes the consensus function takes $\mathcal{O}(m+n)$ time to execute as it needs to check the decision of $m+n$ nodes from blockchain before coming to a decision.
Scalability: Scalability is the property of a system to adapt to the growing load by utilizing the network resources properly. Our model is scalable as it adjusts to growing load of news by letting the nodes to vote for multiple news simultaneously. Thus, the news classification is done in a parallel manner by the available nodes, which enhances the performance of our network. Figure 7 demonstrates the scalability of our proposed model. As can be seen in the figure, with increase in number of user and validator nodes, our model is able to classify more number of news articles in parallel thereby displaying the scalable nature of our model.
Fairness: Fairness in our proposed model is a measure of the following aspects:

- The consensus outcome should not be concentrated within the hands of a few people.
- There should be impartial selection of nodes.
- The model should not discriminate against correct and honest members.

Our model satisfies the first aspect of fairness by dividing the power of decision making to the two sets of nodes (user and validator) which drive their decision simultaneously. The consensus outcome is decided only when both set of nodes stipulate decision, otherwise the decision is delayed until the next consensus window is reached. Thus, our model does not leverage power of decision making into the hands of few people.

The second aspect is satisfied as well because in our approach there exists a predefined method to select the nodes based on their reputation score. Only the nodes with reputation score greater than the threshold are suitable to become validator nodes. The method of node selection and updation is done dynamically after each news post and it cannot be tweaked by any authority. Our model satisfies the third

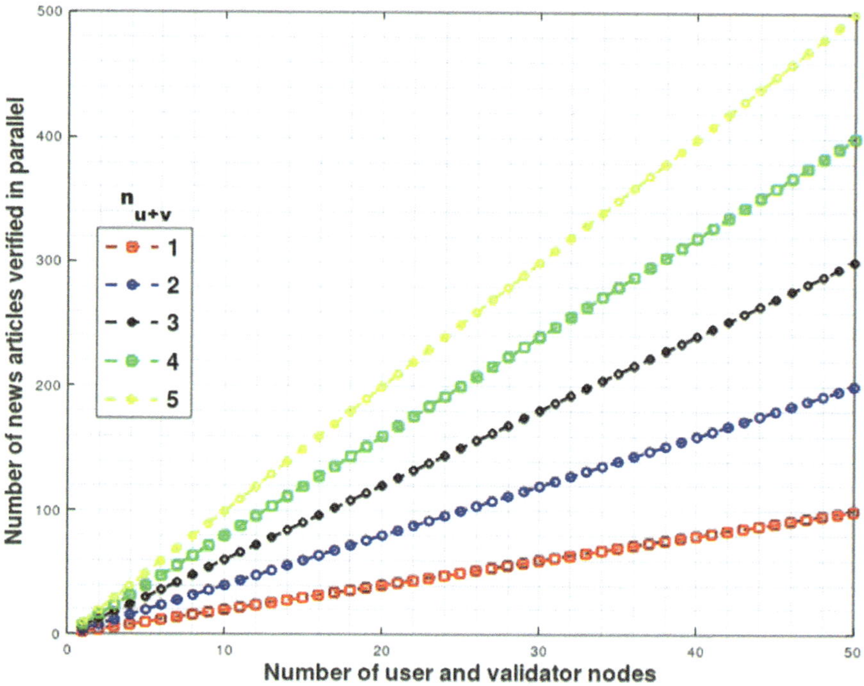

Fig. 7 Scalability of our proposed model

aspect by providing incentives to nodes for their honest behavior and punishing them when they are found to behave in a malicious manner.

Time for news verification: In our experimental settings, we have considered a lease time of 5 min. The verification time for a particular news depends upon the number of users and validators allotted for deciding on its truthfulness. Experimentally it was found when the number of user and validator nodes were less than 6 and 4 respectively for particular slot of sliding window the news took more time to get verified as it had to get verified in the consecutive slots. The more time consumption for less number of users was predominantly due the fact that some nodes after registering themselves for news verification choose not to take part in voting process. However, with increase in the number of nodes these nodes became a fraction of total nodes and the process of news verification became fast. As Fig. 8 shows when the number of user and validator nodes is increased beyond 20 and 12 respectively the news verification time remained constant in the course of time. Thereby, indicating the flexibility of our approach.

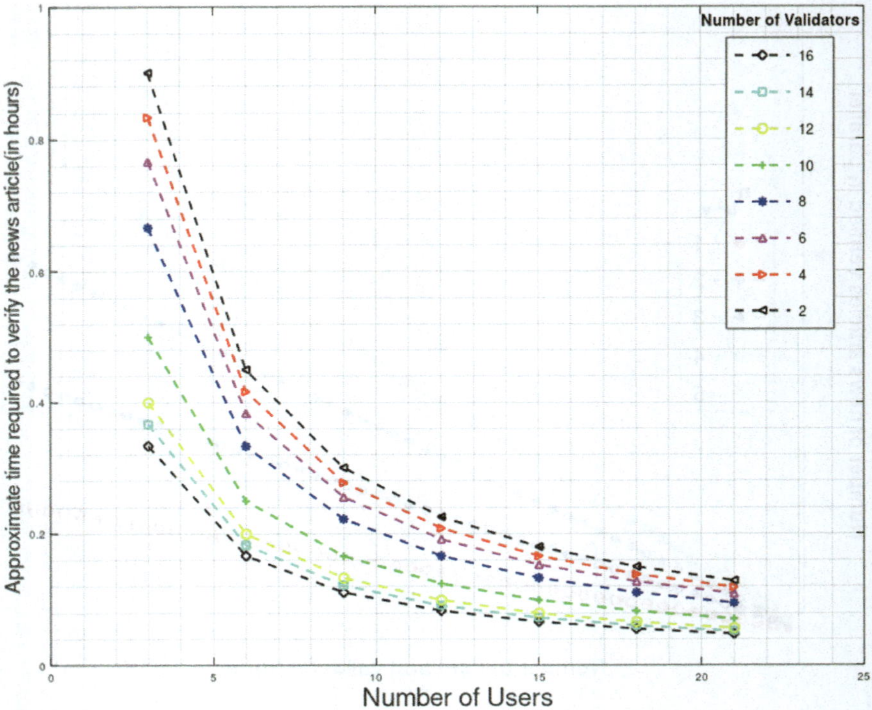

Fig. 8 Time required to verify the news

6.2 Defense Against Attacks

1. Byzantine Generals attack: In Byzantine Generals attack, the nodes vote maliciously in order to disrupt the proper flow of network. Since every distributed system is susceptible to this attack, we devised a consensus mechanism wherein for associating verified tag to a news article $\frac{2}{3}rd$ active users in the network and $\frac{1}{3}rd$ of the active validators should have upvoted the article and for associating fake tag to a news article $\frac{2}{3}rd$ of active users in the network or $\frac{1}{3}rd$ of the active validators should have downvoted the article within a random time slot. As Fig. 9 shows when the number of malicious users or validators are less than 30% the news got verified in the first slot of sliding window. When the number of malicious nodes increased beyond 30% it took more time for news to get verified as the duration of next time slot increased giving nodes more time to agree upon a decision. However, if the percentage of malicious user nodes and validator nodes will increase beyond 70 and 30% simultaneously, our model could misclassify the news. This is a theoretical attack considering the fact that the nodes are selected randomly and probability of selecting colluding malicious nodes decreases with increase in number of users.

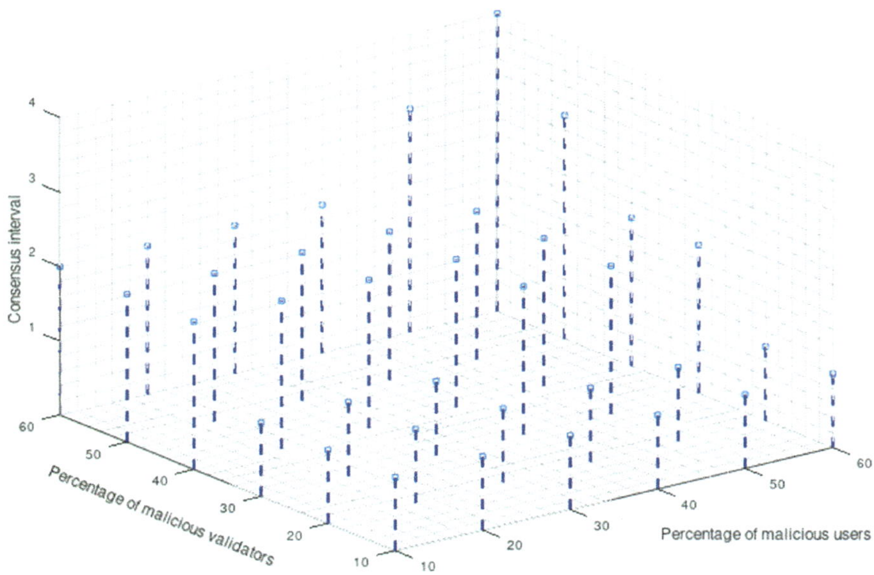

Fig. 9 Contention window over which consensus takes place in presence of malicious nodes

2. Sybil attack: In this type of attack, the attacker forges a large number of identities in the blockchain network and tries to affect the consensus outcome. In our proposed model, since the nodes are selected based on their reputation score, the likelihood for such an attack to succeed is very rare. Even when there are attacker nodes selected among the user nodes, the news voted by the user nodes will simultaneously be voted by validator nodes. Therefore, our consensus algorithm can resist the Sybil attack with an impressive probability.

Efficiency of our model under these attacks: Fig. 10 shows the efficiency of our proposed model under Sybil and Byzantine Generals attack. Initially, when there are few number of user and validator nodes, these attacks lead to decrease in efficiency of our proposed model. This happens primarily due to the fact that probability of malicious nodes and nodes with multiple fake identities occupying a single slot of sliding window and outdoing the desired percentage of 30% for validators and 70% for users to change the decision in their favor is more when their are few number of nodes. However as the number of nodes increase the efficiency of our proposed model is found to increase progressively. After the number of user and validator nodes are found to surpass 35 and 15 respectively our model is found to classify 99% of news accurately.

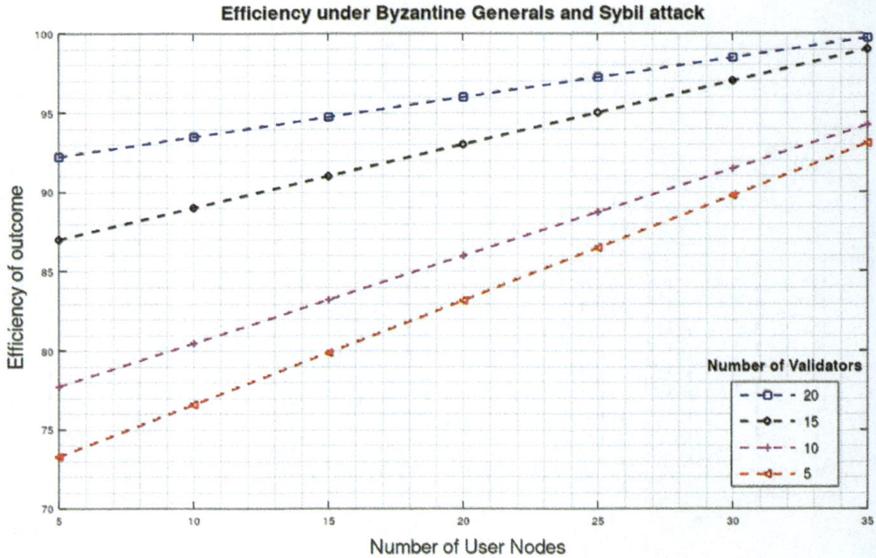

Fig. 10 Efficiency in presence of byzantine generals and sybil attack

7 Relevance of Our Proposed Model in Current World

Combating fake news has been a prime research area in recent years after the impact it caused in national and international sociopolitical relationships challenging the ethics of journalism as well as the authenticity of social media. Previous work shows that spreading fake news can be tackled but cannot be eliminated. And the same can be inferred from the fact that people thrive on novelty and fake news always tends to provide some novel information [21]. Therefore, people are more likely to share fake news without any background check of the fact. They become a chain through which misinformation is propagated as they are neither penalized nor rewarded for their actions on the current social media platforms. Although several researchers have addressed the menace but a few have focused on the characteristic feature of human behaviour associated with the issue. Humans are more susceptible to falling prey to fake news, which is correlated with the sensitivity of the news.

While a user has the freedom of anonymity, it abets bad behaviour. We have aimed to provide a platform where users can freely express themselves but at the same time be liable for their actions. Considering the above-mentioned fact, humans are more vulnerable in spreading fake news and being the biggest challenge in creating a balanced platform where the fine line between freedom and abusive behaviour is maintained [21]. Therefore, we have aimed to address the problem by rewarding the users for good behaviour. Since good reputation confers several advantages, the rewards correspond to their reputation on the platform. However, a possible disadvantage to this may be that the actual behaviour of a person may vary from the reputation he has maintained over the platform, therefore decreasing the effectiveness of the system.

Apart from using blockchain as an immutable ledger we have specified a proper consensus mechanism between users and validators to assert the authenticity of any news article. However, different categories of news such as political, sports, entertainment impact people with different intensities. The consensus mechanism specified can be improved if the factor of sensitivity of the news categories is included in achieving the consensus, thereby reducing the possibility of 51% attack [5] in the system and also increasing the authenticity of the platform in distinguishing fake from fact.

We aim to build a society where the menace of fake news will cease to exist. Our model is designed so that the users will trust the integrity and authenticity of news stories published through our dApp. Our model promotes the right to free speech but at the same time it curtails the spread of misinformation. It guarantees the provenance, accountability, transparency and traceability of data by using blockchain as a secure peer-to-peer platform for storing and exchanging information. These features together with the use of Smart Contracts play an effective role in combating fake news as transactions cannot be altered with once they have been accepted and validated by the consensus algorithm. However, in the initial stages the platform relies on few trusted news networks for validity, which makes it centralized. The platform will be more efficient as the time passes and more users join the network. Also, the current model can be enhanced by including the user preferences for news categories.

8 Conclusion and Future Work

We have proposed a framework for incentivizing user behavior on social networks to achieve two objectives: *combating fake news* and *accountability for sharing data*. While the current architecture of social networking does not offer much in terms of privacy, security, and trust, the intrinsic architecture of blockchain ensures that the user can tweak the sharing criterion in the proposed framework for a secure, trusted, and rewarding networking experience. Our proposed prototype is uniquely capable of recording metadata about digital media on blockchain so that it becomes trivial to prove their authenticity in a manner that can be trusted. Our model's ultimate aim is to make content creator's accountable for what they create and reward/penalize them for their behavior. This is achieved through our crowd-sourced and blockchain-based incentive mechanism, wherein we provide incentives to nodes after following their actions through blockchain. The experimental results of our proposed model indicate its scalability and flexibility. Moreover, under Sybil and Byzantine Generals attack as the number of user and validation nodes increase beyond the desired threshold value, our proposed model is able to classify 99% of news articles accurately.

Moving this work ahead, our proposed model can be expanded to crawl over different web sources and categorize their news articles based upon the metadata structure associated with the current news article which has been classified either as verified or fake. Further, our model can be used as a framework/API by web based news portals to authenticate their news articles and provide appropriate rewards and punishments to their content writers for their fitting behavior. They could likewise stop traffic flow to those portals which result in propagation of fake news.

References

1. Ahmed, H., Traore, I., & Saad, S. (2017). Detection of online fake news using n-gram analysis and machine learning techniques. In *International Conference on Intelligent, Secure, and Dependable Systems in Distributed and Cloud Environments* (pp. 127–138). Springer.
2. Christodoulou, P., & Christodoulou, K. (2020) . Developing more reliable news sources by utilizing the blockchain technology to combat fake news. In *2020 Second International Conference on Blockchain Computing and Applications (BCCA)* (pp. 135–139). IEEE.
3. Conroy, N. K., Rubin, V. L., & Chen, Y. (2015). Automatic deception detection: Methods for finding fake news. *Proceedings of the Association for Information Science and Technology, 52*(1), 1–4.
4. Dwivedi, A. D., Singh, R., Dhall, S., Srivastava, G. & Pal, S. K. (2020). Tracing the source of fake news using a scalable blockchain distributed network. In *2020 IEEE 17th International Conference on Mobile Ad Hoc and Sensor Systems (MASS)* (pp. 38–43). IEEE.
5. Eyal, I., & Sirer, E. G. (2014). Majority is not enough: Bitcoin mining is vulnerable. In *International Conference on Financial Cryptography and Data Security* (pp. 436–454). Springer.
6. Fraga-Lamas, P., & Fernández-Caramés, T. M. (2019). Leveraging distributed ledger technologies and blockchain to combat fake news. arXiv preprint arXiv:1904.05386.
7. Gilda, S. (2017). Evaluating machine learning algorithms for fake news detection. In *2017 IEEE 15th Student Conference on Research and Development (SCOReD)* (pp. 110–115). IEEE.
8. Huckle, S., & White, M. (2017). Fake news: A technological approach to proving the origins of content, using blockchains. *Big data, 5*(4), 356–371.
9. Karame, G. O., Androulaki, E., Roeschlin, M., Gervais, A., & Čapkun, S. (2015). Misbehavior in bitcoin: A study of double-spending and accountability. *ACM Transactions on Information and System Security (TISSEC), 18*(1), 1–32.
10. Lamport, L., Shostak, R., & Pease, M. (2019). *The byzantine generals problem* (pp. 203–226). In Concurrency: The Works of Leslie Lamport.
11. Mikeln, M., & Perović, L. (2018). Eventum: Platform for decentralized real-world data feeds.
12. Murimi, R. M. (2019). A blockchain enhanced framework for social networking. *Ledger, 4*.
13. Ozbay, F. A., & Alatas, B. (2020). Fake news detection within online social media using supervised artificial intelligence algorithms. *Physica A: Statistical Mechanics and its Applications, 540*, 123174.
14. Paul, S., Joy, J. I., Sarker, S., Ahmed, S., Das, A. K., et al. (2019). Fake news detection in social media using blockchain. In *2019 7th International Conference on Smart Computing & Communications (ICSCC)* (pp. 1–5). IEEE.
15. Qayyum, A., Qadir, J., Janjua, M. U., Sher, F. (2019). Using blockchain to rein in the new post-truth world and check the spread of fake news. arXiv preprint arXiv:1903.11899.
16. Rubin, V. L., Chen, Y., Conroy, N. J. (2015). Deception detection for news: Three types of fakes. In *Proceedings of the 78th ASIS&T Annual Meeting: Information Science with Impact: Research in and for the Community* (pp. 83). American Society for Information Science.
17. Shae, Z., & Tsai, J. (2019). Ai blockchain platform for trusting news. In *2019 IEEE 39th International Conference on Distributed Computing Systems (ICDCS)*, pp. 1610–1619. IEEE.

18. Shang, W., Liu, M., Lin, W., & Jia, M. (2018). Tracing the source of news based on blockchain. In *2018 IEEE/ACIS 17th International Conference on Computer and Information Science (ICIS)* (pp. 377–381). IEEE.
19. Shu, K., Sliva, A., Wang, S., Tang, J., & Liu, H. (2017). Fake news detection on social media: A data mining perspective. *ACM SIGKDD explorations newsletter, 19*(1), 22–36.
20. Tacchini, E., Ballarin, G., Della Vedova, M. L., Moret, S. & de Alfaro, L. (2017). Some like it hoax: Automated fake news detection in social networks. arXiv preprint arXiv:1704.07506.
21. Vosoughi, S., Roy, D., & Aral, S. (2018). The spread of true and false news online. *Science, 359*(6380), 1146–1151.
22. Yimin Chen Niall, J. C. & Rubin, V. L. (2015). News in an online world: The need for an "automatic crap detector", using blockchains. In *Proceedings of the 78th ASIS&T Annual Meeting: Information Science with Impact: Research in and for the community* (pp. 82). American Society for Information Science.

Framework for Fake News Classification Using Vectorization and Machine Learning

Yogita Dubey, Pushkar Wankhede, Amey Borkar, Tanvi Borkar, and Prachi Palsodkar

Abstract Fake news are widely offered in digital media to raise the visitors hit and in an offbeat, it acts on users emotions. The foremost ordinary example of such fake news throughout this pandemic, are the various remedies to cure covid. As a result of which individuals are unable to acknowledge any kind of genuine news. People try and attempt numerous things which will never help in curing this contagious disease. Moreover, it might lead to some other major health issues. In this paper, a framework is provided for the classification of news as fake vs real. Text data is pre-processed using Natural Language Processing (NLP) by performing tokenization, text cleaning and vectorization. N-gram and TF-IDF vectorization is used. Seven Machine Learning (ML) algorithms are then applied for classification. Two different datasets Kaggle and ISOT is used for experimentation and evaluated on the same scale using different evaluation metrics to demonstrate the efficacy of the proposed framework.

Keywords Fake news · Classification · Natural Language Processing · Machine learning · Model evaluation

1 Introduction

There is a massive amount of data generated on social networks with various social media formats. Fake News is news, stories, hoaxes, or articles created deliberately to mislead readers through different types of communication mediums. These goes simultaneously with the new world which is also to handle lots of information which is generated every second [1, 2].

The global fact is that world is aware about fake news. The internet and thus the social media changed however it's produced and unfolded. Fake news is formed with a multistep methodology. This basically involves creation of the content by someone else, talking about that content to some third person and passing or posting it through

Y. Dubey (✉) · P. Wankhede · A. Borkar · T. Borkar · P. Palsodkar
Yeshwantrao Chavan College of Engineering, Nagpur, India

M. Lahby et al. (eds.), *Combating Fake News with Computational Intelligence Techniques*, Studies in Computational Intelligence 1001,
https://doi.org/10.1007/978-3-030-90087-8_16

social media platform as real news. Another approach is to make fake news websites and steal the content for a tricky domain name and a well-organized design. These fake news website owners steal from an ironic website or "Clickbait" whose objective is to earn more website hits and profit generation. Once the fake news owners have content, the next step is to monetize it [3].

Why do individuals fall prey to such kinds of fake news? We have got such straightforward access to the real facts regarding everything on the internet. Unfortunately, individuals start believing these fake stories and real facts can't facilitate rectify this problem. Individuals don't notice some things that will be apparent. This is due to cognitive bias in remembering, reasoning, or evaluating some issues which leads to mistaken conclusions. We focus more on headlines and tags and not read the entire article. Social media plays a catalyst role in passing those signals which affects more on our sense of recognition of knowledge with more acceptance. Fake news takes works on the principal of discrimination with chances of false information to circulate [4].

How will we tend to defend ourselves from Fake News? Besides the growing quantity of fake news on-line, there are a various platform on the internet which checks the authenticity of the news and articles. The restrictions in their effectiveness in eliminating fake news justify an excellent deal regarding why individuals don't cash in fact-checking systems [5]. Varied info resources became accessible for example- news agencies, fact-checking organizations that facilitate effectively investigate and convey the unfold of false information and fake news. Individuals should also check the authenticity of news of social media before sharing, like or commenting. Individuals got to question the dependability of facts even after they apprehend them. Despite the availability of fact-checking sites and tools, there are still problems associated with fake news spread as these sites might not be reliable as expected [6].

There are many algorithms used by researcher for the detection of fake news. Language approaches mainly utilize linguistics information. These approach takes into consideration all the words in a sentence and also the letters in a word. The structure of the sentence and how it fits together in a paragraph are also examined. The main basis is grammar and syntax [7, 8]. The Topic-agnostic approach focuses on the important topic of the article instead of the entire article. Linguistic options and internet mark-up language to distinguish fake and real news. Topic-agnostic examples are an outsized range of advertisements, attractive headlines with attention-gaining watchwords and different text patterns [9]. In the recent years, ML algorithms are widely used for fake news detection using training of datasets. Rumour identification framework has been developed using machine learning approach to differentiate faux and real news by reducing the ambiguity of the posts [10]. Knowledge-based approach utilizes the exterior sources to verify if the news is fake or real and spot the news before it becomes viral. Recently ML and knowledge-based approach is combined to detect fake news [11–14] and also hybrid approach with the use of recurrent neural network are also used [15, 16].

The objective of this paper is proposed a framework for fake news detection using the concept of NLP and ML algorithms. The rest of the paper is organized as follows. Material and method used is discussed in Sect. 2. Different evaluation metrics to demonstrate the efficiency of proposed framework are discussed in Sect. 3. Results and discussion are described in Sect. 4 followed by conclusion in Sect. 5.

2 Materials and Methods

Here, we have used two datasets: Kaggle [17] and Information Security and Object Technology (ISOT) [18]. Kaggle fake news dataset contains the information about the news with the parameters as author, text and news articles. ISOT fake news dataset consist of various fake news and reliable articles, acquired from distinct genuine news sites. We have used 2000 samples from each dataset. Kaggle dataset contains 971 (49%) unreliable news and 1029 (51%) reliable news. ISOT contains 991 (50%) unreliable news and 1009 (50%) reliable news. In both the dataset, we have approximately 50–50% data, so that results should not bias towards one category, and we can have compared the results on the same scale. The block diagram representation of the proposed system for the detection of fake news is shown in Fig. 1.

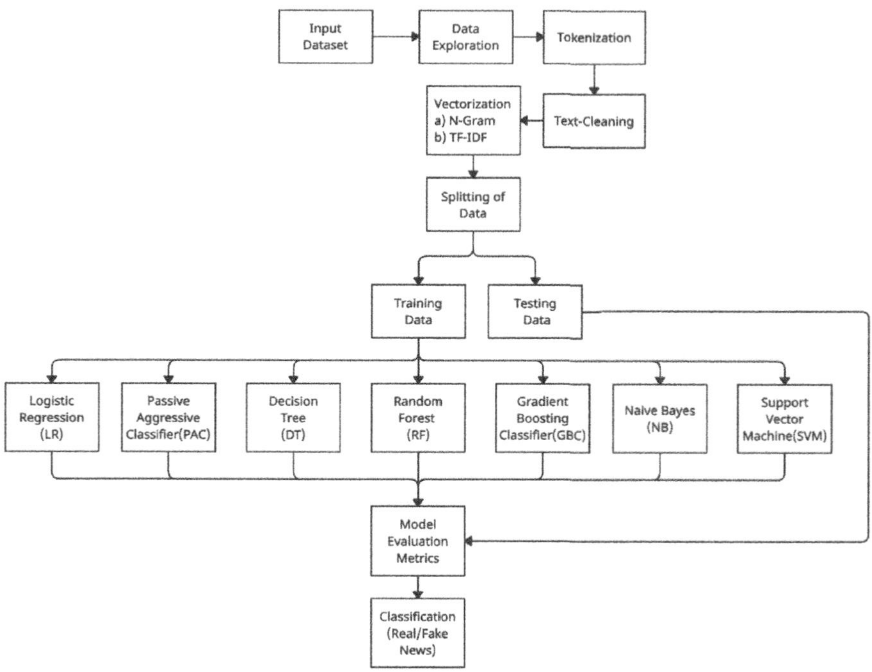

Fig. 1 Proposed framework for fake news classification using NLP and ML algorithms

2.1 Natural Language Processing (NLP)

NLP is widely used for an email spam filter, autocomplete, autocorrect applications. The areas of NLP are sentiment analysis, topic modelling, text classification, parts of speech tagging, sentence segmentation. Pre-processing on the dataset is done by NLP where we have use NLTK toolkit of python [19]. The step involved are described below.

Text Pre-Processing
Pre-processing of text data will be required in NLP if we are using ML algorithms for classification of news as fake vs real. The methods are classified into different types depending on the nature of the task. They are described below.

2.1.1 Tokenization

Tokenization is the fundamental step in NLP which involve separating a piece of text into smaller unit called as Tokens. Here, tokens can be word, sub-words or, characters. Let us take an example, let text = "I am learning and playing." After tokenizing ['I', 'am', 'learning' 'and', 'playing', '.'] [19].

2.1.2 Text Cleaning

Using Natural language processing, there are many operations and products which are being developed. The main input for any machine learning model like classification, Q&A model, Sentiment are clean tokens. So, consider the text which contains symbols and words which convey meaning to the model while training. So, we will remove unwanted symbols or words before efficiently feeding them to the model. This method is called Text Cleaning. Text cleaning involve following steps [20, 21]

(a) Removing Punctuations: The meaning of the sentence doesn't change due to punctuation. Hence by removing them doesn't affect our classification. The conversion of all characters to lowercase, removing punctuation marks, stopwords and typos can avoid unhelpful and unimportant parts of the data. This is helpful for text analysis on pieces of data like comments or tweets.

(b) Removing Stopwords: The removal of stopwords means getting rid of commonly used words which do not add much meaning such as am, is, the, there, etc. The processing time and space are two valuable aspects of efficient database, and we would not want our words to create unnecessary problems with processing time and space. This is achieved by storing a list of words that we consider to stop words. The list of stopwords of 16 different languages are stored in NLTK in python.

(c) Lemmatization: This is the process of gathering the modified forms of a word to be analysed as a single root word or lemma. A lemma is a canonical form, dictionary form, or citation form of a set of words. It reduces the modified words properly ensuring root word belongs to the language. WordNet Lemmatizer is used. It is a collection of verbs, adjectives, nouns, adverbs and these are group together on synonyms of words.

2.1.3 Vectorization

Vectorization map words or phrases from vocabulary to a corresponding vector of real numbers to create feature vectors. This feature vector is used to find word predictions and word resemblances [22]. The methods of vectorization are described below.

(a) N-Gram Vectorization: N-Gram vectorization initiates with the matrix creation with columns of length N. It detects cooccurring words within a given window. In N –Gram computation typical movement is of one word in forward. Let's take an example text = I am studying NLP then after vectorizing if N = 1 which is also called as Count Vectorization ['I', 'am', 'studying', 'NLP']. If N = 2 it is Bi-gram Vectorization ['I am', 'am studying', 'studying NLP'].

(b) Term Frequency Inverse Document Frequency (TF-IDF): Creates a matrix whose columns are unique words only. The cells contain a weight that signifies how important a word is for a particular text message. It is a scoring measure widely used in information retrieval (IR) or summarization [23]. The weight of a word is calculated by the following formula given in Eq. (1)

$$w_{i,j} = t_{i,j} \times \log\left(\frac{N}{df_i}\right) \tag{1}$$

where, $t_{i,j}$ denotes number of times sample i occur in j divided by the total number of samples in *j*. N is the number of articles in the dataset and df_i is the number of articles containing term *i*. The output of vectorizers are sparse matrices which are defined as the matrix whose most of the elements are 0. After the data has been processed by NLP, machine learning algorithms are applied to classify the news as fake versus real [23, 24].

2.2 Machine Learning Algorithm

2.2.1 Logistic Regression (LR)

LR model provide intuitive equation to classify the data into two classes or multi-classes [25] using the hypothesis $h_\theta(x) = g(\theta^T x)$, here, θ is the parameter which is required to be tuned to get the best accuracy for the features x and g is the sigmoid function or logistic function which can be given as-

$$g(\theta^T x) = \frac{1}{1 + e^{-\theta^T x}}$$

which results in the range of 0–1.

The cost used in LR model is maximum likelihood function given by Eq. (2)

$$J(\theta) = \frac{1}{N}\left(-y^T \log(h) - (1-y)^T \log(1-h)\right) \tag{2}$$

The objective is to reduce this cost function with respects to parameters θ.

2.2.2 Passive-Aggressive Classifier (PAC)

We have used Passive Aggressive Algorithm as the Kaggle dataset was having 20,000 sample and it was creating memory issue while training. PAC do not require a learning rate but include a regularization parameter [26].

2.2.3 Decision Tree (DT)

DT is used to classify the news articles as our dataset was having target output. We have used features as the internal nodes [27]. The main terminologies used in decision tree are root node which represent the entire dataset, leaf which give the output which is fake news or real news in our case. Data splitting is done as per some rules and conditions. Branch or Sub Tree is formed after splitting. Unwanted branches are pruned from the tree. The Decision Tree algorithm works on the following way. It begins with root node which is decided by information gain of the feature calculated by using:

$$IG = Entropy(N) - weighted\,avearge \times Entropy\,of\,each\,feature$$

Here, Entropy is the measure of randomness in data and can be calculated by

$$E(N) = -\sum_{i=0}^{1} p_i \log_2 p_i \tag{3}$$

where N is the total numbers of articles present in the dataset. We have use 2000 articles from Kaggle dataset and 1000 articles from ISOT dataset (For DT only).

2.2.4 Random Forest (RF)

The building blocks for Random Forest algorithm is Decision tree. The bagging method is used to generate the required prediction in Random Forest. Predictions are made by using training dataset which includes observations and features. The training data fed to the random forest algorithm is responsible for the different outputs of decision trees. These are ranked outputs, in which the highest is selected as the final output [28].

2.2.5 Gradient Boosting

Gradient boosting algorithm [29] is a weak learner decision tree algorithm, which generally outperforms random forest algorithm. Gradient boosting is used on fixed size decision trees as base learners. Generally gradient boosting at the mth step would fit a decision tree $h_m(x)$ to pseudo-residuals. Let the number of its leaves be J_m. The tree partitions the input space into $J_m$disjoint regions R_{1m}, ...$R_{J_m}m$ and predicts a constant value in each region. The output for input x can be given as:

$$h_m(x) = \sum_{j=1}^{J_m} b_{jm} 1 R_{jm}(x) \tag{4}$$

where b_{jm} is the value predicted in the region R_{jm}.

2.2.6 Naive Bayes (NB)

Naive Bays algorithm is a probabilistic classifier which predicts the output on the basis of the probability and make quick predictions that helps in building the fast machine learning models. Bayes' theorem is used to determine the probability of a hypothesis with prior knowledge. Bayes' theorem is mathematically modelled as

$$P(A/B) = \frac{P(B/A)P(A)}{P(B)} \tag{5}$$

Naïve Bayes' Classifier generate likelihood table by finding the probabilities of given features by converting dataset into frequency table. Then calculate the posterior probability which is used for the prediction or classification [30].

2.2.7 Support Vector Machine (SVM)

SVM algorithm can be used for binary as well as multi-class classification. The aim of the SVM algorithm is to generate the best line or decision boundary that can

Table 1 Confusion Matrix for binary classification

		Actual output	
Predicted output		Real news	Fake news
	Real news	TP	FP
	Fake news	FN	TN

isolate n-dimensional space into classes so as to put the new data point in the accurate groups. SVM pick out the extreme points/vectors that creates the hyperplane which is called as support vectors [31].

3 Evaluation Metrics

To evaluate the performance of the proposed framework for the classification of fake news and real news, different quantitative indices are used. These are discussed below.

3.1 Confusion Matrix

Confusion matrix is basically a tabular representation of classification model performance on the test data. For binary classification as in our case, this is 2×2 matrix with rows indicating predicted output and column indicating the output obtained by proposed framework. This matrix contains four parameters, with diagonal entry as True Positive (TP) and True Negative (TN) and cross diagonal entry as False positive (FP) and False Negative (FN). The example of Confusion Matrix for binary classification is shown in Table 1.

The remaining indices are mainly based on these four parameters which are discussed below.

Accuracy: Proportion of true results may be real or fake with respect to total number of news sample	$Accuracy = \frac{\lvert T_P + T_N \rvert}{N}$
Precision: Gives the number of articles that are marked as real out of all the positively predicted real news	$Precision = \frac{\lvert T_P \rvert}{\lvert T_P + F_P \rvert}$
Recall: Represents the number of articles predicted as real out of the total number of true positive and false negative cases	$Recall = \frac{\lvert T_P \rvert}{\lvert T_P + F_N \rvert}$
F1-Score: Harmonic mean of recall and precision	$F_1 - Score = 2 \times \frac{Precision \times Recall}{Precision + Recall}$

(continued)

(continued)

ROC Curve: Performance of a classification model at all classification thresholds and plots two parameters True Positive Rate (TPR) and False Positive Rate (FPR)	$TPR = \frac{\lvert T_P \rvert}{\lvert T_P + F_N \rvert} \; FPR = \frac{\lvert F_P \rvert}{\lvert F_P + T_N \rvert}$
Mathews Corelation coefficient (MCC): MCC measures the differences between the actual sets and predicted set	$MCC = \frac{(T_P \times T_N) - (F_N \times F_P)}{\sqrt{(T_P + F_P)(T_P + F_N)(T_N + F_P)(T_N + F_N)}}$

3.2 Jaccard Index

The Jaccard index give the similarity between two sets. with 0 means no similarity and 1 mean complete match of predicted output $\hat{y}$ and actual output y. Jaccard similarity between two sets $\hat{y}$ and y can be calculated using.

$$J(y, \hat{y}) = \frac{\lvert y \cap \hat{y} \rvert}{\lvert y \cup \hat{y} \rvert} = \frac{\lvert y \cap \hat{y} \rvert}{\lvert y \rvert + \lvert \hat{y} \rvert - \lvert y \cap \hat{y} \rvert} \tag{6}$$

3.3 Kappa Index

Kappa index (KI) gives the percentage of overlap between two sets. It compares an observed accuracy with an expected accuracy. KI is calculated using

$$KI = \frac{Observed\ Accuracy - Expected\ Accuracy}{1 - Expected\ Accuracy} \tag{7}$$

3.4 Log-Loss

Log-loss indicates what is the similarity between the prediction probability p_i and corresponding actual/true value y_i (0 which is fake news or 1 which is real news in our case). If predicted probability differs from the actual value, the log-loss value will be higher. Less value of log loss is the measure of good classification. Log loss is calculated using

$$\text{Log loss} = -\frac{1}{N} \sum_{i=1}^{N} y_i \log p_i + (1 - y_i) \log(1 - p_i) \tag{8}$$

4 Results and Discussion

This section describes the result obtained using proposed framework on Kaggle [17] and ISOT dataset [18] using seven different ML algorithms with count and TF-IDF vectorization. Table 2 summarises the overall accuracy, Precision, Recall and F1-score for each of the algorithms on Kaggle dataset. It can be seen that maximum accuracy of 99% is achieved by decision tree (DT) and gradient boosting (GB) algorithm using count vectorization with N = 1 and Term frequency inverse document frequency (TF-IDF) vectorization approach. Logistic Regression (LR), Passive aggressive classifier (PAC) and random forest (RF) reported accuracy of 90%, 91% and 91% respectively using count vectorization and accuracy is increased to 94%, 95% and 96% respectively with TF-IDF vectorization approach. Support vector machine (SVM) reported accuracy of 92% and 99% respectively with count and TF-IDF vectorization approach. Naive bayes (NB) algorithm reported accuracy of 93% and 0.98% respectively with count and TF-IDF vectorization approach. Maximum Precision of 99% is reported by GB and DT algorithm with both count and TF-IDF vectorization. LR, PAC, RF and NB algorithm reported recall of 90%, 92%, 92% and 93% respectively with count vectorization, which increases to 94%, 95%, 96% and 98% with TF-IDF vectorization. And Finally Maximum F1-score of 99% is reported by GB and DT algorithm with both count and TF-IDF vectorization approach. LR, PAC, RF and NB algorithm reported precision of 90%, 92%, 92% and 93% respectively with count vectorization, which increases to 94%, 95%, 96% and 98% with TF-IDF vectorization.

Table 3 summarises the overall accuracy, Precision, Recall and F1-score for each of the algorithms on ISOT dataset. LR, PAC, RF, NB, SVM, GB and DT reported accuracy of 90%, 84%, 88%, 88%, 89%, 74% and 85% respectively with count vectorization and with TF-IDF vectorization method the accuracy reported is 92%, 91%, 85%, 83%, 90%, 69%, 90% with lowest using GB algorithm. Precision rate of 91%, 85%, 89%, 89%, 89%, 86%, 74% respectively with count vectorization

Table 2 Overall accuracy, Precision, Recall and F1-score obtained using various algorithms on Kaggle dataset

Model	Accuracy		Precision		Recall		F1-Score	
	Count	TF-IDF	Count	TF-IDF	Count	TF-IDF	Count	TF-IDF
LR	0.90	0.94	0.91	0.94	0.90	0.94	0.90	0.94
PAC	0.91	0.95	0.92	0.96	0.92	0.95	0.92	0.95
RF	0.91	0.96	0.92	0.97	0.92	0.96	0.92	0.96
NB	0.93	0.98	0.94	0.98	0.93	0.98	0.93	0.98
SVM	0.92	*0.99*	0.92	0.99	0.92	0.99	0.92	0.99
GB	*0.99*	*0.99*	0.99	0.99	*0.99*	*0.99*	0.99	0.99
DT	*0.99*	*0.99*	0.99	0.99	*1.00*	*0.99*	0.99	0.99

Table 3 Overall accuracy, Precision, Recall and F1-score obtained using various algorithms on ISOT dataset

Model	Accuracy		Precision		Recall		F1-Score	
	Count	TF-IDF	Count	TF-IDF	Count	TF-IDF	Count	TF-IDF
LR	0.90	0.92	0.91	0.92	0.91	0.92	0.91	0.92
PAC	0.84	0.91	0.85	0.92	0.85	0.92	0.85	0.92
RF	0.88	0.85	0.89	0.86	0.89	0.86	0.89	0.86
NB	0.88	0.83	0.89	0.86	0.89	0.82	0.89	0.82
SVM	0.89	0.90	0.89	0.92	0.89	0.91	0.89	0.91
GB	0.74	0.69	0.86	0.90	0.86	0.90	0.86	0.90
DT	0.85	0.90	0.74	0.69	0.74	0.69	0.74	0.69

and 92%, 92%, 86%, 86%, 92%, 90% and 69% respectively with TF-IDF vectorization is obtained using these seven algorithms. Recall percentage obtained using these seven algorithms are 91%, 85%, 89%, 89%, 89%, 86%, 74% respectively with count vectorization and 92%, 92%, 86%, 82%, 91%, 90% and 69% with TF-IDF vectorization. F1score obtained is similar to recall for all the algorithms.

Table 4 summarizes Jaccard Score (JS), Kappa Index (KI) and Matthew Correlation Coefficient (MCC) obtained using all seven algorithms with count and TF-IDF vectorization on Kaggle dataset. Maximum JS value, KI and MCC of 97% and 98% is obtained using DT with count and TF-IDF vectorization. JS value of 83% and 88%, 84% and 92%, 84% and 93%, 86% and 96%, 86% and 98% and 97% and 98% respectively with count and TF-IDF vectorization is reported by LR, PAC, RF, NB, SVM and GB algorithms respectively. KI of 80%, 83%, 83%, 86%, 84% and 98% respectively with count vectorization which increases to 88%, 91%, 93%, 96%, 99% and 98% respectively is obtained using LR, PAC, RF, NB, SVM and GB algorithms. Similarly, MCC value of 81%, 84%, 84%, 87%, 85% and 98% is reported either count vectorization which increases to 88%, 91%, 93%, 96%, 99%, 98% respectively with TF-IDF vectorization.

Table 4 Jaccard Score, Kappa Index and ECC obtained using various algorithms on Kaggle dataset

Model	Jaccard Score		Kappa Index		MCC	
	Count	TF-IDF	Count	TF-IDF	Count	TF-IDF
LR	0.83	0.88	0.80	0.88	0.81	0.88
PAC	0.84	0.92	0.83	0.91	0.84	0.91
RF	0.84	0.93	0.83	0.93	0.84	0.93
NB	0.86	0.96	0.86	0.96	0.87	0.96
SVM	0.86	0.98	0.84	0.99	0.85	0.99
GB	0.97	0.98	0.98	0.98	0.98	0.98
DT	*0.99*	*0.98*	0.99	0.98	0.99	0.98

Table 5 Jaccard Score, Kappa Index and ECC obtained using various algorithms on ISOT dataset

Model	Jaccard Score		Kappa Index		ECC	
	Count	TF-IDF	Count	TF-IDF	Count	TF-IDF
LR	0.81	0.85	0.81	0.85	0.81	0.85
PAC	0.72	0.86	0.69	0.83	0.69	0.83
RF	0.80	0.76	0.78	0.71	0.78	0.71
NB	0.80	0.64	0.78	0.65	0.78	0.68
SVM	0.79	0.81	0.78	0.80	0.78	0.80
GB	0.76	0.82	0.72	0.81	0.72	0.81
DT	0.61	0.5	0.48	0.38	0.48	0.38

Table 5 summarizes Jaccard Score (JS), Kappa Index (KI) and Matthew Correlation Coefficient (MCC) obtained using all seven algorithms with count and TF-IDF vectorization on ISOT dataset. It is observed from Table 4, DT reported very less JS, KI and ECC value on ISOT dataset. But another algorithm performs well on ISOT dataset. JS score of 81%, 72%, 80%, 80%, 79%, and 76% with count vectorization and 85%, 86%,76%, 64%,81% and 82% with TF-IDF vectorization is obtained using LR, PAC, RF, NB, SVM and GB algorithms, respectively. Similarly, KI values of 81%, 69%, 78%, 78%, 78% and 72% with count vectorization and 85%, 83%, 71%, 65%, 80%, and 81% with TF-IDF vectorization is obtained. ECC values are same as that of KI values except using NB algorithm with TF-IDF vectorization which is 68%.

Figure 2 shows the log-loss values obtained using all seven algorithms with count and TF-IDF vectorization on Kaggle and ISOT dataset. As illustrated in Fig. 2a LR, RF, SVM, GB and DT reported less than 0.4 log-loss with count and DT-IDF vectorization expect for PAC and NB algorithm which reported slightly high log-loss in the range of 0.8–1.2. Figure 2b shows that log loss values obtained on ISOT dataset is slightly high from the range of 0.5–4. For DT algorithm its value is more nearly 10.

It is observed in AUC curve in Fig. 3a, b, approaching to 1 shows a perfect classification for Kaggle dataset approaching to ideal. This also indicates data is specified more towards the right wing whereas in Fig. 4a, b, indicates AUC curve spread between $0.5 < AU < 1$ gives high chance of classification and indicates spread of the data on right as well as in left wing.

5 Conclusion

A framework for the classification of fake news is proposed using NLP with vectorization and ML algorithm. Pre-processing of text data from the two dataset is done using NLP and for classification seven various machine learning algorithms are used.

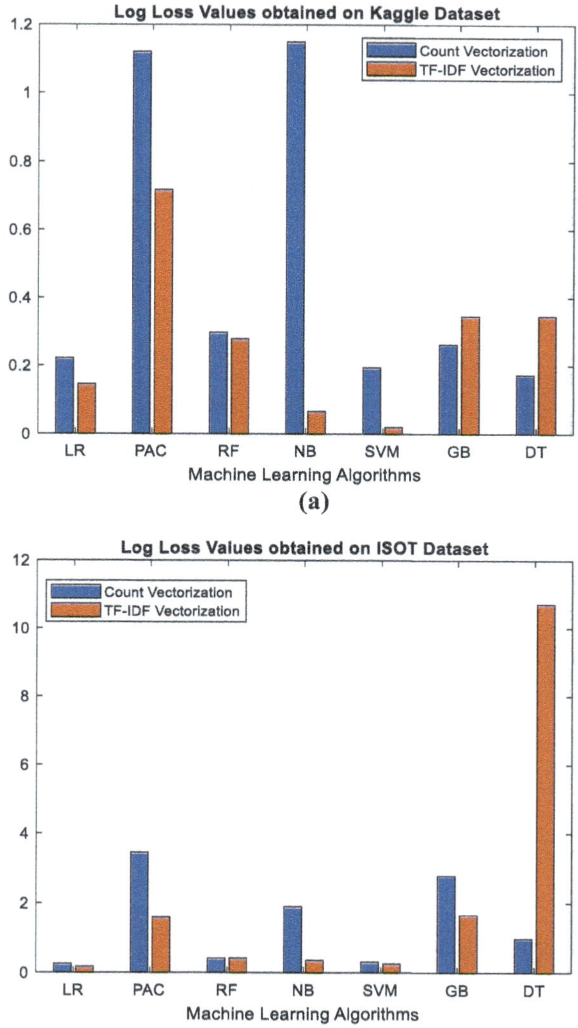

Fig. 2 The log-loss values obtained using all seven algorithms with count and TF-IDF vectorization on **a** Kaggle and **b** ISOT dataset

To demonstrate the performance of proposed framework, various evaluation metrics are used. It is observed that Decision Tree and Gradient bosting algorithm works better than other algorithms on Kaggle dataset. TF-IDF vectorization give more accuracy than count vectorization for all seven-machine learning algorithm. It is also observed that, simple logistic regression works better after the data is pre-processed using NLP on ISOT dataset.

Fig. 3 ROC curve obtained using seven machine learning algorithms on **a** Kaggle dataset with count vectorization, **b** Kaggle dataset with TF-IDF dataset

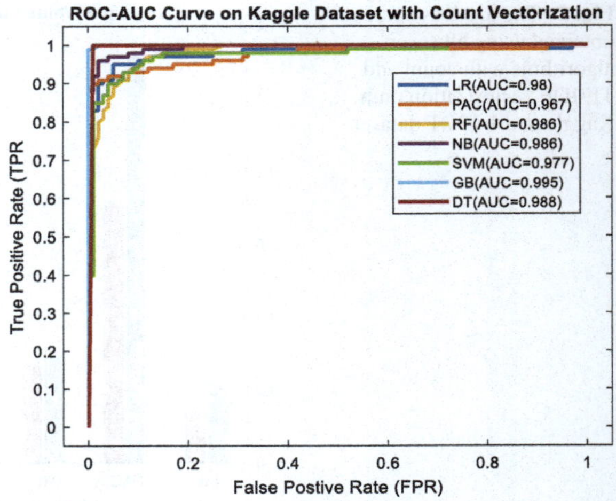

(a)

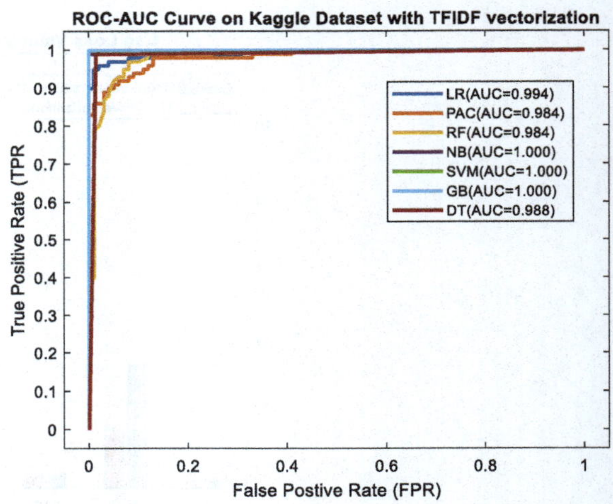

(b)

Fig. 4 ROC curve obtained using seven machine learning algorithms on **a** ISOT dataset with count vectorization, **b** ISOT dataset with TF-IDF vectorization

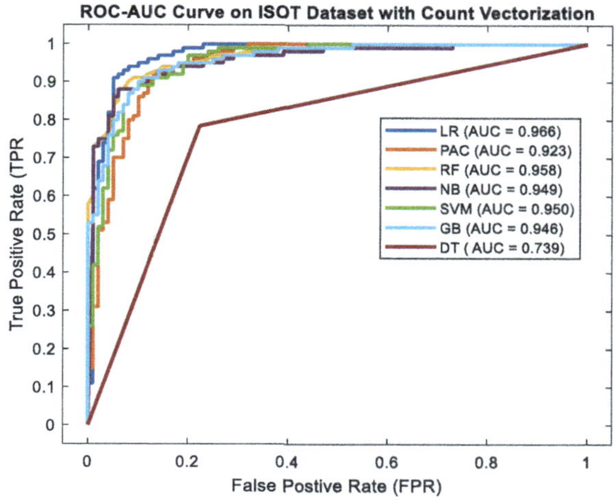

(a)

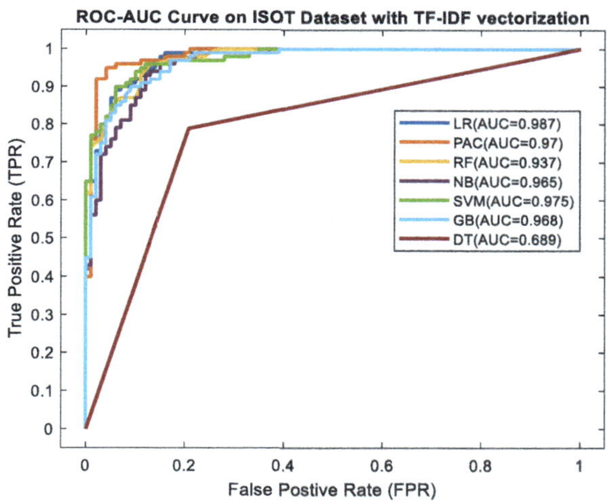

(b)

References

1. Allcott, H., & Gentzkow, M. (2017, May). Social Media and fake news in the 2016 election. *Journal of Economic Perspectives, 31*(2), 211–236.
2. Soll, J. (2016, Dec). The long and brutal history of fake news. *POLITICO Magazine, 18*
3. Gabielkov, M., Ramchandran, A., Chaintreau, A., & Legout, A. (2016). Social Clicks: What and Who Gets Read on Twitter? In *International conference on measurement and modelling of computer science* (pp. 179–192)
4. Fiske, S. T., & Taylor, S. E. (2013). Social cognition: From brains to culture (2nd ed.). CA: SAGE.
5. Lim, C. (2017). Checking how facts- checkers check. 16 May 2017

6. Shao, C. (2018). Anatomy of online misinformation network. *PLOS ONE, 13*(4), 196087
7. Klyuev, V. (2018). Fake news filtering: Semantic approaches.
8. Horne, B. D., & Adali, S. (2017). This just in: Fake news packs a lot in title, uses simpler, repetitive content in text body, more similar to satire than real news. In *International AAAI conference on web and social media* Vol. 8.
9. Castelo, S., Almeida, T., Elghafari, A., Santos, A., Pham, K., Nakamura, E., & Freire, J. (2019). A topic-agnostic approach for identifying fake news pages. *World Wide Web Conference, 2019*, 975–980.
10. Qazvinian, V., Rosengren, E., Radev, D. R., & Mei, Q. (2011). Rumor has it: Identifying misinformation in microblogs. *Conference on Empirical Methods in Natural Language Processing, EMNLP, 2011*, 1589–1599.
11. Pérez-Rosas, V., Kleinberg, B., Lefevre, A., & Mihalcea, R. (2017). Automatic detection of fake news. In *International Conference on Computational Linguistics* (pp. 3391–3401). Santa Fe, New Mexico, USA.
12. Yang, Y., Zheng, L., Zhang, J., Cui, Q., Li, Z., & Yu, P. S. (2018). TI-CNN: Convolutional neural networks for fake news detection.
13. Sivasangari, V., Anand, P. V., & Santhya, R. (2018). A modern approach to identify the fake news using machine learning. *International Journal* of *Pure* and *Applied Mathematics, 118*(20).
14. Ahmed, S., Hinkelmann, K., & Corradini, F. (2019). Combining machine learning with knowledge engineering to detect fake news in social networks—A survey. In *Proceedings of the AAAI 2019 spring symposium* Vol. 12.
15. Ruchansky, N., Seo, S., & Liu, Y. (2017). CSI: A hybrid deep model for fake news detection. *ACM on Conference on Information and Knowledge Management, CIKM, 2017*, 797–806. https://doi.org/10.1145/3132847.3132877
16. Okoro, E. M., Abara, B. A., Umagba, A. O., Ajonye, A. A., & Isa, Z. S. (2018). A hybrid approach to fake news detection on social media. The Nigerian *Journal* of *Technology, 37*(2).
17. Kaggle, (2018). Fake news. San Francisco, CA, USA: Kaggle. https://www.kaggle.com/c/fake-news
18. ISOT, Fake News. University of Victoria, Engineering, Canada. https://www.uvic.ca/engineering/ece/isot/datasets/fake-news/index.php
19. Bird, S., Klein, E., & Loper, E. (2009) Natural language processing with Python—Analyzing text with the natural language toolkit.
20. Patil, S. M., & Malik, A. K. (2019). Correlation based real-time data analysis of graduate students behaviour. In K. Santosh, R. Hegadi (eds.), *Recent trends in image processing and pattern recognition. RTIP2R 2018*. Communications in Computer and Information Science Vol. 1037. Springer.
21. Shetty, B. (2018). Natural language processing (NLP) for machine learning. at towardsdatascience, Medium.
22. Ultimate guide to deal with Text Data (using Python) – for Data Scientists and Engineers by Shubham Jain, February 27, 2018
23. Ranjan, A. (July 2018) Fake news detection using machine learning. Department Of Computer Science & Engineering Delhi Technological University.
24. Mitchell, T. M. (2006). *7e Discipline of machine learning*. Carnegie Mellon University.
25. Chao-Ying Joanne Peng, Kuk Lida Lee, Gary M. Ingersoll, "An Introduction to Logistic Regression", Indiana University-Bloomington, September 2002
26. Crammer, K., Dekel, O., Shalev-Shwartz, S., & Singer, Y. (2006, March). Passive-Aggressive Algorithms. School of Computer Science & Engineering, The Hebrew University, Jerusalem 91904, Israe.
27. Patel, H. H., Prajapati, P. (2018, Oct). Study and analysis of decision tree based classification algorithms. Dept. of Information Technology, CSPIT, Charotar University of Science and Technology, Changa, Gujarat, India.
28. Breiman, L. (2001, Jan). Random forests. Statistics Department University of California Berkeley, CA 94720.

29. Natekin, A., Knoll, A. (2013). Gradient boosting machines. Department of Informatics, Technical University Munich, Garching, Munich, Germany.
30. Kaviani, P. (2017, Nov). Mrs. Sunita Dhotre, Short survey on naive bayes algorithm. *International Journal of Advance Engineering and Research, Department of Computer Engineering*, Bharati Vidyapeeth University, College of Engineering, Pune.
31. Thandar, M., & Usanavasin S. (2015). Measuring opinion credibility in Twitter. In H. Unger, P. Meesad, S. Boonkrong (eds.), *Recent Advances in Information and Communication Technology 2015*. Advances in Intelligent Systems and Computing Vol. 361. Springer.

Fact Checking: An Automatic End to End Fact Checking System

Sajjad Ahmed, Knut Hinkelmann, and Flavio Corradini

Abstract Fact checking is an important topic that needs to be studied scientifically to determine how fake news is spread. Previous work in this area has primarily focused on document- level fact checking. In this paper, however, we will focus on individual statements and the relationship between target statements and the overall news text. In larger context, we will compare statements to known facts, which we will tag within the statement. For dual verification, we will compare our findings to forty mainstream news sources as well as the online encyclopedia (Wikipedia). If a news is detected as fake the existing techniques should block it immediately due to its function, as we cannot replace it. However if a news is detected as fake, we need at least an expert opinion or review before blocking that particular news. This process helps third-party fact-checking organizations to solve the issue; but it is also a time-consuming process. We will attempt to solve the problem of automatically identifying factual claims at the sentence level. Despite its importance, this is a relatively under-studied problem. Existing fake news systems are based on predictive models that simply classify whether a news item is fake or not. Some models use source reliability and network structure so the major challenge in these cases is to train the model. But due to the unavailability of corpora, this is impossible to accomplish. We created a new corpus for social media claims, containing statements that have been fact checked by three reputable sources, and then trained a machine learning model to predict the facts of the news. We presented a fact checking system that takes news as input and then produces an output with an aggregation such as fake, non fake or unclear. To the best of our knowledge it is the only system that has such capabilities.

Keywords Fake news detection · Fact checking · Check worthy statements · Classification techniques

---

S. Ahmed (✉) · K. Hinkelmann · F. Corradini
Department of Computer Science, University of Camerino, Macerata, Italy
e-mail: ahmed.sajjad@unicam.it

K. Hinkelmann
FHNW University of Applied Sciences and Arts Northwestern Switzerland, Windisch, Switzerland

M. Lahby et al. (eds.), *Combating Fake News with Computational Intelligence Techniques*, Studies in Computational Intelligence 1001,
https://doi.org/10.1007/978-3-030-90087-8_17

1 Introduction

Fake news contains information that may be false or inaccurate [1] and separating false from true text is a challenging and difficult task [2]. The important issue is the speed of dissemination, which refers to the information on social media networks. This is a challenging problem that requires attention and alternative solutions. The existing fake news systems are based on the predictive models that simply classify that the news is fake or not fake. Crowdsourcing model a complex problem that can be solved by open calls [3]. In this model, a crowd of indeterminate size is involved in solving the problem and the work is divided among the participants to achieve a cumulative result. There is a possibility that a large number of groups of non-experts can collaborate well on tasks rather than small groups of experts with large efforts [4]. Besides Wikipedia, the news aggregation site—Reddit.com [5] is another example of this application with similar methodology. Crowdsourcing leads to high disagreement among contributors [6]. It has been shown that disagreement is not noise but a signal, and that crowdsourcing can not only be cheaper and scalable, but also have higher quality with more information. Thus, the representation of disagreement can be used to detect low-quality workers. Figure 1 shows the proposed diagram for the fact checking model.

Some models use source reliability and network structure, so the major challenge in these cases is to train the model, but this is impossible due to the unavailability of corpora. There is a need for an alternative approach that combines knowledge with data and requires automation of fact checking that looks at the news content in depth with expert opinion in the same place to detect the fake news. An important motivation for my research is the efforts to introduce an automated fact checking application. There are various ways to check the credibility of news that is fake or not. Popat et al. [7] proposed an approach to check the fact of the claim using credibility check. They check the credibility from the social media websites/news and then pass it to a classifier for credibility check. They conducted different experiments with fact checking websites e.g. snopes.com, politifact.com. Most of the automated methods were based on supervised learning. According to [8, 24], in order to check the truth of

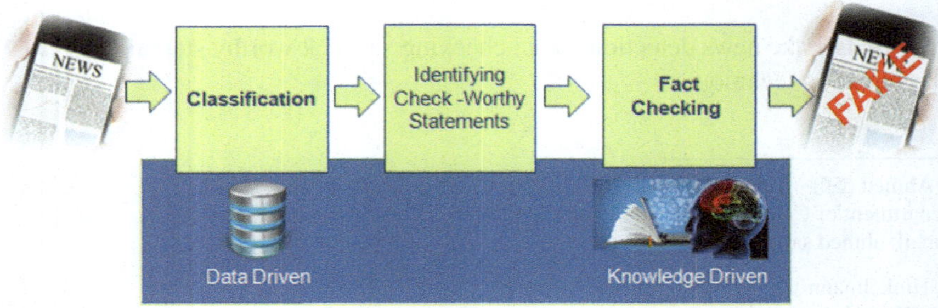

Fig. 1 Proposed diagram for fake news detection

the news through fact checking, the main limitation of the text classification approach is that fact checking requires world knowledge [9].

There is no flexible definition of "fact" for different contexts of information dissemination. Fact is considered by us in this paper from a conventional and qualitative perspective. We understand it as a statement that usually corresponds to a specific event or a state of knowledge. A fact is defined by Wikipedia as being consistent with objective reality or something that can be proven with evidence; thus, if a statement can be shown to be consistent with experience, it can be defined as a fact [10]. However, we recognize that this may not be the end of the story and significantly more research is needed to get a more concrete idea about fact. Some claims contain facts, but this is unimportant as general public will not be interested in knowing these claims. Some other claims contain facts and the public would want to know about those facts. These facts could be useful for fact checking. In our dataset, we tag the location (country), name of the person, organization, event etc. Our automated system compares these tags extracted from the claim with the main body of the news; based on this comparison it generates the percentage and finally the verdict of this claim with evidence and aggregation as output. At the end, it shows the status of the news as fake, true or unverified (Fig. 11). A comprehensive global list of fact checking websites is provided by Duke Reports,[1] where two hundred and ninety fact checking websites are available to date in many countries and in many languages. The following table provides details of the first ten known fact checking websites and how they operate (Table 1).

In addition, to improve the results of its search engine, Google also uses a knowledge graph to gather information from a variety of sources. The extracted information is presented to users in an information box that is displayed next to the search results. Figure 2 shows an example of the Google knowledge graph (Figs. 1, 2, 3, 4 and 5).

In this paper, we focus on the first step: predicting fact-checking statements. Our contributions can be summarized as follows:

- **New dataset:** We create a new dataset of manually-annotated claims extracted from three reputable news organizations after the 2016 US presidential election, which we make available to the research community.
- **Context Modelling:** We developed a new approach that is a combination of text classification and fact checking of check-worthy statements. We model not only the text content, but also the context, i.e., how the target sentence relates to the neighboring sentences.

2 Literature Review

The studies [11–13] focus on classification and fact checking of check worthy statements to address the problem of detecting fake news. Text classification mainly

[1] https://reporterslab.org/fact-checking/.

Table 1 Fact checking websites comparison

Name	Topic	Content	Labels
Snopes[a]	Political and social issues	– News Articles – Videos	True, false, mixture, unproven, outdated, scam, mostly true, half true
FactCheck[b]	American politics	– Debates – Speeches – Interview – TV ads	True, false, no evidence
PolitiFact[c]	American politics	– Statements	True, mostly true, half true, false, mostly false, pants on fire
The Washington Post[d]	American politics	– Statements – Claims	One pinocchio, two pinocchio, three pinocchio, four pincocchio, verdict pending
FullFact[e]	Economy, health and education	– Articles	Not clear
TruthOrFiction[f]	Politics, religion, nature, food, medical	– Email Rumors	Truth, fiction
HoaxSlayer[g]	Not specific	– Articles – Messages	Hoaxes, scams, malware, fake news, true, humor, spams
RealClearPolitics[h]	Politics Defense Energy Heath	– News	Not specify
Our.news[i]	Politics	– Articles – News	Accepts, rejected, left spin, no spin, etc
Media Bias[j]	Politics media	– News	Bias, least biased, right, right center

[a] https://www.snopes.com/
[b] https://www.factcheck.org/
[c] http://www.politifact.com/
[d] https://www.washingtonpost.com/news/fact-checker
[e] https://fullfact.org/
[f] https://www.truthorfiction.com/
[g] http://hoax-slayer.com/
[h] https://www.realclearpolitics.com/
[i] https://our.news/
[j] https://mediabiasfactcheck.com/

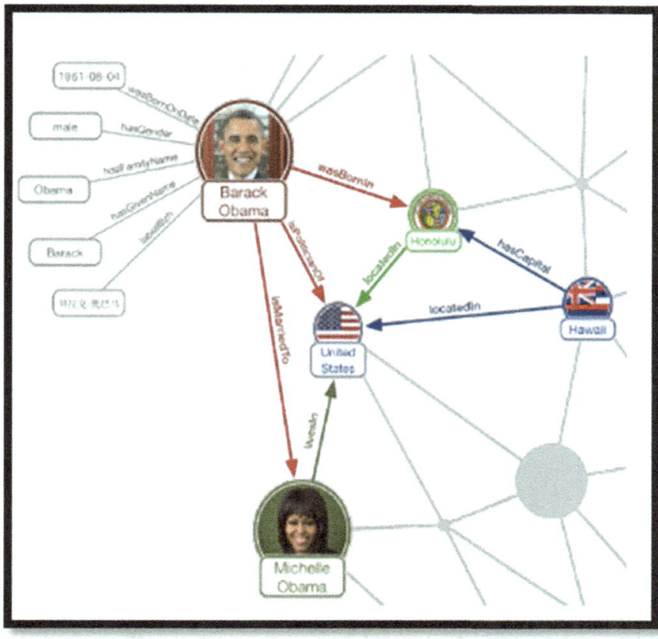

Fig. 2 Example of knowledge graph [5]

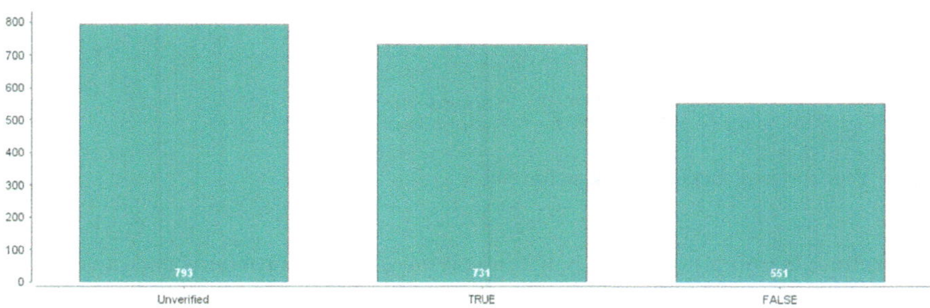

Fig. 3 Class distribution of sentences

focuses on extracting text features and incorporating these features into classification models e.g. Decision tree, SVM, logistic regression, K nearest neighbor. A state of the art presented by authors is the combination of knowledge (Fact-checking) and data (Text classification, Stance detection) [11]. At the end selection of the best algorithm that performs well. Detection of check-worthy statements is a subtask in the fact-checking process, the automation of which would reduce the time and effort required to fact-check a statement. Manual fact-checking is at a disadvantage nowadays, but automated fact-checking can help to reduce the human burden [12].

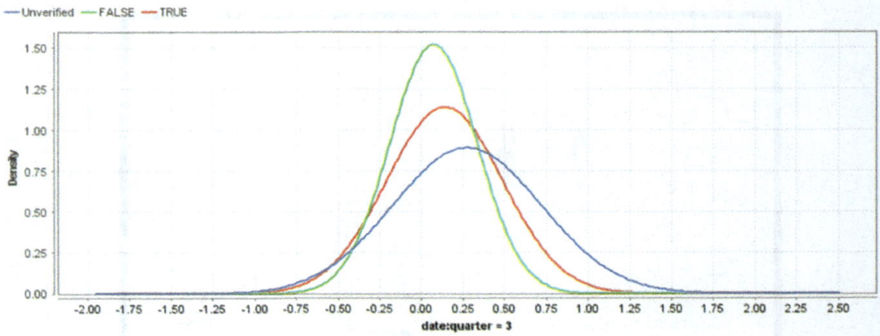

Fig. 4 Dataset class labelling chart

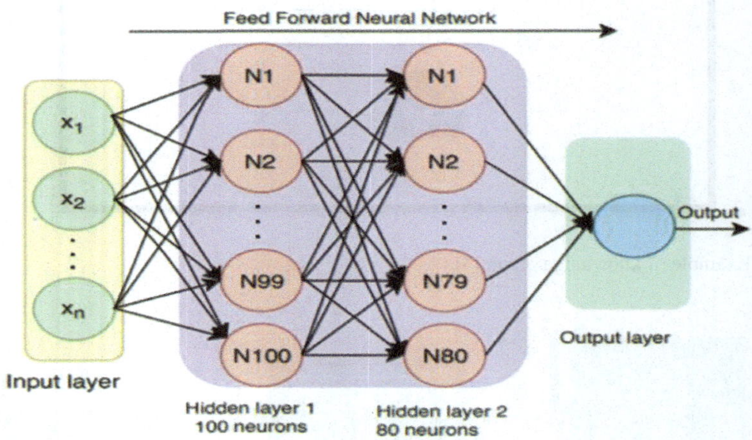

Fig. 5 Feed forward neural network architecture

Machine learning nowadays helps in solving complex problems and developing AI systems, especially in cases where we have tacit knowledge or the knowledge is unknown [13]. Computational fact checking aims to provide the user with an automatic system that can classify true and false content. Predominantly, computational fact checking works on two points that identify check worthy claims and then discriminate the veracity of factual claims. It works on the basis of user's key bases and viewpoints on the content [14]. Open web sources are used as reference which can distinguish the news into true and false [15, 16]. Separation of fake content into three categories: serious fabrication, large scale hoaxes and humorous content was the main goal of this work. They provide a way to filter, vet, and verify the news and discussed in detail the pros and cons of these news [17].

This study is a data-oriented application that simply uses an available dataset and then applies a deep learning method and finally, proposes a new text classifier that can predict whether the news is fake or not [18]. Traditionally, all rumor detection techniques are based on message level detection and analyze credibility using data but in real-time detection using keywords, the system collects related microblogs using a data collection system that solves this problem. The proposed model combines user based, propagation based and content based models and checks the credibility in real time and returns the answer within thirty-five seconds [19]. The above tasks have previously been treated in isolation, as previous work [29] proposed factuality prediction focused only on input claims. Another work proposed by focused only on specific domains e.g., Wikipedia.

3 Methodology

An important task in fact checking is to analyse how different features are involved in result exploration and dataset comparison. Features such as dataset size and the text length are also discussed as part of the analysis. For the classification of texts as real or fake news, each text was pre-processed and 'cleaned'. Feature extraction techniques are then used before classification is performed. Framework that integrates various components of the fact checking process, starting from check-worthy statements from mainstream news media sites, Text Retrieval, source collection, Fact checking of claims, linguistic analysis and aggregation. This process is outlined in Fig. 6 and will be explained in the next sections.

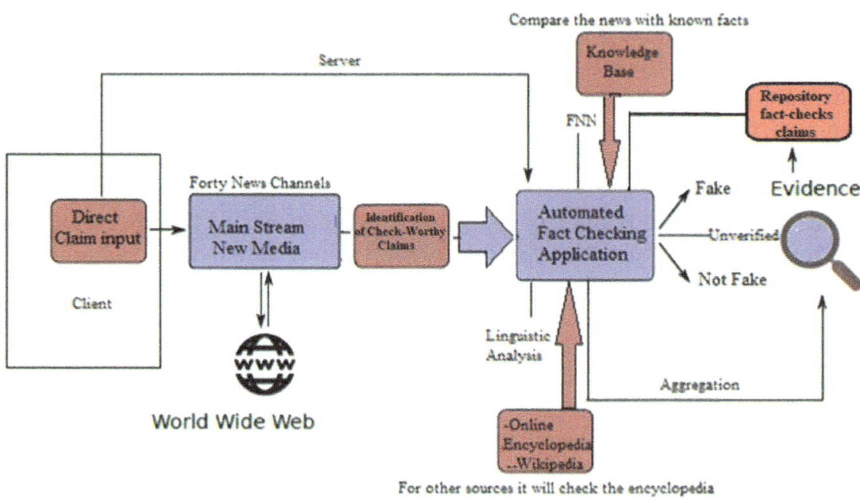

Fig. 6 System Framework for automated fact checking

Table 2 Dataset row structure example set

Claim	Source	Tags	Claim Label
An oil pipeline exploded in Saudi Arabia	Dailymail.co.uk	Pipeline, Saudi + Arabia	Fake
Microsoft is going to acquire Mojang AB	Avsforum.com	Microsoft, Mojang	Non-Fake
A fourth-grade student from Texas was suspended after threatening another student with magic	Dailymail.co.uk	Magic, Texas, Hobit, Lord + of + the + rings	Unverified

3.1 Data Exploration

To create an effective automated application, it is important to examine the datasets used and the techniques used to extract features from the data. It is to analyze and compare how different features are included in the dataset. Features such as the size of the datasets and the length of the texts will also be discussed as part of the analysis. For the classification of texts as real or fake news, each text was pre-processed and cleaned. Feature extraction techniques are then used before classification is performed. This process is outlined in Fig. 4. For this task, we collected news articles from different websites. The dataset was separated by different attributes: web page, claim, description, label, tags, domain, and date. We analyzed the dataset and examined how the articles differed from each other, e.g., by content and feature. We also sorted the data by different result indicators, such as how often it was shared. All articles were then labelled as fake, true, and unverified (which is ambiguous). The corpus contained 2146 news articles, of which 731 were true claims, 793 were unverified claims and 551 were false claims. In the next step, we identified the features that can help us distinguish the claims as fake or not when we compare them with known facts. For each claim we tagged known entities, such as name, location, country, organization name and any other item that can help us in fact checking. The organizations considered are Politifact,[2] Emergent,[3] and Daily Mail.[4] Figure 3 shows the distribution of the classes. We used RapidMiner[5] a powerful Machine Learning tool for data exploration.

Table 2 shows an example of how a row in the data set is constructed. Each row in a file consists of the claim of article, source, tags and claim label.

As can be seen in Fig. 4, almost half of the articles are true; this is due to proper source and evidence. We have tagged these claims because some people are more interested to know about the public figures or politicians.

[2] https://www.politifact.com/.

[3] https://www.emergent.info.

[4] https://www.dailymail.co.uk/home/index.html.

[5] https://rapidminer.com/.

We tagged all the claims we have because it was difficult to tag the whole article, so we just tagged the claim because it was easier to manage. We did an 80–20 split of the data for the training and testing sets.

3.2 Model Description

Multiple models were trained, tested and compared to evaluate the effectiveness of different feature extraction and modelling techniques. I experimented with classification models that have proven effective and provided good results on related sentence classification tasks in previous chapters. The Feed Forward Neural Network provided promising results and I continued to experiment with it.

3.2.1 Feed Forward Neural Network

Feed Forward Neural Networks (FNN) take a fixed input and feed it forward through the network to produce an output without generating cycles. Feed Forward Neural Network (FFN) are classification models that learn from the input data through their weights [20]. The FNN and LSTMs work particularly well on data that is sequential, such as natural language, because they allow a longer context to be represented in the prediction. In contrast, only the output is considered immediately. This is done by training LSTM layers within the network when to keep or forget information. This means that the network can retain information at a variable rate. This long memory, allows for more accurate prediction. Within an LSTM cell in a node, lies the capability of storing and forget previous input. The LSTM contains three gates to decide whether to forget information about a previous input: an input gate, an output gate, and a forget gate. Each time data is passed through the cell, information relating to previous input is passed from the hidden layers and then produces an output. Exploring relationship entities, specifically news articles, publishers and users, and user contributions. The diagram below illustrates this process as this system has been helpful in the developing automated fact checking applications (Fig. 5).

4 Web Application Development Task

Manual fact checking is not only time consuming but also a challenging process. Therefore, the increasing demand for automation has encouraged researchers to look into automatic fact checking. We have presented a fact checking system that combines text classification and fact checking to detect fake news. Our developed model incorporates various components of machine learning and knowledge engineering, such as retrieving documents from mainstream media sources with different

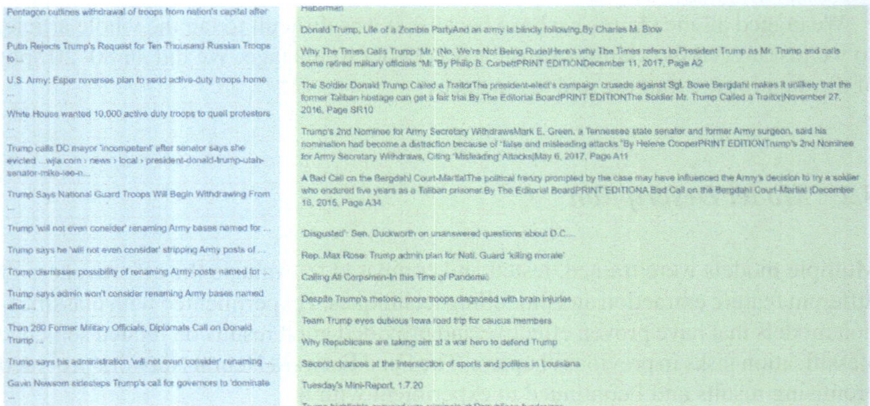

Fig. 7 Source collection from mainstream media and top search

types of reliability, classification, evidence extraction, linguistic analysis and aggregation. The general architecture of the proposed fact checking system is shown in Fig. 6. This system is accessible through a web browser and has two sides: Client and Server. The first step in this process is that the user on the client side sends a request to the server in the form of a textual claim. This request is processed by the server which forwards the request data to the document retrieval component, which then retrieves a list of documents (shown in Fig. 6) from four different sources: Wikipedia, mainstream news media (forty news organizations). The retrieved result is further refined by bypassing the document retrieval module. The perspective of each relevant document with respect to the claim is detected by the fact checking component, which is typically modeled by using tags and comparing those tags to the claim of the news. Further explanations of model predictions are rationalized at the sentence level using the same component. Linguistic comparison also takes place in the fact checking component to analyze the language of each document after it has passed through the linguistic component. Finally, the aggregation component makes the final decision about the factuality of the claim by aggregating the predictions of the classification and fact checking about the claim. It can predict the factuality of a given claim with proper evidence at document and sentence level in support of its prediction. We have described all four components below (Figs. 7, 8, 9 and 10).

The Python micro-framework Flask[6] was used to create the application using HTML[7] and CSS.[8] Flask was chosen due to its lightweight nature which was suitable for the purpose of this application. The user-interface consists of a text area for text input and a button to run the search. Figure 9 shows the main user interface, while Fig. 11 shows the result of a prediction.

---

[6] https://pypi.org/project/Flask/.

[7] https://html.com/.

[8] https://getbootstrap.com/.

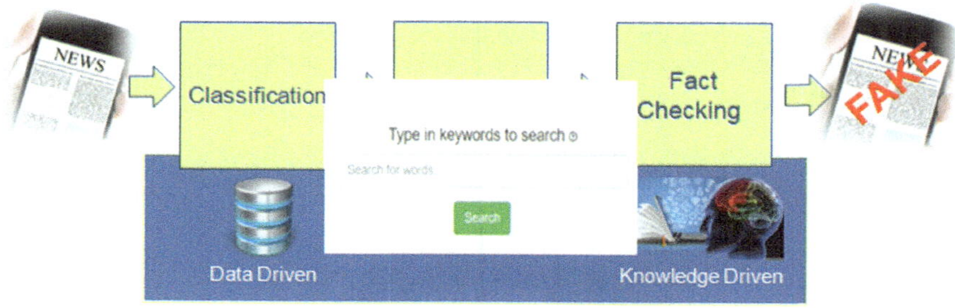

Fig. 8 The web-applications main interface

Claim input panel:

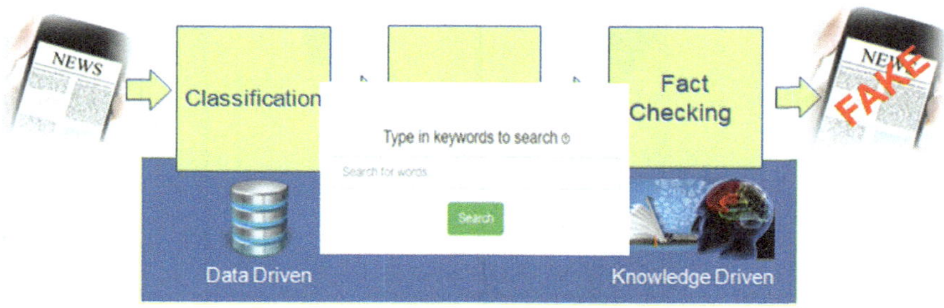

Fig. 9 Claim input panel for users

Our developed fact checking system comprises of the following three approaches. The first phase of evidence extraction takes place based on the fact checked given claims through the user's text input window. Then to check the reliability of the given claims and retrieved media sources [21]. At the end, a fact checking module takes place to check the claim through FNN algorithms and also double check the results through linguistic check. The above three steps correspond to different Natural Language Processing (NLP) and Information Retrieval (IR) tasks, that involve information extraction. Existing approaches were mostly used for text classification problems and utilized different linguistic, stylistic and semantic features [28] and few used information from external sources [27]. All of these steps were typically carried out individually and the results were then proposed. For example, looking at recent work on the fact extraction and verification (FEVER) [23], the focus is on a specific domain (e.g., Wikipedia) and according to [24, 29] algorithms have been proposed to predict the factuality of the claims by focusing mainly on the input claims and their metadata information (e.g., the speaker of the claim). To the best of our knowledge, there is to date no end-to-end fact checking system that is a combination of data and knowledge with the ability to search Wikipedia and mainstream media sources across the web to perform fact checking for specific claims. We attempted to fill

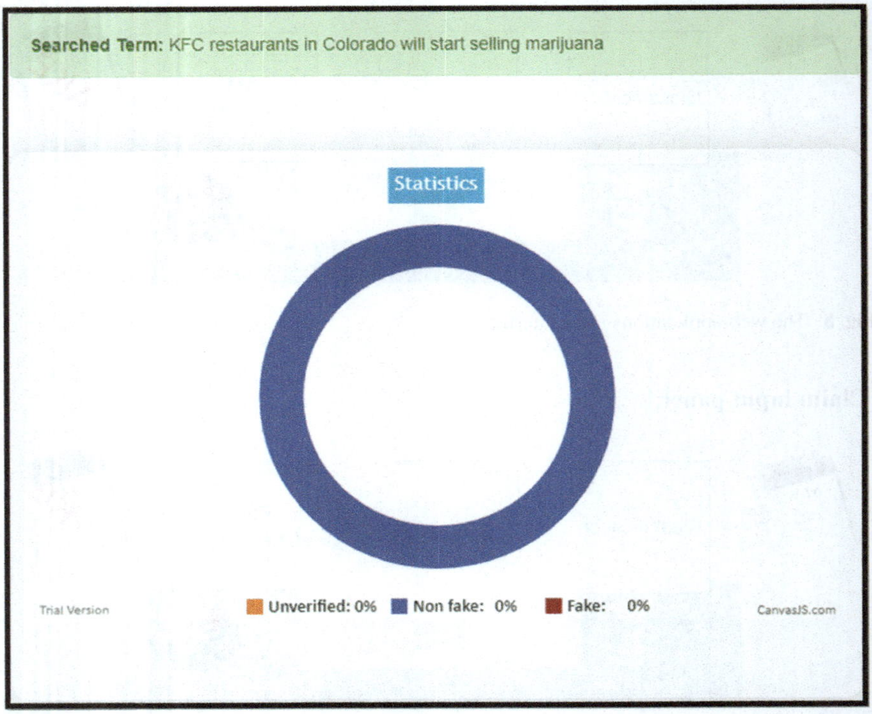

Fig. 10 An example of a fake prediction (General)

these gaps and developed a proposed fact checking system consisting of different fact checking steps (Fig. 9) that is not only capable of searching different sources, but also predicting the factuality of claims, and presenting a set of evidence with explanations to support the prediction. It gave the results based on fake, non fake and unverified with aggregation of factuality. An example is shown in Fig. 11, where the claim with 90% factuality is labelled as fake. In our work, we present the proposed fact checking system as an online application for automatic fact checking of claims. Our developed system is helpful for individuals and professionals to check the facts of claims in one place as it not only has the ability to check the factuality of a claim with aggregation after multiple checks, but also presents relevant documents as evidence to support its prediction for given claim. In the future, we plan to further extend the system and make it even more advanced and user-friendly by focusing on advancement of underlying components (e.g., stance detection), topic detection, and credibility comparison, and source, author based cross-language settings.

After Sentence Level Comparison:

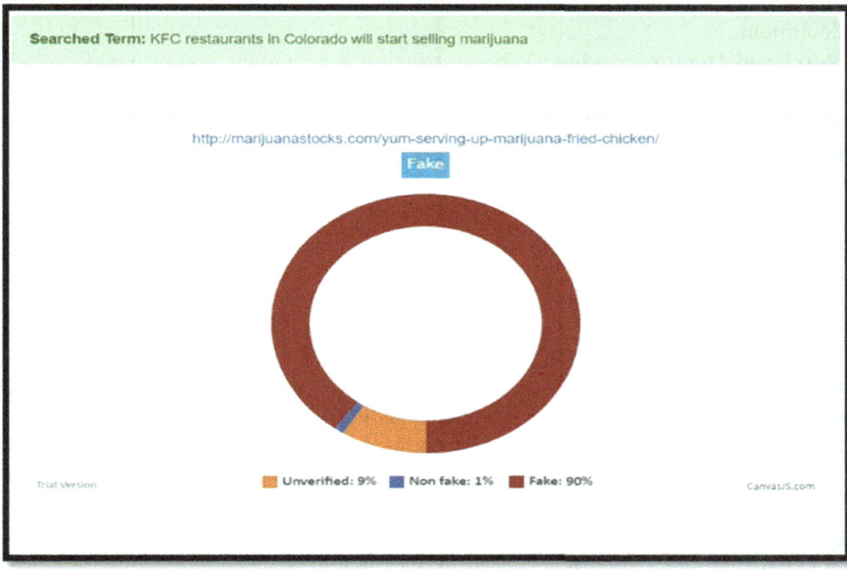

Fig. 11 An example of a fake prediction with claim and evidence

4.1 Source Collection

Currently, search engines (e.g., Google, Bing, and Yahoo) are used to retrieve the relevant documents for a given query from any media source. Four types of sources: Wikipedia and high, mixed and low factual media are used to retrieve relevant documents. Usually, journalists spend a considerable amount of time verifying their sources of information [25, 26]. Sometimes, a list of unreliable online news sources given by journalists from some fact checking organizations was also created. We extracted the information from news sources with high accuracy using available libraries that provide parsers for information extraction (Stanford NLP). In our work, we used the above three categories of media sources to retrieve documents using the document retrieval component. In addition to the above forty mainstream media sources and open web search documents, we also used Wikipedia, which contains the most of the factually accurate information. Figure 7 shows the top search documents after entering the query and based on targeted mainstream media sources collected against the given claim.

Front end comprises of three views:-

- **Claim input view**: For entering a claim to be checked for factuality (Fig. 8).
- **Result view**: This contains lists of retrieved documents from factuality type sources: Wikipedia, Open browser Search and mainstream news media (Forty

Organizations). The final score for the entered claim for each list is displayed at the top of the page, and the fact checking score appears next to it for each document.

- **Retrieved Document view:** When retrieving a document, the proposed system not only displays the text of the document but also displays the important sentences based on their score to the claim in the highlighted form (Fig. 8). The results are further displayed in the form of pie chart and the result is shown in Fig. 11.

4.2 Text Retrieval

This step is feasible because we only need to retrieve the data using different APIs from different news agencies. Our developed tool has both functionalities e.g. entering own keywords or selecting fact checked claims that the system updates from mainstream media. In the first step, an input claim is directly converted into a query by considering its verbs, nouns and adjectives [26]. The initial query checked from the claim text and knowledge base, if there are matches the claims are checked. For this purpose we used Natural Language Toolkit (NLTK)[9] which is suitable for linguistically related tasks. It extracts relevant documents from mainstream news media sites and also from open search. It also checks the relevant documents from Wikipedia. Finally, it determined the forty links with the highest match to the given claim. This approach is a good change from the existing approach where human fact checkers mainly focus on multiple sources rather than relying on one source (like Wikipedia). The user view or claim input window is shown in Fig. 8.

Some researchers approach this problem with a specific text processing and are based on the separation of fake and non-fake text. On the other hand, some previous researchers have separately studied the individual components of this multi-step process which includes—(i) retrieving potentially relevant documents for a given claim [28, 29], (ii) verifying the reliability of the media sources from which documents are retrieved, (iii) predicting the stance of each document with respect to the given claim (Mohtarami et al., 2018), and then predicting the factuality of claims [27]. In our work, we present an automated web based fact checking tool that combines all four components into one framework and has the potential to predict the factuality of a given claim along with evidence for its document and sentence level predictions (Fig. 12).

4.3 Aggregation

FFNs linguistic analysis and fact checking were performed in parallel with the given claims and based on the claim the retrieved documents from all sources. After fact

[9] https://www.nltk.org/.

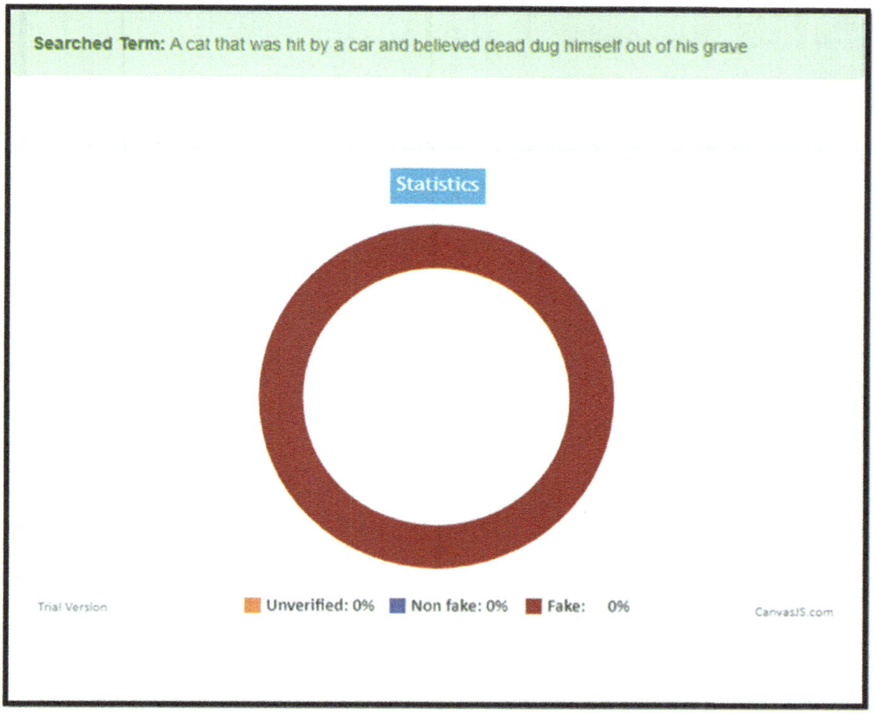

Fig. 12 An example of a non-fake prediction

checking, each claim was assigned an average score and then aggregate score, which is shown at the top of the list of retrieved documents in Fig. 13. A higher agrees score means the claim is true and a higher disagree score means false.

5 Evaluation Results

For demonstration purposes, a simple web application was created. The web application takes a text as the users input claim and classifies the text as fake, not-fake or unverified based on a pre trained model in the fact checking module. For dual verification of the results, a linguistics analysis is performed before coming to the verdict and aggregation of the given claim. The text is cleaned in the same way as in the training and testing phase. Stop words are removed along with punctuation before lemmatization is performed. The output of this is converted into TF-IDF values which are passed into a pre-trained FNN model trained on the given dataset.

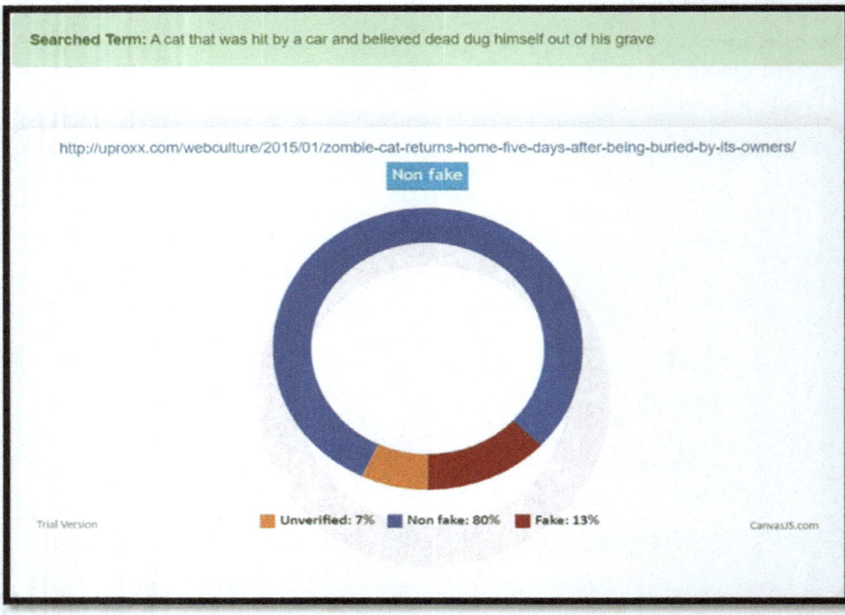

Fig. 13 An example of a non-fake prediction with claim and evidence

5.1 Fake Claim Example

Claim input panel

Overall result

As Fig. 10 shows, the overall factuality result of the claim is non fake (True) because different media channels have reported this claim, so the systems initial response is false based on different news media reported this news. Our developed system tests the factuality of the claim at sentence level after comparing the claim with different checks highlighted in Fig. 6.

Finally, if we check the claim through the fact checking module and compare it with known facts, you can see this in Table 3 in the tags section. In the table you can also see that Racket Report is an unreliable source, and this was a fake news claim. Our system suggested 90% factuality on the basis of comparison with mainstream news media organizations and Wikipedia. The final results can be seen in Fig. 11.

After Sentence Level Comparison.

Table 3 A sample fake claim with assessment and explanation

Claim	KFC restaurants in Colorado will start selling marijuana
Headline Date Description	KFC restaurants in Colorado will start selling marijuana 03/04/2017 KFC Gets Occupational Business License To Sell Marijuana In Colorado Restaurants KFC Gets Occupational Business License To Sell Marijuana In Colorado Restaurants
Tags	KFC, Marijuana, Hoaxes, Fake + News, Colorado
Evidence Source	The Racket Report is an unreliable source, and this was a fake news article. Snopes provided a debunking Emergent
Label	Fake

5.2 *Non Fake Claim Example*

Another claim published by Emergent with the headline that the cat claws its way out of the grave after five days, is exactly the true claim. The details of the news are mentioned in the (Table 4).

When we check the factuality the claim, the initial results based on the other media sources suggest that this claim is fake on the results, but we need further sentence level investigation to verify the fact of the news, so we compare it to more with known facts. The initial results can be seen in Fig. 12.

After Sentence Level Comparison

After comparing it to various facts drawn from Tampa's human society, the system has concluded that the claim is 80% non fake, 7% unverified, and 13% fake. So, based

Table 4 A non-fake claim with assessment and explanation

Claim	A cat that was hit by a car and believed dead dug himself out of his grave
Headline Date Description	Cat claws out of grave 5 days later 1/26/2017 Bart the cat showed up in his neighbour's yard five days after being buried. He should make a full recovery, according to the Humane Society
Tags	Cat, Animals, Florida, Zombies
Evidence	The Humane Society in Tampa provided images and background on the cat and believes the cat's injuries are consistent with the story. Bart's owner, Ellis Hutson, said that one neighbour helped him bury the cat, and another neighbour found Bart. "I open the door and my neighbour's standing there with the cat in her hand," Hutson told ABC. "She said, 'Bart is not dead.' I said, 'That impossible. We buried Bart.'" The involvement of the humane society combined with the other people in this story leads us to consider it true
Source	Emergent
Label	Non-fake

Table 5 Unverified claim with assessment and explanation

Claim	A fourth-grade student from Texas was suspended after threatening another student with magic
Headline Date Description	Parent: Fourth-grader suspended after using magic from 'The Hobbit'. Interview 2/2/2017 Allegedly, the 9-year-old told a classmate his magic ring would make them disappear. The boy had recently seen "The Hobbit" with his family and was supposedly inspired by that and the powerful ring in "The Lord of the Rings
Tags	Magic, Texas, Hobit, Lord + of + the + rings
Evidence Source	The Odessa American was the first with the story Jan. 30, interviewing the boy's father, Jason Steward. They reported the child was suspended "for allegedly making a terroristic threat," though Kermit Elementary School Principal Roxanne Greer declined to comment. Until the school confirms the incident, we will keep this as Unverified Daily mail
Label	Unverified

on the majority, the system predicts that the overall result of the claim is not fake. The overall result can be seen in Fig. 13.

5.3 *Unverified Claim Example*

Our final claim is the headline of a fourth grade student who was suspended from school after threatening his classmate. The status of this claim is unclear, as there has been no comment from the school. You can see the full story of the news in the (Table 5).

Overall Result

Initial findings based on the other media sources, that only gave the student father's point of view, not the other side's point of view. The system shows that claim is 100% unverified (Fig. 14).

After Sentence Level Comparison

After reviewing the factuality at the sentence level, which involves comparing the text with different known facts such as the location of the incident, the father's attitude towards the media etc. We found that 65% of the status of the claim is unverified, but on the other hand like the location of the school and the student's father stance, so the system predicts 13% true and 20% false status (Fig. 15).

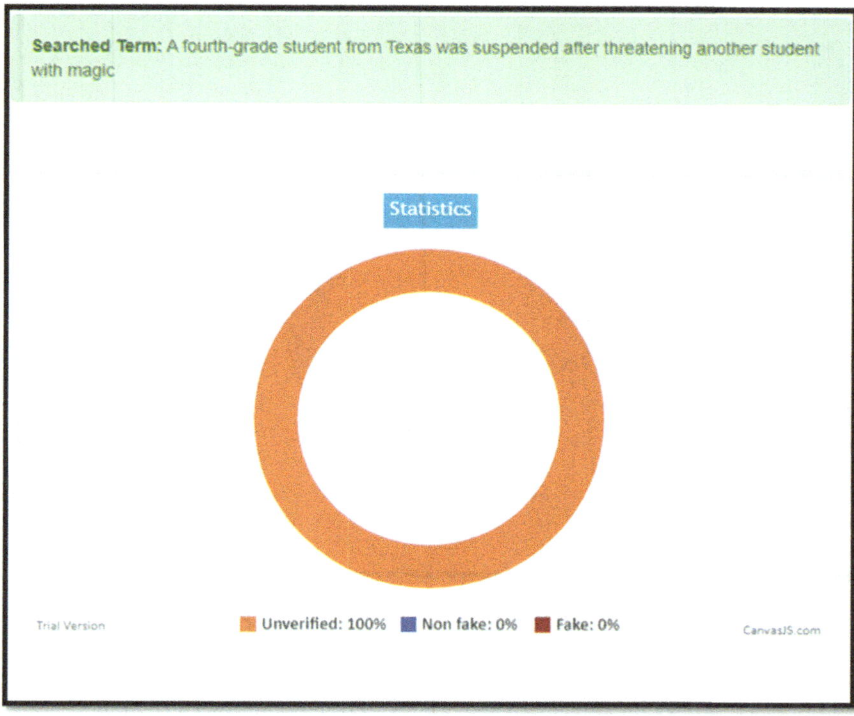

Fig. 14 An example of an unverified prediction (General)

6 Conclusion and Future Directions

Fake news detection is a real problem for various sectors of society, which I have discussed in detail in this paper. A framework was developed that facilitated the evaluation of different classification and feature extraction techniques, as well as the creation of a simple web application that could classify text input from users as false, true or unverified after a combination of machine (text-based) and fact checking (human-based). The results are limited due to the small size of the dataset, i.e., there were not enough texts to effectively train and test the model. Ultimately, the classification techniques analyzed in this project are not substantial enough to effectively combat fake news, but the results have proven valuable for the potential that fact checking can also be done in a method by incorporating knowledge engineering with current methods. Some of the features discussed in this work will definitely be part of future research in fact checking e.g. to check the news when it's published online according to the source, credibility, time, location, etc. We were able to demonstrate that with the integration of text classification and fact checking of check worthy statements, fake news can be detected. We hope that this system will

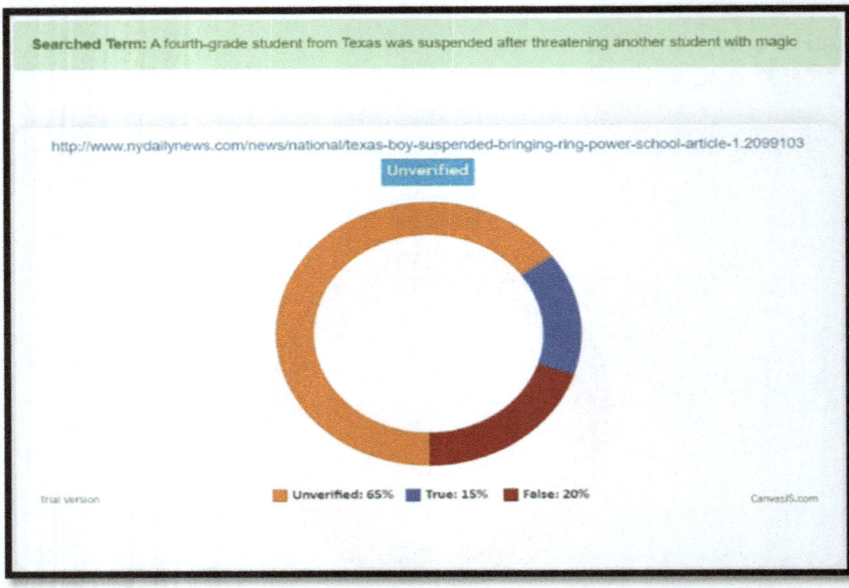

Fig. 15 An example of an unverified result prediction with claim and evidence

provide a strong base for future research as due to its potential, it can help individuals and society.

References

1. Zannettou, S., Sirivianos, M., Blackburn, J., & Kourtellis, N. (2019). The web of false information: Rumors, fake news, hoaxes, clickbait, and various other shenanigans. Journal of Data and Information Quality, 11(3). https://doi.org/10.1145/3309699
2. Lazer, D. M. J., Baum, M. A., Benkler, Y., Berinsky, A. J., Greenhill, K. M., Menczer, F., & Zittrain, J. L. (2018). The science of fake news. *Science, 359*(6380), 1094–1096. https://doi.org/10.1126/science.aao2998
3. Chatzimilioudis, G., Konstantinidis, A., Laoudias, C., & Zeinalipour-Yazti, D. (2012). Crowdsourcing with smartphones. *IEEE Internet Computing, 16*(5), 36–44. https://doi.org/10.1109/MIC.2012.70
4. Howe, J. (2006). The rise of crowdsourcing. *In Wired Magazine, 14*, 1–4.
5. Zhou, Z., Guan, H., Bhat, M. M., & Hsu, J. (2019). Fake news detection via NLP is vulnerable to adversarial attacks. arXiv preprint arXiv:1901.09657.
6. Dumitrache, A., Inel, O., Aroyo, L., Timmermans, B., & Welty, C. (2018). CrowdTruth 2.0: quality metrics for crowdsourcing with disagreement. arXiv preprint arXiv:1808.06080.
7. Popat, K., Mukherjee, S., Strötgen, J, & Weikum, G. (2016). Credibility assessment of textual claims on the web. In *Proceedings of the 25th ACM International on Conference on Information and Knowledge Management*. ACM, 2173–2178.
8. Rashkin, H., Choi, E., Jang, J. Y., Volkova, S., & Choi, Y. (2017). Truth of varying shades: Analyzing language in fake news and political fact-checking. In *EMNLP 2017 - Conference on*

Empirical Methods in Natural Language Processing, Proceedings, (pp. 2931–2937). https://doi.org/10.18653/v1/d17-1317
9. Nakashole, N., & Mitchell, T. M. (2014). Language-aware truth assessment of fact candidates. In *52nd Annual Meeting of the Association for Computational Linguistics, ACL 2014 - Proceedings of the Conference* (Vol. 1, pp.1009–1019). https://doi.org/10.3115/v1/p14-1095
10. Weber, K., & Alcock, L. (2004). Semantic and syntactic proof productions. *Educational studies in mathematics, 56*(2–3), 209–234.
11. Ahmed, S., Hinkelmann, K., & Corradini, F. (2019). Combining machine learning with knowledge engineering to detect fake news in social networks-a survey. In *Proceedings of the AAAI 2019 Spring Symposium* (Vol. 12, p. 8).
12. Ahmed S., Balla K., Hinkelmann K., & Corradini F. (2021). Fact checking: Detection of check worthy statements through support vector machine and feed forward neural network. In: K. Arai (Ed.), *Advances in information and communication. FICC 2021* (Vol. 1364) Advances in Intelligent Systems and Computing . Springer. https://doi.org/10.1007/978-3-030-73103-8_37.
13. Ahmed, S., Hinkelmann, K., & Corradini, F. (2020). Development of fake news model using machine learning through natural language processing. *International Journal of Computer and Information Engineering, 14*(12), 454–460.
14. Houvardas, J., & Stamatatos, E. (2006). N-gram feature selection for authorship identification. Lecture Notes in Computer Science (Including Subseries Lecture Notes in Artificial Intelligence and Lecture Notes in Bioinformatics), 4183 LNCS, 77–86. https://doi.org/10.1007/11861461_10.
15. Banko, M., Cafarella, M. J., Soderland, S., Broadhead, M., & Etzioni, O. (January, 2007). Open information extraction from the web. In IJCAI (Vol. 7, pp. 2670–2676).
16. Magdy, A., & Wanas, N. (2010). Web-based statistical fact checking of textual documents. In *International Conference on Information and Knowledge Management, Proceedings* pp. 103–109. https://doi.org/10.1145/1871985.1872002
17. Rubin, V. L., Chen, Y., & Conroy, N. J. (November, 2015). Deception detection for news: three types of fakes. In *Proceedings of the 78th ASIS&T Annual Meeting: Information Science with Impact: Research in and for the Community* (p. 83). American Society for Information Science.
18. Bajaj, S. (n.d.). The pope has a new baby! Fake news detection using deep learning. Retrieved from https://web.stanford.edu/class/cs224n/reports/2710385.pdf.
19. Zhou, X., Cao, J., Jin, Z., Xie, F., Su, Y., Chu, D., & Zhang, J. (2015, May). Real-Time News Cer tification System on Sina Weibo. In *Proceedings of the 24th International Conference on World Wide Web* (pp. 983–988). ACM.
20. Bengio, Y. et al. (2003) A neural probabilistic language model. *Journal of Machine Learning Research.* https://doi.org/10.1162/153244303322533223
21. Baly, R., Mohtarami, M., Glass, J., M'arquez, L., Moschitti, A., & Nakov, P. (2018b). Integrating stance detection and fact checking in a unified corpus. In *Proceedings of the 16th Annualw Conference of the North American Chapter of the Association for Computational Linguistics.* New Orleans, LA, USA: NAACL-HLT '18.
22. Nakov, P., Mihaylova, T., Marquez, L, Shiroya, Y., & Koychev, I. (2017). Do not trust the trolls: Predicting credibility in community question answering forums. In *Proceedings of the International Conference Recent Advances in Natural Language Processing (RANLP)* (pp. 551–560).
23. Thorne, J., Vlachos, A., Christodoulopoulos, C., & Mittal, A. (2018). Fever: A large-scale dataset for fact extraction and verification. In *Proceedings of the 2018 Conference of the North American Chapter of the Association for Computational Linguistics (HLT-NAACL)* (pp. 809–819).
24. Wang, W. Y. (2017). Liar, liar pants on fire: A new benchmark dataset for fake news detection. In *Proceedings of the 55th Annual Meeting of the Association for Computational Linguistics (Volume 2: Short Papers)* (pp. 422–426). Association for Computational Linguistics.
25. Nguyen, A. T., Kharosekar, A., Lease, M., & Wallace, B. C. (2018). An interpretable joint graphical model for fact-checking from crowds. In *32nd AAAI Conference on Artificial Intelligence, AAAI 2018*, ((3), pp. 1511–1518).

26. Potthast, M., Hagen, M., Gollub, T., Tippmann, M., Kiesel, J., Rosso, P., & Stein, B. (2013). Overview of the 5th international competition on plagiarism detection. CEUR Workshop Proceedings (Vol. 1179).
27. Mihaylova, T., Nakov, P., Màrquez, L., Barrón-Cedeño, A., Mohtarami, M., Karadzhov, G., & Glass, J. (2018). Fact checking in community forums. In *32nd AAAI Conference on Artificial Intelligence, AAAI 2018* (pp. 5309–5316).
28. Karadzhov, G., Nakov, P., Màrquez, L., Barrón-Cedeño, A., & Koychev, I. (2017a). Fully automated fact checking using external sources. In *International Conference Recent Advances in Natural Language Processing, RANLP, 2017-Sept (February 2018)* (pp. 344–353). https://doi.org/10.26615/978-954-452-049-6-046
29. Obrien, N., Latessa, S., Evangelopoulos, G., & Boix, X. (2018). The language of fake news: Opening the black-box of deep learning based detectors. In *Proceedings of the Thirty-second Annual Conference on Neural Information Processing Systems (NeurIPS)–AI for Social Good.*

Fake News and COVID-19 Pandemic

False Information in a Post Covid-19 World

Mohiuddin Ahmed, Chris Martin, Tristram Walker, and James Van Rooyen

Abstract The Covid-19 pandemic has greatly impacted the landscape of modern society. This holds implications when considering how false information may be used in a post Covid-19 world. When considering scenario that false information could affect such as medical information, vaccinations, and propaganda there is a clear need for deepened understanding and reaction as it happens, to combat potential catastrophic outcomes. Combining this with the stamping out of false information on social media platforms, as well as more traditional media outlets such as newspapers, reliable medical advice can be distributed in a concise and effective way being of great benefit to the wider public health. Understanding how false information could impact the post Covid-19 world through examining specific scenario and by extension the resulting impact, it is possible to prepare and potentially reduce the effect false information may have leading to a safer post Covid-19 society for all. In this chapter, we have highlighted different aspects of false information in the context of covid-19.

1 Introduction

When considering the future it is important to note that whilst we are facing and working through a pandemic, an infodemic is also prevalent [1]. The term infodemic is a phrase that was used first during the SARS epidemic and refers to the rapid distribution of information that is often false or misleading [1]. This infodemic is something of great importance when considering solutions to the current pandemic. Globally the infodemic and the pandemic must be stopped together otherwise there will not be a positive outcome for either. By stopping the rapid dissemination of false information, we can reliably trust professional health directives and public health will successfully strengthen. However, the infodemic is not something that is easy to stop and thus, later this paper will discuss misinformation detection techniques, and suggestions will be made as potential solutions to this problem.

M. Ahmed (✉) · C. Martin · T. Walker · J. Van Rooyen
School of Science, Edith Cowan University, Perth, Australia
e-mail: m.ahmed.au@ieee.org

M. Lahby et al. (eds.), *Combating Fake News with Computational Intelligence Techniques*, Studies in Computational Intelligence 1001,
https://doi.org/10.1007/978-3-030-90087-8_18

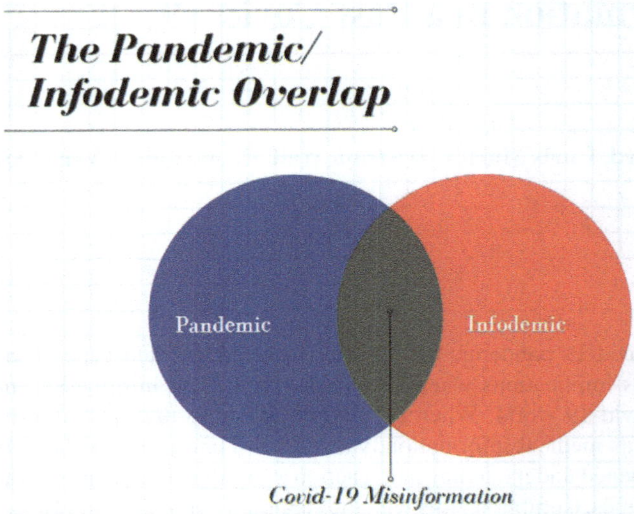

Fig. 1 A Venn diagram displaying the pandemic/infodemic overlap

Figure 1 shows the overlap between the infodemic and pandemic, demonstrating where Covid-19 misinformation can run free and cause havoc amongst the community. Understanding this overlap allows for one to focus on both separate issues to have a direct impact on the middle issue. Whilst efforts against Covid-19 misinformation are important, it can equally be argued that if focus is given to both the pandemic and infodemic separately it will naturally reduce Covid-19 misinformation. This could be in the form of vaccines reducing the risk and thus prevalence of the virus or considering the infodemic, the reduction of misleading information available to people so that correct sound information can be heard.

Medical misinformation as part of the ongoing infodemic surrounding Covid-19 is a large problem with fatal consequences. One such example of these fatal consequences occurred in Iran where over three hundred people died due to receiving information that drinking alcohol cures Covid-19, leading people to drinking counterfeit alcohol that contained the toxic substance methanol, leaving many others with serious health issues [2]. Furthermore, rumours about the curative effects of chloroquine and hydroxychloroquine have also had deadly consequences as people act on this information as they would sound medical advice. Medical misinformation can be so destructive due to the nature that health advice, or in this case false health advice created by cyber criminals or the like, can have a direct impact on someone, as it informs people on the way to act in the best interests of their health. Thus, with the panic that has ensued due to the Covid-19 pandemic, medical misinformation has had a breeding ground where rumour can circulate quickly (enhanced by modern day tools such as social media), reaching many people in a short time. Here it will become an opposition to correct health advice that may become lost in a sea of rumour. This is clearly not the desirable outcome as it has a large negative impact on society as people are unable to determine what is correct reliable information. When considering the outreach that medical misinformation has had during the Covid-19 pandemic one can look at key political and social figures and their spread of misinformation as a key driving force. Medical Doctor Nahid Bhadelia when speaking with American news company NBC states that Donald Trump seems to be creating discord regarding the development of vaccines and wreaking havoc due to his wide outreach [3]. A person with Trump's power now a former US President can have a huge impact on public health going forward if he continues to engage in this form of misinformation. Bhadelia further states how Trump is using misinformation to spread the message that the worst of the pandemic is over something considered by medical professionals to not be false [3]. This could be disastrous to the post Covid-19 world, where people to believe they are safe, when the reality is the pandemic still poses a great threat to the wider public health. If this kind of medical misinformation continues by people of high significance, there is no denying it will have an impact on the post Covid-19 world. In this case there is large concern for the post Covid-19 world as vaccines are the most effective way of fighting this type of large public health issue and misinformation that opposes this firstly puts people who do not get vaccinated as a result at immediate risk, however, also lowers the whole public health's immunity as it reduces the overall resilience making all people more vulnerable. In a post Covid-19 world we need to have the best possible immunity and anti-vaccination misinformation is a hurdle that must be overcome for the good of all public health.

1.1 Chapter Organization

Rest of the chapter is organized as follows. Section 2 highlights the impact of fake test results followed by false vaccination in Sect. 3. Propaganda, social media and off-line outlets are discussed in Sects. 4, 5 and 6. The chapter is concluded in Sect. 7.

2 Fake Test Results

Fake Covid-19 test results are becoming a large-scale issue as the misinformation being presented is creating growing concern for the flow on effect that may occur after presenting fake test results. This issue is happening in worldwide and examples can be seen in countries such as Nigeria where over thirty nine thousand travellers shunned getting proper COVID-19 tests putting countless lives at risk [4]. The Nigerian government showed particular concern over this matter as the false negative Covid-19 tests has a great impact on the countries health as people who have the virus can slip by unnoticed and spread it to the wider community once they leave the airport. For whatever reason they are presenting false test results is not of great importance, rather the fact that they have presented them and the wider health implications that their singular actions may result in.

Another example of this can been seen in Lagos, where the government has stated that they will not hesitate to prosecute people selling fake test results due to their destructive nature [5]. This clear plan to take severe action against the potential consequences of selling or buying false test results shows just how big the threat they pose is. It can be seen a somewhat flow on effect where one person with the virus buys a false test and then in the process is able to further spread the deadly disease without consequences, leading to more people becoming sick and potentially dying as a result.

Furthermore, a man travelling in Mississauga, Canada was charged after presenting false Covid-19 test results at the airport in hopes to travel [6]. This scenario is not uncommon with many countries reporting instances of fake test results at airports as people seek to travel at whatever cost it has to public health. Border Force services in the UK have reported that they are intercepting up to one thousand fake tests a day as people seek to enter the country [7]. Parsley also goes on to raise the issue that without guidelines and a standardised approach to what is and is not acceptable as a valid negative test result, there will always be ambiguity and people will gain access to countries without presenting valid test results [7]. This leads to the discussion of creating a unified test verification system that can be recognised and validated, however this poses its own challenges as different countries have different standards and us total uniformity is ultimately unreachable. Furthermore to this issue, there is nothing to make valid methods obtained verified by doctors completely fool proof, as doctors may collude with family members or friends (amongst other people) to produce fully verified fake tests that will pass through the system completely unnoticed. Thus, it is further realised that a state of total validation when it comes to fake test results is likely unrealistic.

This issue holds significant ramifications for a post Covid-19 world. Reliable test results are vital when considering how to reduce the spread of the virus and reach a point where society can return to a somewhat normal. Fake test results pose a large threat since the people presenting them clearly do not care about the public health risk they are creating by supplying them. Furthermore, it is not possible to create a system where all fake results can be immediately identified and thus it is likely that some will pass through validation processes unnoticed. This will lead to potential virus cases being unnoticed and by extension the virus spreading unknowingly amongst society and area of great concern. As we approach a post Covid-19 world it is important to understand the impact that fake test results have as mentioned above so nations can best prepare for their inevitable presentation.

3 False Vaccination

Vaccinations are a key defence against many diseases and the development of a vaccine to help contain the Covid-19 pandemic has been of great global interest and importance. However, misinformation regarding vaccines can have great negative effect on the potential rollout of these vaccines as people become misinformed regarding vaccine advice, thus making decisions based on said misinformation putting themselves and others at risk.

Figure 2 represents current COVID-19 vaccine acceptance from a survey conducted by Lazarus on a wide demographic, incorporating many nations and types of people [8]. Perhaps the most alarming statistic that can be gleamed from this is the change in the "Completely Agree" category. Such a sharp drop in this category when changing from "Generally Available" to "Employer Recommends" demonstrates how people's free will is important and they are much more likely to be accepting if they do not feel like they are being pushed into it. Furthermore, every other category saw an increase when recommended by an employer showing that it can in fact do more harm than good. It is important to understand how people accept vaccine information as shown above, as it allows a tailored solution to be developed, where vaccine information can be delivered in a way that improves public acceptance.

Looking at past examples clarifies what can become reality when misinformation regarding vaccines gains traction. This has been shown in the United States in 2015 where there were multiple measles outbreaks in schools in (but not limited to) California and Texas, as a result of students receiving vaccination exemptions after

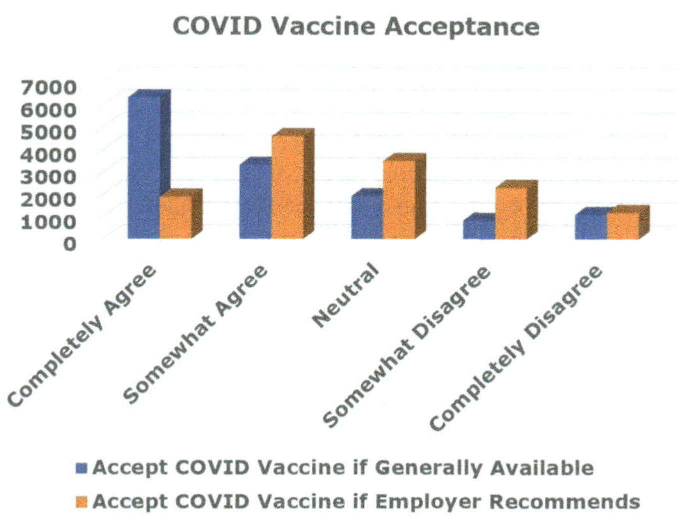

Fig. 2 Statistics showing current COVID Vaccine acceptance if available freely or recommended by employer (Survey conducted in 2020)

protesting school vaccination admittance policies [9]. The outbreaks occur due to one (or a number) of people who are not vaccinated contracting the virus and then from this person it spread to immunised people. Cases of measles outbreaks such as this are not limited to the United States with countries such as Samoa in 2019 where the measles virus was introduced to the community by a traveller from New Zealand causing an epidemic due to the low vaccine uptake [10].

A similar situation was seen in Japan with relation to HPV vaccines where initially an uptake of 70% was seen in 2010. However, due to misinformation rumours spreading in 2013 uptake fell to just 1% and the Japanese Government decided to stop recommending the vaccine all together [10]. In this case all the information leading to the drop in uptake within the community was misinformation that had no foundation in medical fact. Issues with the HPV vaccine have also been seen in countries such as Ireland and Denmark where similar antivaccination protests and the spread of misinformation have cause great concern for public health [10]. Therefore, when considering Covid-19 it is important from the perspective of the greater public health that every person is vaccinated to avoid this kind of scenario.

Due to misinformation surrounding vaccines however recent survey show that 40% of US citizens would choose not to have the vaccine [11]. This is particularly concerning as community uptake of the vaccine is vital to it being successfully effective. Vaccine misinformation has been circulated since the very being of the pandemic with people claiming the whole pandemic has been created to profit from selling a vaccine in the future [12]. This puts the fight against vaccine misinformation on the back foot with methods such as debunking myths and false information not always proving useful as people continue to believe misinformation they have been told whilst acknowledging the fact that is truly false [11]. It has been found however, that emphasizing medical consensus around vaccine safety appears to put people at rest and reduces the anxieties people have regarding vaccines resulting in a stronger stance to reject vaccine misinformation [11] a stance that would be greatly beneficial to public health.

A study conducted in Israel on 1941 subjects relating to Covid-19 has found some interesting results when regarding vaccine uptake. The study found doctors not directly associated with patients who have Covid-19 are likely to be more sceptical regarding vaccines, that nurses are more likely than physicians when considering vaccines and that males are more likely to be open to the uptake of vaccines (this is somewhat based on research indicating males have more violent reactions to the virus) [13]. Dror states that whilst vaccines are important from a herd immunity perspective, they are especially important in frontline workers due to their more direct contact with the virus, thus nurses posing resistance to vaccines is an issue of high priority [13]. If it is possible to better educate these people so they are not as susceptible to misinformation and antivaccination movements then positive change may occur. Otherwise, dangerous trends such as these will continue and herd immunity may become something that is not achievable, a large problem for the post Covid-19 world as herd immunity is a vital aspect of being able to have a liveable future with the virus. Furthermore, this article [13] goes on to state that the main issue people seem to have with the current vaccines is that people believe they are dangerous and unreliable due to being developed so quickly, however, this vaccine whilst being developed quickly was based on other models of SARS virus which greatly expedited the process [14] leading to a vaccine based somewhat on previous knowledge, thus reducing development time. This once again demonstrates the need for greater education to combat the sea of misinformation people are currently believing to progress into a safe post Covid-19 world.

A sign of positive action against the spread of misinformation regarding Covid-19 vaccines can be seen by examining social media companies' actions relating to user accounts and the information (or misinformation) they are spreading on their platforms. This has been evidenced in the fact that Instagram banned and removed the account of Robert Kennedy a prominent vaccine sceptic in February over misleading information regarding Covid-19 vaccines [15]. Kennedy had over three hundred thousand followers and thus, the spread of the misinformation he could produce was vast effecting large quantities of people and by extension the greater public health (Herrera, 2021). Furthermore, Twitter is now implementing warnings on posts that may be misleading ("Twitter Tackles COVID-19 Vaccine Misinformation," 2021) a sign of positive change on a platform that is rife with misleading information on a myriad of topics. Perhaps though the most significant step by an online company has come from Google, who have funded a number of projects and firms around the world as part of the GNI Vaccine Counter-Misinformation Open Fund. This fund seeks to help broaden fact checking software especially amongst populations that may be targeted with disproportionate amounts of misinformation. This fund is a step in the right direction looking to stop misinformation at its source and help the people targeted by it particularly the overexposed. Positive action like this is a must for all companies as we deal with the Covid-19 infodemic and Google is showing a leading example for others to follow.

Overall, vaccines are undoubtedly the most effective measure the post Covid-19 world can take to reduce the threat of the virus and improve public health. However, compliance is key and thanks to misinformation surrounding the safety and potential side effects of said vaccines uptake may not be welcomed by as many people as needed for effective herd immunity. Thus, it is vital that education programs are put in place to counteract this rapidly spreading misinformation so that vaccines become effective, otherwise it is clear to see what may happen if this is not the case, show above with outbreaks of measles in communities with low vaccination rates. Whilst some social media companies are taking action against the spread of misinformation, they cannot combat it alone and thus a group effort is needed to have a lasting positive result.

4 Covid-19 Propaganda

Misinformation lends itself well to the very essence of propaganda which Oxford Dictionary defines as information especially of a biased or misleading nature, used to promote a particular point of view. Thus, the dissemination of most misinformation is propaganda as it adheres to this definition. Propaganda surrounding Covid-19 is of great concern as once misinformation becomes weaponised to promote and persuade people to believing misleading and false information the implications can be catastrophic. The distinction that not all propaganda is misinformation and furthermore, not all misinformation is propaganda is important to consider when moving forward.

Covid-19 propaganda has taken many forms whether it be pertaining to the fact the virus is a hoax, that 5G signals are spreading the virus or that vaccinations are dangerous etc. These forms of propaganda can be weaponised and directed at target audiences for greater impact. In British Columbia fliers have been placed inside delivered newspapers promoting Covid-19 denial agenda a form of direct propaganda using misinformation to extend influence [16]. The culprits behind this specific propaganda attack unknown however links were provided amongst the fliers to websites that seek to disseminate misinformation regarding the pandemic. This could be seen as an indirect targeted propaganda attack; the culprits here are able to hide in the shadows as the fliers have no direct links back to them thus making this method a somewhat safe criminal practice due to the case there is no trail leading directly back to them. Publisher Chris Mackie states that unfortunately once the paper is dropped off to its location there is no way to please people picking it up or in this case fill it with propaganda [16], thus tracking and apprehending culprits is almost impossible.

More alarming than this however are the charges made by the Rome Doctors Guild to Italian doctors who have been found guilty of promoting anti-vaccination propaganda [17] a community that is of vital importance to the fight against this kind of propaganda. When considering how to best protect against the Covid-19 pandemic doctors are at the forefront as healthcare professionals and need to advocate for sound medical information the exact opposite of what these doctors have done. Due to the position of power doctors hold due to their vast knowledge and medical expertise the wider community trusts what they have to say. Thus, it is possible for people in this position of power to abuse it and support and/or disseminate misinformation propaganda regarding the pandemic to people who trust and in turn will believe them. This is particularly dangerous since people are more likely to act on information, they have received by people they trust and thus it is of utmost importance that medical professionals advocate for correct medical information. In this case the Rome Doctors Guild has taken precaution due to the destructive nature of some doctor's current behaviour and acted quickly to stop any further spread of propaganda a positive outcome for the greater public health. Another alarming occurrence of propaganda in relation to the Covid-19 pandemic has been conducted by the Central Propaganda Department of the Communist Party of China a group dedicated to the spread of propaganda to promote Chinese communist agenda. The publishing of "A Battle Against Epidemic: China Combating covid-19 in 2020" a book overseen by this organisation is a prime example of using propaganda to impact masses of people and influence their opinions. The text states how China is combating the pandemic effectively however has been pulled from online publishing sources as members of the propaganda party reconsider their approach to misinformation dissemination to garner the best possible public response [18]. Noting that China does not have free speech media it is an interesting country to observe as almost all public outreach from the government is propaganda used to press the countries agenda. It is important to note here that the Chinese government is under pressure due to how bad the pandemic has become and the fact that it is the believed origin of the virus. Thus, propaganda is being used as a powerful tool to promote that the country is successfully combating and will defeat the virus and that medical workers are warriors battling these harsh conditions when in reality the situation may be the complete opposite [18]. Seconding this claim that the reality of the situation may be completely different to the actuality Goldstein states that China has steeped to new lows with YouTube videos targeting the US response to the pandemic sprouting large amounts of misinformation regarding in relation to statistics regarding the pandemic, a direct propaganda attack meant to either confuse the view or oppose the current correct information [19]. It seems that Chinese agenda will remain the same as it always has even during the worldwide pandemic and the Central Propaganda Department of the Communist Party of China will go to great lengths through propaganda to achieve their goals.

Examining propaganda relating to the pandemic is of great importance when considering the fight for reliable information in the post Covid-19 world. By understanding the agenda of said propaganda it is possible to launch anti-propaganda campaigns that seek to undo the damage that has been done. Furthermore, through examination it is possible to discover trends amongst Covid-19 propaganda that may be useful to identify new propaganda as it begins to spread and thus, the level at which it can be disseminated can be reduced. Lastly, it is important to note that China will always pose threats regarding propaganda not only related to Covid-19, therefore it is of paramount importance that misinformation and by extension propaganda originating from China is seen as a threat and treated accordingly as it appears (as evidenced above) that the Central Propaganda Department of the Communist Party of China will continue its endeavours without consideration for the validity of what they are saying.

5 False Information in Social Media

Social media has become a breeding ground for misinformation particularly when pertaining to the Covid-19 pandemic. In a post Covid-19 world careful attention is needed into how information spreads and effects people through social media to improve not only public health, but people's general wellbeing. Social media is a fantastic tool and service, however, in the hands of cyber criminals it can be exploited and weaponised making it dangerous without many user's knowledge. The following section will aim to demonstrate the harm misinformation can have when paired with social media by first demonstrating instances of misinformation in social media that pertain to Covid-19 and then the resulting fallout. Previous examples have been listed in the sections above that are applicable to false information in social media, however this section will not cover these specific instances again.

Thinking of the main forms of social media that are currently used worldwide: Facebook, Twitter, Instagram and TikTok are the most used and thus have the widest outreach. An agreement could be made that YouTube also has a place amongst these however due to the limited social interaction aspects when compared to the others it will not be discussed here rather the focus will be on Facebook and Twitter as these two platforms have had the most prevalence in society during the time of the Covid-19 pandemic.

Websites who spread Covid-19 misinformation gained almost half a billion views from Facebook links in April 2020 alone [20] an extremely alarming statistic that demonstrates the current outreach misinformation is having through Facebook specifically. Furthermore, research conducted in Bangladesh suggests that even one piece of medical misinformation could lead to as many as eight hundred deaths a staggering figure when considering just how much medical misinformation is currently circulating [20]. Reports show that the highest viewed pieces of medical misinformation prevalent on Facebook have as many views as current reliable medical information an alarming prospect [20].

Due to the severity of the situation Facebook CEO Mark Zuckerburg along with CEOs for both Google and Twitter were brought before American congress in March 2021 to discuss what they as companies were doing to deal with this large-scale misinformation issue (Cao, 2021). Discusses at this meeting lead to all companies stating how they had successfully removed thousands of sources of misinformation from their services; however, this effort is clearly not resounding with huge success as the amount of misinformation still on these services is of great concern. It leads to one asking the question is it possible to completely block out misinformation pertaining to the pandemic in a post Covid-19 world which would a clear no, however with greater efforts by all parties it may be possible to reduce the impact that such misinformation has.

As a way of combating misinformation on Twitter, the company is trialling a service know as Birdwatch. Birdwatch allows users to write notes directly to tweets they believe are misleading with the hope people will read them and understand that the nature of the original post might be misleading [21]. Whilst still being in the testing phase and thus not in full community circulation, this solution poses a strong sustainable solution due to the fact it is community driven meaning work to debunk or remove misinformation is not solely a task for the service provider (as evidenced above, the issue is not something that can be managed entirely by the company alone due to its scale) allowing for more posts to be highlighted for potential readers. Another tool being developed currently called Bluesky is attempting to create open-source, decentralised standards for social media services [21]. If Bluesky is a success it could lead to the development of moderation algorithms which could be groundbreaking in the fight against misinformation present on social media post Covid-19.

An alarming prospect can be how algorithms employed by social media services may be worsening the problem. Facebook's current algorithms for disseminating information have contributed to health misinformation being viewed an approximate 3.8 billion times during 2020 alone [22]. When social media algorithms are exacerbating the problem to such a high degree, where does one draw the line and say these algorithms are causing more harm than good? Further to this does this make the social media services somewhat responsible for the infodemic and should they be held accountable? Whilst social media services such as Facebook are removing as much medical misinformation as they possibly can to combat the situation, are they making the infodemic situation worse by their own doing, where the number of posts removed are heavily outweighed by those given greater outreach by their own algorithms. It becomes a situation where the platform might be fighting a losing battle removing posts where they are creating more without their direct knowledge. This issue needs to be addressed with the highest importance and algorithms modified to prevent the further dissemination of medical misinformation, otherwise all the hard work that is currently being demonstrated by social media is for naught.

5.1 Evidence of Misinformation in Social Media

Thinking of the main forms of social media that are currently used worldwide: Facebook, Twitter, Instagram and TikTok are the most used and thus have the widest outreach. An agreement could be made that YouTube also has a place amongst these however due to the limited social interaction aspects when compared to the others it will not be discussed here rather the focus will be on Facebook and Twitter as these two platforms have had the most prevalence in society during the time of the Covid-19 pandemic.

Websites who spread Covid-19 misinformation gained almost half a billion views from Facebook links in April 2020 alone [20] an extremely alarming statistic that demonstrates the current outreach misinformation is having through Facebook specifically. Furthermore, research conducted in Bangladesh suggests that even one piece of medical misinformation could lead to as many as eight hundred deaths a staggering figure when considering just how much medical misinformation is currently circulating [20]. Reports show that the highest viewed pieces of medical misinformation prevalent on Facebook have as many views as current reliable medical information an alarming prospect.

Due to the severity of the situation Facebook CEO Mark Zuckerburg along with CEOs for both Google and Twitter were brought before American congress in March 2021 to discuss what they as companies were doing to deal with this large-scale misinformation issue [21]. Discusses at this meeting lead to all companies stating how they had successfully removed thousands of sources of misinformation from their services; however, this effort is clearly not resounding with huge success as the amount of misinformation still on these services is of great concern. It leads to one asking the question is it possible to completely block out misinformation pertaining to the pandemic in a post Covid-19 world which would a clear no, however with greater efforts by all parties it may be possible to reduce the impact that such misinformation has.

As a way of combating misinformation on Twitter, the company is trialling a service know as Birdwatch. Birdwatch allows users to write notes directly to tweets they believe are misleading with the hope people will read them and understand that the nature of the original post might be misleading [21]. Whilst still being in the testing phase and thus not in full community circulation, this solution poses a strong sustainable solution due to the fact it is community driven meaning work to debunk or remove misinformation is not solely a task for the service provider (as evidenced above, the issue is not something that can be managed entirely by the company alone due to its scale) allowing for more posts to be highlighted for potential readers. Another tool being developed currently called Bluesky is attempting to create open-source, decentralised standards for social media services [21]. If Bluesky is a success it could lead to the development of moderation algorithms which could be ground-breaking in the fight against misinformation present on social media post Covid-19.

An alarming prospect can be how algorithms employed by social media services may be worsening the problem. Facebook's current algorithms for disseminating information have contributed to health misinformation being viewed an approximate 3.8 billion times during 2020 alone [22]. When social media algorithms are exacerbating the problem to such a high degree, where does one draw the line and say these algorithms are causing more harm than good? Further to this does this make the social media services somewhat responsible for the infodemic and should they be held accountable? Whilst social media services such as Facebook are removing as much medical misinformation as they possibly can to combat the situation, are they making the infodemic situation worse by their own doing, where the number of posts removed are heavily outweighed by those given greater outreach by their own algorithms. It becomes a situation where the platform might be fighting a losing battle removing posts where they are creating more without their direct knowledge. This issue needs to be addressed with the highest importance and algorithms modified to prevent the further dissemination of medical misinformation, otherwise all the hard work that is currently being demonstrated by social media is for naught.

6 False Information in Media Forms Outside of Social Media

Whilst social media is becoming a leading force in the dissemination of false information in modern day society, it is important to also consider other media forms to gain further perspective on the issue. In fact, broadcast news has been seen to reach more people when regarding news content (shown in Fig. 3) Print news media is also an important source of information particularly for the older generation even though it ranks towards the bottom of most statistical analysis. A survey completed in Australia in 2020 on news media consumption [23] produced the following results:

Of high interest here is the fact that over 60% of people surveyed consumed free to air news media the highest of any category, followed second by radio media. This is of particular interest as these are traditional forms of media rather than digital media. Note that social media still comes in very high at over 40% but not as high as one would maybe expect. These results show the varied demographic of media

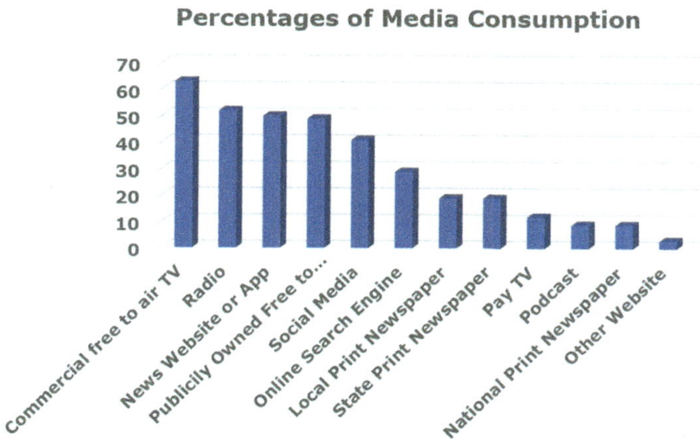

Fig. 3 Statistics on news media consumption in Australia in 2020

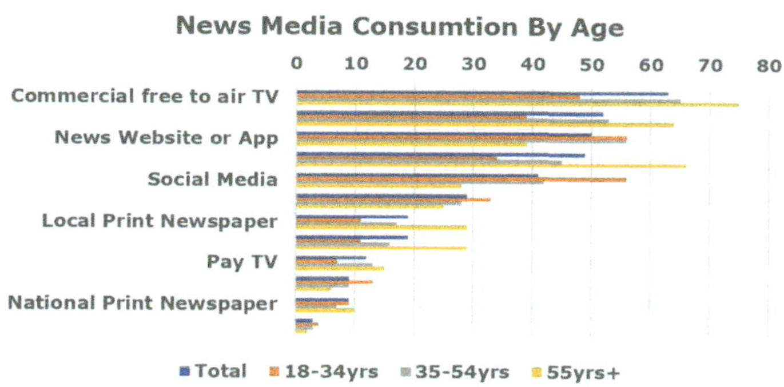

Fig. 4 News media consumption by age

consumption, demonstrating how people do not rely on just one media form for their news content [23]. Further to this Fig. 4 demonstrates the usage of these media types by age demographic:

Important aspects of this age demographic breakdown show how traditional media forms such as commercial free to air TV, Radio and in particular print news media are consumed the most by the over 55yrs bracket. In contrast to this digital media forms such as social media, online search engine and podcast are heavily dominated by the 18–34 age bracket. Thus, the differences between what types of news media are consumed by different generations can be seen. These differences highlight the need for a multifaceted approach to false information detection and reduction.

Readfearn states that in the fight against misinformation one must ensure people base their knowledge off expert advice and not succumb to rumour, however, determining what is expert advice can sometimes be misleading [24]. This pertains to all forms of media above and thus, attention to a broad range of information types is vital. As an example, it has been shown that the elderly in particular are far less likely to be open to mobile phone technology let alone social media [25]. Considering this, it is important to treat false information in all media forms to be equally damaging as they effect different groups, and the impact can be much larger than first anticipated. Understanding the demographics that consume each type of media is a positive step in the right direction to combat false information in all its forms.

6.1 Print Media

Newspapers are a form of print media outside the realms of social media. Whilst most news media is presented in digital formats today, print versions still circulate within communities and thus this physical media can be used to spread false information. Who is to stop an editor from writing an opinion piece within the paper rife with misinformation, pushing their views onto countless others who read it? Furthermore, often eye-catching headlines are subtle forms of false information as the actual article has little information pertaining to it, or it has been manipulated in such a way where it is taken out of context to mean something entirely different [26]. Understanding that newspapers print or digital are constructions of the companies who make them [24] also allows one to understand that their motives may not always align with dissemination of accurate reliable information. A key example of this would be print news media in China where the Central Propaganda Department of the Communist Party of China has a large influence on what is published, potentially changing it into nothing more than political propaganda. Developing across the board standards regarding information published in newspapers would be a positive step in the fight against this issue post Covid-19, however, this provides its own challenges as large scale, unified standards are hard to achieve within countries let alone across them. Thus, careful policing of what is published by authorities dealing with news media is perhaps the most suitable solution, although still flawed.

Another issue with print media such as newspapers has been touched on above, where once the paper has been delivered there is no regulating what others may be able to add to it. As seen in Canada, false information can be planted in a paper that has been delivered by the delivery person, or passers-by at random. This makes physical papers volatile to misinformation as criminals can plant false information after papers have potentially passed regulations regarding misinformation. Understanding this shows how easy it is for a positive checking system to be worked around and where there is a way criminals usually find it, thus making it a hard issue to comprehensively solve. Therefore, print media can be seen as volatile and dangerous even in the best of scenario a situation the reader should be aware of.

6.2 *Broadcast Media*

Television broadcast media is another form where propaganda and false information can spread with ease. Like social media and print newspapers, it has a wide outreach and viewers generally trust what they are consuming, thus leaving themselves vulnerable to false information. Here the responsibility lies with the producers of broadcast media such as news and talk shows, with positive steps being taken in instances such as the Sinclair Broadcast Group editing television show "America This Week" to not include pandemic misinformation included by the host [27]. Here, the producers took action against an individual who used his privileged position to spread false information, editing out sections to not subject users to thoughts that were not inline with expert advice and the companies views [27].

Taking another perspective China has banned many social media sources to allow for credible information to circulate through broadcast media even going as far to make Twitter only available through a government regulated VPN [28] a process that has seemingly hidden information from people in order to avoid chaos and ensure reliable information does not get lost (even if there broadcast media is driven by political interest.) Similarly this can be seen in Spain where information has been shielded from the population in order to reduce panic and the spread of rumour (both results of large scale misinformation spreading) [28]. However, this does lead one to question whether the censoring of information available to people crosses ethical boundaries, with the rights to obtain information. Furthermore, it leaves the government's free to disseminate the information they see fit, a dangerous prospect if they were to use that power to spread false information rather than that from reliable sources. Overall, an interesting approach that has seen some success when reducing the spread of false information, however there is potential for it to backfire if people are unable to access expert advice when needed.

Understanding that broadcast news media is usually trusted even though it is a construct of the people producing it [24] leads to further discussion, where people can become too trusting if false or not entirely true information is presented. It is the responsibility of those producing it (as evidenced in the "America This Week" example above) to ensure the information is sound and companies should be held legally accountable for behaviour that intentionally attempts to spread false information.

7 Conclusion

Traditional forms of media (that being outside of social media) present their own challenges when attempting to stem the spread of false information in a posit Covid-19 world. A tailored approach that targets each form of media is hence needed. Whether this be strict regulations, policing, and governance over what can and cannot be broadcasted or printed, or a more situational approach as things arise, it is of utmost importance to consider the demographic that still consumes these media forms, this being for example, that social media campaigns on the threats of false information within newspapers are likely to be less effective as the audiences do not always overlap. Targeted campaigns within the media type are thus arguably the best approach, where those who are vulnerable are already engaged thus their having attention is already achieved. Already having their attention, being able to draw one to understanding the dangers of false information is easier through advertisements and warnings thus exposing them to the threat they may not have previously perceived.

References

1. Niemiec, E. (2020). Covid-19 and misinformation. *EMBO Reports, 21*(11), e51420.
2. Love, J. S., Blumenberg, A., & Horowitz, Z. (2020). The parallel pandemic: Medical misinformation and covid-19: Primum non nocere. *Journal of General Internal Medicine, 35*(8), 2435–2436.
3. Nahid, B. (2021). Trump's covid misinformation is now mainstream. and winter is coming. https://www.nbcnews.com/think/opinion/trump-s-covid-misinformation-now-mainstream-winter-coming-ncna1247484. Accessed July 10, 2021.
4. Friday, O., & Adelani, A. (2021). Fg threatens sanctions as 39,070 travellers shun coronavirus test. https://punchng.com/fg-threatens-sanctions-as-39070-travellers-shun-coronavirus-test/. Accessed July 10, 2021.
5. Covid-19: Lagos raises the alarm over fake test results, inaugurates 10 oxygen centres. https://infoweb.newsbank.com/apps/news/. Accessed July 10, 2021.
6. Ashley Newport. Man charged with allegedly using fraudulent covid-19 document at pearson airport in mississauga. https://www.insauga.com/man-charged-with-allegedly-using-fraudulent-covid-19-document-at-pearson-airport-in-mississauga. Accessed July 10, 2021.
7. David, P. (2021). Passengers faking covid test results: Travel, Feb 13 2021. Copyright—Copyright 2021 I, All Rights reserved; Last updated—February 14, 2021.
8. Lazarus, J. V., Ratzan, S. C., Palayew, A., Gostin, L. O., Larson, H. J., Rabin, K., & ...& El-Mohandes, A. (2020). A global survey of potential acceptance of a COVID-19 vaccine. *Nature Medicine, 27*(2), 225–228.
9. Hotez, P., Batista, C., Ergonul, O., Peter Figueroa, J., Gilbert, S., Gursel, M., Hassanain, M., Kang, G., Kim, J. H., Lall, B., Larson, H., Naniche, D., Sheahan, T., Shoham, S., Wilder-Smith, A., Strub-Wourgaft, N., Yadav, P., & Elena Bottazzi, M. (2021). Correcting COVID-19 vaccine misinformation. *EClinicalMedicine, 33*, 100780.
10. Helen, P.-H., & Lisbeth, A. (2020). Impact of antivaccination campaigns on health worldwide: Lessons for Australia and the global community. *Medical Journal of Australia, 213*(7), 300.
11. van der Linden, S., Dixon, G., Clarke, C., & Cook, J. (2021). Inoculating against COVID-19 vaccine misinformation. *EClinicalMedicine, 33*, 100772.
12. Warren, C. (2020). Officials gird for a war on vaccine misinformation. *Science, 369*(6499).
13. Dror, A. A., Eisenbach, N., Taiber, S., Morozov, N. G., Mizrachi, M., Zigron, A., & ...& Sela, E. (2020). Vaccine hesitancy: the next challenge in the fight against COVID-19. *European Journal of Epidemiology, 35*(8), 775–779.
14. Li, Y. D., Chi, W. Y., Su, J. H., Ferrall, L., Hung, C. F., & Wu, T. C. (2020). Coronavirus vaccine development: From SARS and MERS to COVID-19. *Journal of Biomedical Science*, 27(1).
15. Instagram bans Robert f Kennedy Jr over COVID vaccine posts. https://www.bbc.com/news/world-us-canada-56021904. Accessed July 10, 2021.
16. Covid-19 denial propaganda littered around the north Okanagan. https://www.vernonmorningstar.com/news/covid-19-denial-propaganda-littered-around-the-north-okanagan/. Accessed July 10, 2021.

17. Day, M. (2020). Covid-19: Italian doctors are disciplined for anti-vaccination propaganda. *BMJ*, m4962.
18. China's propagandists are trapped by their own rhetoric. https://www.economist.com/china/2020/03/05/chinas-propagandists-are-trapped-by-their-own-rhetoric. Accessed July 10, 2021.
19. Goldstein: Propaganda video about COVID-19 a new low for China. https://torontosun.com/opinion/columnists/goldstein-propaganda-video-about-covid-19-a-new-low-for-china. Accessed July 10, 2021.
20. Facebook funnelling readers towards covid misinformation—study. https://www.theguardian.com/technology/2020/aug/19/facebook-funnelling-readers-towards-covid-misinformation-study. Accessed July 10, 2021.
21. Congress Grills Zuckerberg and Big Tech CEOs for Spreading COVID Misinformation. https://observer.com/2021/03/covid-vaccine-misinformation-facebook-tech-companies-testify/. Accessed July 10, 2021.
22. Health misinformation pages got half a billion views on Facebook in April. https://www.technologyreview.com/2020/08/19/1007383/health-misinformation-pages-got-half-a-billion-views-on-facebook-in-april/. Accessed July 10, 2021.
23. Media Content Consumption Survey. https://www.communications.gov.au/what-we-do/television/2020-media-content-consumption-survey. Accessed July 10, 2021.
24. Coronavirus overload: Five ways to fight misinformation and fear. https://www.theguardian.com/world/2020/mar/22/coronavirus-overload-five-ways-to-fight-misinformation-and-fear. Accessed July 10, 2021.
25. Fernández-Ardèvol, M. (2016). An exploration of mobile telephony non-use among older people. In *Ageing and technology* (pp. 47–66). Transcript Verlag.
26. Ecker, U. K., Lewandowsky, S., Chang, E. P., & Pillai, R. (2014). The effects of subtle misinformation in news headlines. *Journal of Experimental Psychology: Applied, 20*(4), 323–335.
27. Former fox news host spreads virus misinformation on his sinclair show. https://www.nytimes.com/2020/10/16/business/media/eric-bolling-sinclair-coronavirus.html. Accessed July 10, 2021.
28. Cegarra-Navarro, J. -G., Vătămănescu, E. -M., Martínez-Martínez, A. (2020) A knowledge hiding approach to cope with covid-19: A comparison between spain and china. In *2020 IEEE International Conference on Technology Management, Operations and Decisions (ICTMOD)*, pp. 1–6.

Applying Fuzzy Logic and Neural Network in Sentiment Analysis for Fake News Detection: Case of Covid-19

Bahra Mohamed, Hmami Haytam, and Fennan Abdelhadi

Abstract The pandemic we witnessed starting from December 2019, was accompanied by a significant rise in internet usage, and the social media, in particular, people were asked to stay at home to limit the spreading of the Covid-19 virus, this isolation made fake news a dangerous weapon that can directly harm people's wellbeing and encourage antagonism and racism. Considering the re-al danger of this misinformation and disinformation, fake news research witnessed a surge of contributions that apply machine learning models, deep learning, and sentiment analysis, but among these models and especially those that use sentiment analysis, we found that there is a lack of the integration of the fuzzy aspect of our language, which may give more details and accuracy to the detection of fake news. In this work, we extend the classification model from our previous work by combining a deep learning algorithm LSTM with fuzzy logic for sentiment-aware classification of fake news. We experiment with a dataset that contains over than 13 K of covid-19 text content al-ready labeled as being real or fake, and we compared the results of our model with the state-of-the-art methods that do not incorporate fuzzy logic and sentiment for fake news classification. and we observed that our approach yields better results.

Keywords Fake news · Fuzzy logic · Sentiment analysis · Long-shirt-term-memory (LSTM) · Natural language processing (NLP)

1 Introduction

Starting from December 2019, the date of which the first case of coronavirus (COVID-19) was identified in Wuhan, China. This virus has been announced as being a world pandemic on March 11, 2020 [1], and as of June 03, 2021, the statistics show that 236 countries and territories have been affected, and over 171 million cases that have been confirmed, and more than 3.69 million confirmed deaths [2]. Reason for

B. Mohamed (✉) · H. Haytam · F. Abdelhadi
LIST Department of Computer Science, Faculty of Sciences and Techniques, UAE, Tangier, Morocco

M. Lahby et al. (eds.), *Combating Fake News with Computational Intelligence Techniques*, Studies in Computational Intelligence 1001,
https://doi.org/10.1007/978-3-030-90087-8_19

which the Covid-19 virus is considered one of the most dangerous viruses seen by humanity.

Along with the recent flourish of information and communication technologies, and the surge of using social media platforms, fake news has found fertile ground for spreading across the globe. Also with this pandemic, harmful campaigns have exploited the public fear and the lack of information related to the virus to influence people's opinions and spread hatred and negativity, leading the world to fight against both, Covid-19 pandemic and the 'Infodemic', as stated by Tedros Adhanom Ghebreyesus [3], "We're not just fighting a pandemic; we're fighting an infodemic".

Fake news dilemma has started a long before the rise of information technologies and social media [4], but the researches that aim to tackle this problem has gained a lot of attention back to the 2016 US presidential election [5, 6], many contributions have opted for applying machine learning classification algorithms, account analysis, and sentiment analysis, however, the used models in sentiment analysis ignore to consider the fuzzy linguistic character of our language, which can bring more insight and valuable information, to this end we propose a model based on our continuous work [7], in which we combine the LSTM algorithm with fuzzy logic to apply sentiment analysis classification of fake news. we experiment with a dataset of over 13,000 labeled content as real or fake news.

In our work we introduce a new approach that incorporate fuzzy logic linguistic functions [8] as an important feature to get more insight about the expressed sentiment, and applied a specific type of recurrent neural network called LSTM, to tackle fake news classification problem. Also, we evaluate the benefits of incorporating fuzzy logic and sentiment scoring for fake news classification, against a series of machine learning and deep learning models used to treat the same task, and we found that our model yields to enhancing the classification results.

2 Related Work

The term of fake news can be related to both misinformation, e.g., information shared by well-meaning by misguided or misinformed individuals, and disinformation, that is false information intentionally created to mislead people to believe in false information or lead them to act a certain way which may cause tangible threats to the society. Kalsnes [9] has extended the definition of fake news, their work includes the concern of fake news, its origins, circulation, and countering disinformation spreading. A survey of fake news by Zhou and Zafarani [10], categorize fake news researches as two fundamental theories: the news-related theory in which researchers analyze the characteristics of fake content compared to real one [11, 12], and user-related theory that focus on analyzing the accounts posting misleading information [13, 14].

To help researchers tackle fake news, a well-known dataset was released by Wang [15], which contain around 12.8 K short statements collected from POLITIFACT.COM and manually labeled, another dataset was proposed by Nakamura et al. [16], providing over 1 million of samples from multiples categories (i.e., text, image, metadata, and comments data). Also, the authors of FakeNewsNet [17], introduced a data repository containing multi-dimension information from news content, social context, and dynamic information, annotated by journalists and domain experts. The ISOT fake news dataset [18, 19] is another well-known dataset containing more than 23 k of fake news collected from the various unreliable website, and over than 21 k of truthful articles curated from "Reuters",[1] a crowd-sourcing helped with fact-checking BREDBANK, a twitter-specific dataset of real-world events, released by Mitra and Gilbert [20]. Amidst the Covid-19 pandemic, Patwa et al. [21], released a dataset that includes 10,700 social media posts and articles of real and fake news related to coronavirus, the data was manually annotated based on verified Twitter accounts and fact-checking websites respectively, another work by Vijjali et al. [22], curated a Covid-19 related dataset of true and false claims with their explanation pairs, the authors collected about 5500 false-claim and explanation pairs from "Poynter",[2] and then manually rephrase these false-claims to have a true claims, CoAID [23] a Covid-19 healthcare misinformation dataset that includes fake news, collected from websites and social platforms, with users' social engagement about such news.

Along with fake news datasets, various approaches were applied to detect and deter the propagation of unreliable information, Granik and Mesyura [24] used a Naïve Bayes classifier to classify fake news, and their results conclude that artificial intelligence techniques are good for tackling fake news detection issue, the authors of Thota et al. [25], experiment with three variations of neural network architectures, that are: (a) Term Frequency-Inverse Document Frequency (TF-IDF) Vectors with Dense Neural Network, (b) Bag of Words Vector with Dense Neural Network, (c) a pre-trained word embedding's "Google Word2Vec" with a neural network, their results shows that using the approach (a) lead to better results, the work presented in Balwant [26] describes the implementation of a hybrid model based on a Bidirectional LSTM and Convolutional neural network (CNN), and incorporate news text content and user profile information, their work experiment with applying Bidirectional LSTM based on Part-Of-Speech tags on news content, and using CNN with user profile information, and finally a hybrid model that use a fusion of news content and user profile information. Sentiment analysis is also another research area used to combat fake news, in Zaeem et al. [27], the authors model the association between the expressed sentiment and the veracity of the content, and they statistically verify the existence of a relationship between negative emotion and fake news also with positive emotion and reliable news. Also Ajao et al. [28] confirm the existence of a relationship between sentiment and fake news/rumors, by using emotional words in the classification feature set given to machine learning classifiers, leading to an improvement over the state-of-the-art algorithms that does not include opinioned words. However, most used sentiment analysis algorithms, does not consider the fuzzy aspect of human language, and conclude that the sentiment expressed by a person is binary (i.e., positive or negative), which is not always the case, in our previous work [1], we propose a model that incorporate fuzzy linguistic hedges in

[1] https://reuters.com

[2] https://www.poynter.org/ifcn-covid-19-misinformation/

sentiment analysis process for a deep extraction of expressed sentiment, and we will incorporate our fuzzy logic model in this present work along with the LSTM algorithm. This combination is to the best of our knowledge, the first time it would be examined in the context of fake news detection.

3 Main Concepts

In this chapter, we will discuss several concepts that we explored and incorporated into our proposed approach.

3.1 Sentiment Analysis

Also known as opinion mining, is a subdomain of natural language processing, which evolves extracting people's expressed sentiment from a given text, this field has developed rapidly due to the growth of data generated by users on the internet, which touch a variety of field such as, technologies, healthcare, business, politics and more. Alongside this rich data generated, a variety if algorithms were developed to automatically explore and identify expressed opinions. Figure 1 describes a hierarchy of some of the state-of-the-art approaches used for opinion mining.

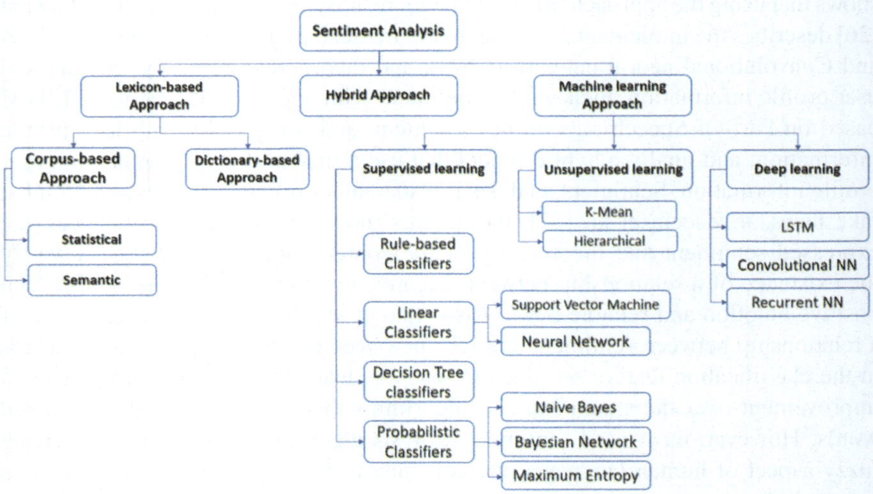

Fig. 1 Hierarchy of some of the approaches used in sentiment analysis fields

3.2 Fuzzy Logic

Fuzzy logic is not a recent theory, but it is back to 1965, when Lotfi Zadeh, also known as the "father" of fuzzy logic, introduced the concept of fuzzy set, in his paper [29], which became a pioneering paper in the development of fuzzy logic theory.

Fuzzy logic is the precise logic of imprecision and approximate reasoning, it aims at modeling human capability of taking rational decisions based on uncertain, inexact, or not totally reliable knowledge from the environment. In the context of sentiment analysis classification, fuzzy logic is passing from the classical bivalence system of absolute truth (i.e., positive or negative sentiment) to considering the degree of positivity or negativity expressed. In our proposed approach we incorporate the use of fuzzy linguistic hedges to tackle sentiment classification. Hereafter, we describe some used vocabulary of fuzzy logic.

Fuzzy set: Is an extension of the classical set, in which every element has a degree of membership ranging between zero and one, assigned by a membership function.

Membership function: As shown in Fig. 2, the membership function defines the fuzziness of the fuzzy set, and assigns a degree of membership to each element of the fuzzy set (Fig. 3).

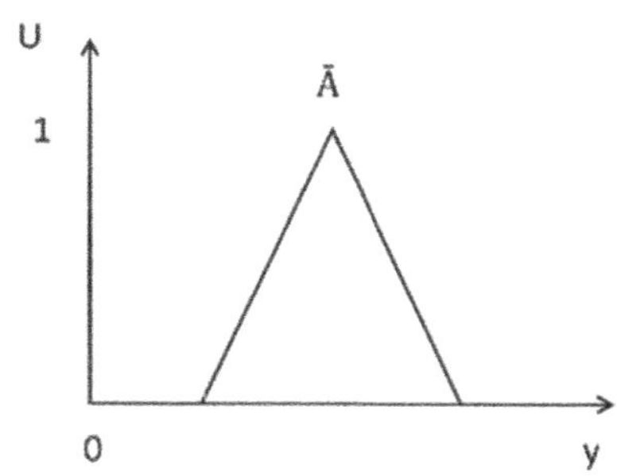

Fig. 2 Membership function for fuzzy set Ā

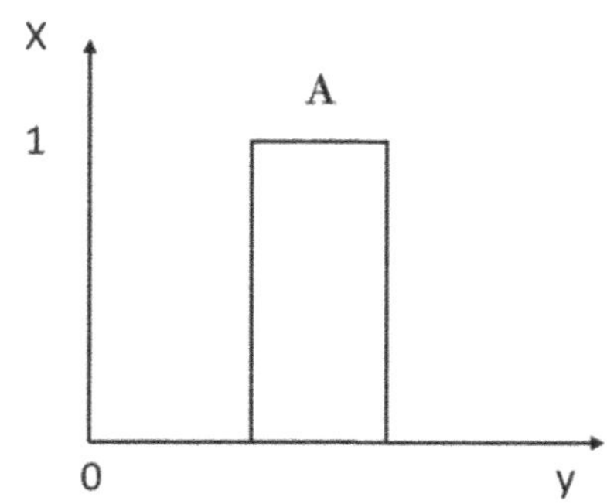

Fig. 3 Membership function for classical set A

$$\left\{ f_A(x) : U \to [0,1], \text{ where } \begin{array}{r} f_A(x) = 1, \text{ if } x \text{ is totally in } A \\ f_A(x) = 0, \text{ if } x \text{ is not in } A \\ 0 < f_A(x) < 1, \text{ if } x \text{ is partly in } A \end{array} \right\} \tag{1}$$

Where f is the membership function that associate to each point x in the universe of discourse U, a real number in the range of [0,1], where the value of $f_A(x)$ at represent the grade of membership of x in A.

Linguistic variable: Is a variable that has words or sentences as values but not a number, this concept was introduced by Zadeh [30–32], in case of a sentiment variable, it is considered a linguistic variable if its values are linguistic (e.g., strongly negative, negative, neutral, positive and strongly positive) rather than numerical.

Linguistic hedges: Are considered as operators that modify the shapes of fuzzy sets by the mean of terms such as "very, more or less", in terms of sentiment extraction, linguistic hedges can modify the strength and the meaning of the opinionated phrase by the presence of a linguistic hedge such as modifiers (e.g., Not), a concentrator like "very" or "extremely" or a dilator (e.g., almost, quite…).

3.3 Long-Shirt-Term-Memory (LSTM)

Introduced by Hochreiter and Schmidhuber [33], is a special type of Recurrent Neural Network (RNN), which was designed with the capability of learning long-term dependencies, by the mean of the cell state that is used for remembering information for a long period of time, which result in context-awareness. This model is widely used in the areas that involve sequential data and the learning of order dependency in a sequence, such as a machine translation, sentiment analysis, speech recognition, text generation, and more.

4 Proposed Approach

In this section we will start by presenting our dataset, and then we will proceed by presenting the proposed approach.

4.1 Dataset Description

In order to test the effectiveness of our approach, we avoid using fake news datasets related to other domains such as politics and economics, this is because we wanted to construct a reliable knowledge base that includes recent and common vocabulary used to describe the Covid-19 pandemic, to this end, we used a combination of labeled

Table 1 Example of real and fake news from the dataset we used as our corpus

Dataset	Content	Label
Infodemic dataset	Current understanding is #COVID19 spreads mostly from person to person through respiratory droplets produced when a person coughs or sneezes similar to how flu spreads	Real
Infodemic dataset	Says to leave objects in the sun to avoid contracting the coronavirus	Fake
Infodemic dataset	Focus on good nutrition as a part of self-care during the #COVID19 pan-demic. Certain vitamins & minerals may have effects on how the immune system works to fight off infections & inflammation. You can obtain these nutrients through #food. Learn more: https://bit.ly/3iONzET	Real
CoAID dataset	The WHO stated that asymptomatic spread of COVID-19 is "very rare" therefore physical distancing and face masks are not necessary	Fake
CoAID dataset	even with new safety guidelines in place health experts say going to movies is very risky during pandemic.	Real
CoAID dataset	The coronavirus pandemic can be dramatically slowed, or stopped, with the immediate widespread use of high doses of vitamin C	Fake

datasets, "Infodemic" dataset collected by Patwa et al. [21] and CoAID [23]. At the time of writing this chapter the "Infodemic" dataset is private, and we managed to communicate with the authors to get access to its content, and as of CoAID dataset, its content has a public access. Table 1 shows some data examples from these datasets.

4.2 *Dataset Statistics*

We gathered a dataset that contains 13,012 of real and fake news posts and articles related to Covid-19, from both labeled datasets [21, 23]. Our corpus is class-wise balanced (i.e., 51.92% of real news samples, and 48.07% of fake news). Table 2 resumes the class-wise distribution of train, validation, and test splits of the dataset.

Table 2 Distribution of data splits used for training, validation and test of our proposed approach

Split used	Fake content	Real content	Total
For training	3718	4018	7736
For validation	1269	1369	2638
For testing	1269	1369	2638
Total	6256	6756	13,012

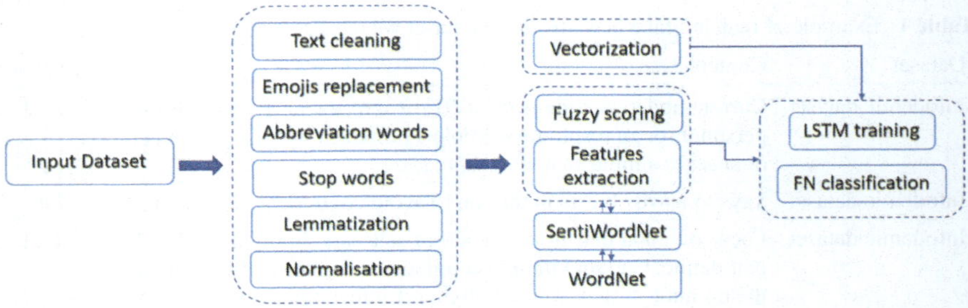

Fig. 4 The architecture of our proposed approach

4.3 Proposed Approach

Our approach consists of three major stages: Text preprocessing, vectorization and fuzzy sentiment scoring, and finally model training and fake news classification. Figure 4 describes the component of our architecture.

4.3.1 Text Preprocessing

In our pipeline we started by a text preprocessing, as a major step for natural language processing, in order to train our model on a clean and noiseless dataset, which will result in getting more meaningful and accurate results, our text preprocessing stage consist of the following phases:

Text cleaning: in this phase, we start by removing URLs and punctuation from the input data, also we further remove special characters to get a cleaned text that will be passed to the following preprocessing phase.

Emojis replacement: emojis are a visual representation of emotions, and as they are more representative than words, their usage has become more popular as a sort of communication, but as natural language deals with words rather than images, we proceed by replacing these emojis by their word's significance.

Abbreviation words: this step consists of replacing abbreviation words like "Gr8, IDK" with "Great, I don't know" respectively and we further process contractions words (e.g., "I'll become I will", "I don't become I do not"), to tackle the principle of least effort used in language expression.

Stop words: we used an English corpus of stop words provided by a natural language toolkit called "NLTK[3]" to remove stop words from the text.

Lemmatization: in which we convert words to their common base form known as lemma, by removing inflectional endings from the words. This operation was done using the same language toolkit "NLTK".

Normalization: in this phase, we process by lowercasing the text, for further analysis.

---

[3] https://www.nltk.org/

4.3.2 Vectorization and Fuzzy Sentiment Scoring

Vectorization: In this phase, we transform our text data into a vector format so it can be used by the LSTM algorithm, for this process we used a python library Gensim [34], to implement word embedding which is a technique capable of learning words relations, and capturing context, semantic and syntactic similarity of words in a document.

We used Word2Vec[4] in order to vectorize our text data, and we used the output as an initial weight for our model.

Fuzzy sentiment scoring: we extract the sentiment degree as a feature that indicates to our model the degree of sentiment expressed within the text, the process of extracting the sentiment degree pass by first identifying opinionated words, and their linguistic hedges using Part-Of-Speech tagger, then we use SentiWordNet[5] and WordNet[6] dictionaries to associate polarity to these opinionated words as an initial score value $f(\boldsymbol{u}_s)$, later on, we modify these values based on the existence and the type of linguistic hedges found, using the fuzzy logic functions proposed by Zadeh to get the final sentiment score as follow:

- For an opinion word has a preceding by a complement hedge (e.g., "Not"), the fuzzy score is modified using (2):

$$f(\boldsymbol{u}_s) = 1 - (\boldsymbol{u}_s) \tag{2}$$

- If the hedge is a concentrator (e.g., "extremely"), we proceed using (3):

$$f(\boldsymbol{u}_s) = [\boldsymbol{u}_s]^2 \tag{3}$$

- If the hedge is a dilator (e.g., "somewhat"), the modified fuzzy score is deduced using function (4):

$$f(\boldsymbol{u}_s) = [\boldsymbol{u}_s]^{1/2} \tag{4}$$

The final fuzzy score is normalized and used as a sentiment feature in our fake news classification model.

---

[4] https://radimrehurek.com/gensim/models/word2vec.html

[5] http://ontotext.fbk.eu/sentiwn.html

[6] https://wordnet.princeton.edu/

5 Results and Discussions

In order to test the effect of incorporating fuzzy logic and sentiment analysis for fake news classification, we test the results of using our model against the state-of-the-art algorithms used in text classification tasks, which are: Support-Vector-Machine (SVM), Naïve Bayes (NB), Random Forest (RF), extreme gradient boosting (XG-Boost), Long-Shirt-Term-Memory (LSTM), and these tests were carried out using the validation as well as test datasets.

The performance of our proposed model is evaluated using: (a) Accuracy, (b) Precision, (c) Recall and (d) F-measure metrics, and considering the results as positive, when the classifier classifies the fake content as fake, then:

- True Positive (TP): is the number of news content correctly classifier as fake.
- False Positive (FP): is the number of news content not correctly classified as fake.
- True Negative (TN): is the number of news content correctly classified as real.
- False Negative (FN): is the number of news content not correctly classified as real.

As of:

(a) *Accuracy:* shows the ratio of the correctly classified news to all the dataset. (i.e., how many news contents was correctly classified), and it is calculated using (5).

$$\frac{TP + TN}{TP + TN + FP + FN} \tag{5}$$

(b) *Precision:* that indicate the cost of FP (i.e., real news that was classified as fake news), and it is found by the following formula.

$$\frac{TP}{TP + FP} \tag{6}$$

(c) *Recall:* which highlights the cost of FN (i.e., fake news that was not correctly classified as fake. The recall of a classifier is given by (7).

$$\frac{TP}{TP + FN} \tag{7}$$

(d) *F-measure:* is a measure that considers both the precision and recall metrics, and it is calculated using (8):

$$\frac{2 * (Recall * Precision)}{(Recall + Precision)} \tag{8}$$

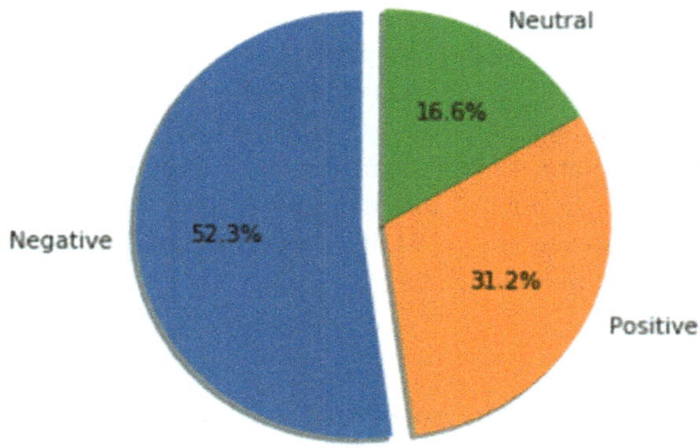

Fig. 5 Sentiment polarity using our fuzzy sentiment scoring

5.1 Results

We started by applying our fuzzy sentiment scoring on the dataset to get some insight related to the characteristics of the content of our fake news dataset, Fig. 5, shows positive, negative, and neutral sentiment percentages, and we observed that neutral content has the lowest value, which can tell that fake and real news tend to contain more opinionated words which can be explained by a way of gaining people engagement.

As stated previously, we used Long-Shirt-Term-Memory (LSTM) algorithm along with fuzzy sentiment scoring as an incorporated feature to tackle fake news classification. we did some experiments on the dataset to define the hyperparameters to be used for our model, and we set the value of 1400 as max_feature for our model based on the length of the text after vectorization, we also pass the fuzzy sentiment feature and word vectors as initial weights, moreover, we used 128 units for our LSTM, and the "sigmoid" as the activation function, and we compile the model using "adam" optimizer and "binary_crossentropy" as the loss function. And we compared the results obtained by our approach with the state-of-the-art algorithms already mentioned. Table 3 provides summary results in terms of accuracy, precision, recall, and F-measure metrics.

Table 3 Summary results of the proposed approach and the state-of-the-art algorithms

Approach\metrics	Accuracy	Precision	Recall	F-measure
Support victor machine	89.56	89.58	89.56	89.56
Naïve bayes	74.80	74.81	74.80	74.80
Random forest	81.24	81.25	81.24	81.24
Extreme gradient boosting	80.13	80.14	80.13	80.13
LSTM (without fuzzy sentiment)	90.21	90.24	90.25	90.24
LSTM (with fuzzy sentiment)	91.40	91.43	91.43	91.43

The experiments were carried on the same test and validation datasets, as shown in Table 3, using the deep learning model (LSTM) without fuzzy sentiment scoring, which perform better than other categories of classifiers, this can be explained by the mean of context-awareness implemented by Long-Shirt-Term-Memory algorithm. Furthermore, when integrating fuzzy linguistics hedges for sentiment scoring as a feature, we noticed that the performance of the proposed approach has improved with an accuracy of 91.40, which can tell that this feature has an impact on the classification model.

6 Conclusion and Future Work

The coronavirus also known as the Covid-19 pandemic has affected the way we live our everyday lives, people were obliged to stay at home in order to limit the spread of this dangerous virus, which results in massive use of internet services, and especially social media platforms like Facebook and Twitter, and news website. This usage has been noticed by malicious users that try by the mean of fake news to deceive and influence people's views. To this end, we proposed a model to tackle the fake news problem, in which we used a context-awareness deep learning model known as Long-Shirt-Term-Memory (LSTM) and we incorporated a fuzzy sentiment scoring feature, calculated using linguistics hedges proposed by Zadeh. And we experiment with a dataset with over 13,000 of fake and real news text content, and the results were compared against the state-of-the-art models. The proposed approach shows that our model gives better results, and therefore may be used for further research.

For our future work, we aim to integrate another source of sentiment extracted from images and implement another deep learning algorithm such as Bidirectional Encoder Representations from Transformers (BERT) to tackle fake news classification.

References

1. World Health Organization. (2020) WHO Director-General's opening remarks at the media briefing on COVID-19—11 March 2020. Available at https://www.who.int/dg/speeches/detail/who-director-general-s-opening-remarks-at-the-media-briefing-on-covid-19---11-march-2020. Accessed 12 Mar 2020.
2. Pan American Health Organization/World Health Organization. Series. "COVID-19 Daily Updates", Collection: "COVID-19 Reports". Available at https://iris.paho.org/handle/10665.2/54169.
3. World Health Organization, Director-General of the World Health Organization (WHO) at a gathering of foreign policy and security experts in Munich. Available at https://www.who.int/director-general/speeches/detail/munich-security-conference.
4. Burkhardt, J. M. (2017). Chapter 1 history of fake news. *Library Technology Report, 53*(8), 5–9.
5. Allcott, H., & Gentzkow, M. (2017). Social media and fake news in the 2016 election. *Journal of Economic Perspectives, 31*(2), 211–36.

6. Howard, P. N., Bolsover, G., Kollanyi, B., Bradshaw, S., Neudert, L. M. (2017). Junk news and bots during the US election: What were michigan voters sharing over Twitter. *Comprop Data Memo, 1*
7. Bahra, M., Fennan, A., Bouktaib, A., & Hmami, H. (2019) Smart city services monitoring frame-work using fuzzy logic based sentiment analysis and apache spark. In *1st International Conference on Smart Systems and Data Science (ICSSD)*, 3–4 Oct 2019. https://ieeexplore.ieee.org/document/9002687.
8. Zadeh, L. A. (1972). A fuzzy-set-theoretic interpretation of linguistic hedges. *Journal of Cybernetics, 2*(3), 4–34.
9. Kalsnes, B. (2018). Fake news. In *Oxford Research Encyclopedia of Communication*. Available at http://dx.doi.org/https://doi.org/10.1093/acrefore/9780190228613.013.809.
10. Zhou, X., & Zafarani, R. (2020). A survey of fake news: fundamental theories, detection methods, and opportunities. *ACM Computing Surveys, 53*, 1–40.
11. Monther, A., & Alwahedi, A. (2018). Detecting fake news in social media networks. *Procedia Computer Science, 5*(141), 215–222.
12. Ahmad, I., Yousaf, M., Yousaf, S., & Ovais Ahmad M. (2020). Fake news detection using machine learning ensemble methods. In *Complexity*.
13. Shu, k., Zhou, X., Wang, S., Zafarani, R., & Liu, H. (2019). The role of user profile for fake news detection. In *ASONAM '19: Proceedings of the 2019 IEEE/ACM International Conference on Advances in Social Networks Analysis and Mining* (pp. 436–439).
14. Espinosa, S., Centeno, M., Rodrigo, R. (2020). Analyzing user profiles for detection of fake news spreaders on Twitter, In *CLEF*
15. Wang, W. Y. (2017) "Liar, Liar Pants on Fire": A new benchmark dataset for fake news detection. In *Proceedings of the 55th Annual Meeting of the Association for Computational Linguistics* (Vol. 2, pp. 422–426).
16. Nakamura, K., Levy, S., & Wang, W. Y. (2020). r/Fakeddit: A new multimodal benchmark dataset for fine-grained fake news detection, In *Proceedings of the 12th Conference on Language Resources and Evaluation (LREC 2020)* (pp. 6149–6157).
17. Shu, K., Mahudeswaran, D., Wang, S., Lee, D., & Liu, H. (2020). FakeNewsNet: A data repository with news content social context and spatialtemporal information for studying fake news on social media. *Big Data, 8*(3), 171–188.
18. Hadeer A., Issa T. & Sherif S. (2018). Detecting opinion spams and fake news using text classification. *Security and Privacy, 1*(1).
19. Hadeer, A., Issa, T., & Sherif, S. (2017) Detection of online fake news using n-gram analysis and machine learning techniques. In *Intelligent, secure, and dependable systems in distributed and cloud environments* (Vol. 10618, pp. 127–138). Springer.
20. Tanushree, M., & Gilbert, E. (2015) CREDBANK: A large-scale social media corpus with associated credibility annotations. In *Ninth International AAAI Conference on Web and Social Media*.
21. Patwa, P., Sharma, S., Pykl, S., Guptha, V., Kumari, G., Akhtar, Md. S., Ekbal, A., Das, A., & Chakraborty, T. (2021). Fighting an infodemic: COVID-19 fake news dataset. In *Communications in Computer and Information Science* (pp. 21–29).
22. Vijjali, R., Potluri, P., Kumar, S., & Teki, S. (2020) Two stage transformer model for COVID-19 fake news detection and fact checking. In *ArXiv.* abs/2011.13253.
23. Cui, L., & Lee, D. (2020). CoAID: COVID-19 healthcare misinformation dataset. In *ArXiv.* abs/2006.00885.
24. Granik, M., & Mesyura, V. (2017) Fake news detection using naive bayes classifier. In *2017 IEEE First Ukraine Conference on Electrical and Computer Engineering (UKRCON)* (pp. 900–903).
25. Thota, A., Tilak, P., Ahluwalia, S., & Lohia, N. (2018). Fake news detection: A deep learning approach. *SMU Data Science Review, 1*(3), 10.
26. Balwant, M. K. (2019) Bidirectional LSTM based on POS tags and CNN architecture for fake news detection. In *2019 10th International Conference on Computing, Communication and Networking Technologies (ICCCNT)* (pp. 1–6).

27. Zaeem, R. N., Li, C., & Barber, K. S. (2020) On sentiment of online fake news. In *2020 IEEE/ACM International Conference on Advances in Social Networks Analysis and Mining (ASONAM)* (pp. 760–767).
28. Ajao, O., Bhowmik, D., & Zargari, S. (2019) Sentiment aware fake news detection on online social networks. In *ICASSP 2019–2019 IEEE International Conference on Acoustics, Speech and Signal Processing (ICASSP)* (pp. 2507–2511).
29. Zadeh, L. A. (1965). Fuzzy sets. *Information and Control, 8*(3), 338–353.
30. Zadeh, L. A. (1975). The concept of a linguistic variable and its application to approximate reasoning—I. *Information Sciences, 8*(3), 199–249.
31. Zadeh, L. A. (1975). The concept of a linguistic variable and its application to approximate reasoning—II. *Information Sciences, 8*(4), 301–357.
32. Zadeh, L. A. (1975). The concept of a linguistic variable and its application to approximate reasoning—III. *Information Sciences, 9*(1), 43–80.
33. Hochreiter, S., & Schmidhuber, J. (1997). *Long Short-Term Memory, 9*(8), 1735–1780.
34. Radim, R., & Sojka, P. (2010) Software framework for topic modelling with large corpora. In *Proceedings of LREC 2010 workshop New Challenges for NLP Frameworks* (pp. 46–50). University of Malta.

Analyzing Deep Learning Optimizers for COVID-19 Fake News Detection

Ayan Chakraborty and Anupam Biswas

Abstract In this time of COVID-19 crisis, the threat posed by the propagation of misinformation leading to mistrust needs to be kept in check. Misinformation related to the vaccines, remedies, false symptoms, etc. are spiraling out of control. We might not be able to directly put a stop to the flow or spread of fake news to a large extent at the moment, but it may be able to identify it as such with the help of Natural Language Processing (NLP) tools and Deep Learning (DL) algorithms. Steps involved in achieving this goal can be narrowed down into collection and analysis of data from various sources, sorting out the articles as covid-relevant and categorizing them as real or fake using DL models. However, DL models use different optimizers in the learning process, which plays an important role in identifying the fake news. This chapter also compares the efficiency of different optimizers in the context of COVID-19 fake news detection using DL models. The newly developed Continuous Coin Betting (CoCoB) Optimizer for DL studied extensively for fake news detection and performed compared with four other widely used optimizers. The comparative analysis shows the CoCoB as well as popularly used Adam optimizers are quite effective in finding optimal classification results for detection of fake news related to COVID-19.

Keywords Deep learning · Fake news · Misinformation · COVID-19

1 Introduction

Over the past decade, social media has climbed up the ladder of connectivity and now carries the title of being one of the major sources of information that people come across on a daily basis. Yet, the news dissipated across social media falls on

---

A. Chakraborty
Department of Electronics and Communications, Tezpur University, Tezpur, India

A. Biswas (✉)
Department of Computer Science and Engineering, National Institute of Technology Silchar, Silchar, India
e-mail: anupam@cse.nits.ac.in

M. Lahby et al. (eds.), *Combating Fake News with Computational Intelligence Techniques*, Studies in Computational Intelligence 1001,
https://doi.org/10.1007/978-3-030-90087-8_20

the relatively unfiltered and unverified end of the spectrum and many would argue it to be quite unreliable due to the spread of rumors and misinformation. According to an article published by MIT News Office on March 8, 2018, [1] false news disperses much faster, deeper and farther in all categories of information and the prime agents are not in-fact bots but the people themselves. On Twitter, false news was found to be retweeted approximately 70% more and traverses six times as fast as legitimate news, and these statistics raise the question of what the scenario must be like with the added pressure of the pandemic looming over everyone's head.

In the era of COVID, the spread of misinformation has had a huge impact on the population. According to a study by WHO, [2] during the first 3 months of the pandemic, nearly 6,000 people around the world were hospitalized due to misinformation related to the COVID-19. This overabundance of information consisting of both reliable as well as unreliable information coupled with the pandemic at hand serves as the breeding grounds for uncertainty and skepticism, which if kept unchecked can be catastrophic. That is where early detection of false news could make a major impact.

For the task at hand, an abundance of news articles, forums as well as data from social media platforms such as Twitter are available, but the availability of labelled data is much more limited and this factor coupled with the fact that contextual references are also required to be considered makes the detection of fake news ever more difficult. DL is an emerging field and has yet to find its proper footing in the realm of NLP. The proper selection of loss functions and optimizers plays one of the most crucial roles in building a DL model as optimizers are the ones responsible for associating the loss function and the model parameters by regularly updating the weights and biases of the nodes. In this chapter, we have used DL techniques for training the tokenized text items using five different optimizers: Adagrad, Adam, RMSProp, Adamax and the relatively lesser known CoCoB (based on the Continuous Coin Betting Algorithm) and at the end, the outcomes of the results produced by all five of these optimizers have been compared for identifying the optimizer that proves to be most efficient for detection of fake news.

The rest of the chapter is organized as follows. Section 2 briefs the prominent works in the domain of fake news detection with DL models. Section 3 details the background details about optimizers in DL. Section 4 elaborates the methodology followed for the analysis. Section 5 details about the experimental setup including dataset, system configuration and DL model. Section 6 presents the results and discusses the performance of optimizers in the context of COVID-19 fake news detection. Lastly, Sect. 7 put forward concluding remarks.

2 Related Work

Detection of fake news is an emerging challenge in the era of social media. DL models are being explored in recent years by several researchers for detecting fake news. This section briefs few of the prominent works where DL models are predominantly explored for fake news detection. Vosoughi et al. [3] investigated the transmission and spread of false and real news distributed across twitter throughout 2006–2017 thoroughly, classified the tweets as false or true by cross-checking the references with six renowned fact-checking organisations and further dug deeper into the reason behind why fake news traverses faster and deeper than real news and what the psychological and conscious factors lead up to it. All-Tanvir et al. [4] demonstrated through their work, the efficiency of five well known machine learning algorithms including SVM, Naive Bayes, Logistic Regression as well as an RNN model in detection of fake news from dataset acquired from twitter, and compared the final accuracies for drawing further conclusions.

Thora et al. [5], performed a detailed study on efficiency of CNNs and RNNs, ensemble methods and attention mechanism models in detection of fake news and came to the conclusion that a CNN+ bidirectional LSTM ensembled network with attention mechanism achieved the best performance and outperformed the existing architectures by a margin of upto 2.5%. Girgis et al. [6] performed a thorough examination on fake news detection using DL approaches by implementing RNN models including GRU and LSTM and compared them with each other and the existing models with proven efficiency. Upon comparing the results, GRU was found to outperform LSTM by an extremely narrow margin. Monti et al. [7] presented a method for detection of fake news using a geometric DL model. They showed that transmission based approaches provide superior outcomes when compared to purely content based approaches. By generalizing CNNs into graphs, they were able to create connections between heterogeneous data such as user profile, activity and influence with the content of the tweets. The trained model was tested and conclusive results were obtained, which were further cross-checked by reliable fact-checking organisations.

With advantages of DL models in fake news detection, it comes up with an inherent challenge of considering appropriate optimizers for the model. Several popular optimizers are in use. Hinton et al. [8] developed RProp as an algorithm that would offer a solution for the Adagrad's shortcomings with diminishing learning rate. The intuition behind RProp was to use only the sign of the gradient while adapting the step size separately for each weight, but RProp failed to work on mini-batches and so along with it, its mini-batch compatible version, RMSProp was introduced. Diederik and Ba [9] introduced Adam, a first-order gradient based optimization technique which calculates individual adaptive learning rate. They essentially presented it as a combination of the RMSProp algorithm and gradient descent with momentum. Few of the proven advantages offered by Adam were found to be that, the magnitudes of parameter updates are not tied to the rescaling of the gradient, it is able to work with relatively less memory and its hyperparameters require very little tuning and yet produce remarkable results. Orabona and Pal [10] explained about Parameter-Free Convex Learning through Coin Betting, about an algorithm which they described as a 'parameter-free algorithm' whose analysis could be represented through an adversarial game of coin-betting, and the goal of the algorithm was to achieve optimum optimization for convex functions without having to tune too many hyper-parameters. Orabona and Tommasi [11] discussed in detail about their approach to the CoCoB or the Continuous Coin Betting algorithm, their aim was to get rid of at least one of the hyperparameters of DL models and in the process, to design a backpropagation procedure that does not have the requirement for any learning rate at all and at the same time, it had to be capable of competing with the renowned and conventionally used optimizers.

3 Background

For implementation of DL approaches in our NLP model, it is essential to look into the suitable optimization techniques as well as the hierarchy of optimizers. Optimization algorithms or strategies which when coupled with the loss functions are responsible for reducing the losses and for providing the most accurate results possible.

3.1 Deep Learning Optimizers

With DL models, our primary objective is to build a model which is capable of offering an optimal solution that fits all of our given data with enough flexibility to predict our future data as well. This involves the process of minimizing the error projected by the loss function throughout the course of the training of the given data, and that's where the role of optimizer comes in. Tied together with the loss function and the parameters, the model is updated repeatedly as the weights are tweaked depending on the response of the loss function.

Based on their learning parameters, different optimizers propose varied results. Given below is the overview of the optimizers used for the comparative analysis:

- Adagrad, which stands for Adaptive Gradient Algorithm, is capable of adapting the learning rates to individual features. In Adagrad, smaller updates are performed for frequent parameters and for the non-frequent parameters, larger updates are performed. It is well suited for sparse datasets. One of the major issues with Adagrad is that the learning rates tend to get smaller over time.
- RMSProp or the Root Mean Square Propagation, deviced by Hinton [8], proposes a solution to the diminishing learning rate problem of Adagrad which is achieved by using a moving average of squared gradient. Here, the gradient is normalized by applying the magnitude of the recent gradient descents.
- Adam or otherwise known as Adaptive Moment Estimation, is another optimizer that provides a solution for reducing the diminishing learning rates observed in Adagrad. Adam is one of the most widely used optimization algorithms, it implements exponentially moving averages in order to scale the learning rate.
- Adamax is a variant of Adam obtained by generalizing the approach to infinity norm and performs better than Adam in certain scenarios.

3.2 CoCoB Optimizer

Modern day DL approaches provide us with cutting-edge performance in a wide range of application scenarios. But the current issue that persists is that those techniques require a decent quantity of hyperparameters to tune with for acquiring efficient results. In particular, tuning the learning rates in the stochastic optimization manner remains one of the problems that holds back the performance of the model. We shall utilize a brand new stochastic gradient descent process for deep networks that doesn't require any adjusting or tweaking of learning rate setting. Unlike the conventional techniques, here, there will be no need for adapting the learning rates nor will there be any need to make use of the assumed curvature of the objective function. Instead, this algorithm interprets the sub-gradient descent as a game of coin-betting. The convergence of this algorithm has already been proven for convex functions and it is known to have shown efficient results.

3.3 CoCoB Optimizer

Considering a gambler making bets on the outcomes of adversarial coin flips repeatedly. The gambler starts with initial money. In each round the bets on the outcome of the coin flip g_t $\{1, -1\}$, where $+1$ denotes heads and -1 denotes tailso assumptions are made about the generation of g_t.

The gambler is free to bet any amount he pleases without asking for any additional money. If the gambler's prediction is incorrect, he loses his betted amount, whereas, if his prediction is correct, he gets to keep his betted amount along with an additional same amount, i.e., an overall gain of the amount he betted that round. His bet is represented by ω_t, where the value of ω_t represents the amount he is betting and its sign represents whether he's betting on heads of tails (+1 for heads and -1 for tails), where t represents any given round. W_t represents the wealth that the gambler acquires at the end of each round and $\mathrm{R_t}$ as the gambler's net reward.

Therefore, Wealth can be expressed as

$$W_t = \varepsilon + \sum_{i=1}^{t} (\omega_i g_i) \tag{1}$$

and the Reward can be defined as,

$$R_t = W_t - \varepsilon + \sum_{i=1}^{t} (\omega_i g_i) \tag{2}$$

Again, a bet in any given round can also be represented as a function of the Wealth in the previous round:

$$\omega_t = B_t W_t \quad (3)$$

Here, the value of B_t represents the fraction of the current wealth being betted and its sign signifies whether the bet is on heads or tails, and since no money can be borrowed, the value of B_t lies within the bounds of [−1, 1], and the outcome of the coin flip g_t is generalized to represent any real number in [−1; 1].

Now, for performing sub-gradient descent through coin betting, we shall take a one-dimensional example. Let us take an example function F(x) = |x − 4|. This function neither has any curvature, nor it is differentiable, and so, application of second order optimization algorithms is not possible. The negative sub-gradient of our function F(x) in ωt is set as equal to the outcome of the coin flip g_t such that, $g_t \in \delta[F(\omega_t)]$. Assuming that there exists a function X such that the betting strategy being implemented will make sure that after T rounds have passed, wealth will be at least

$$X\left(\sum_{t=0}^{T} g_t\right) \quad (4)$$

and the convergence of this sequence was described as the solution to the optimization problem at hand by Orabona and Tommasi [11]. They also described an actual functional betting strategy which ensures optimal convergence rate and adaptivity to gradients, upon which the CoCoB optimizer used in this project is based.

4 Methodology Overview

The flow chart shown in Fig. 1, represents all the steps involved in the analysis methodology to create the experimental environment for generating the results starting with collection of suitable data to generate conclusive plots for comparison.

4.1 Data Collection and Cleaning

The data collected for this work were primarily acquired from pre-existing datasets, but for the portion of data that was acquired from Twitter using the Tweepy [12] library with accurate timelines and relevant hashtags. For the cleaning and preprocessing of the text, standard re [13], nltk [14] libraries have been used. All residual symbols, numbers, stopwords, along with all the '@' mentions are items that have no use in the evaluation and training of our model and hence these were eliminated. Among all the remaining words, the uppercase alphabets were transformed into lowercase and finally the words were stemmed using the 'PorterStemmer' class before being added to the corpus.

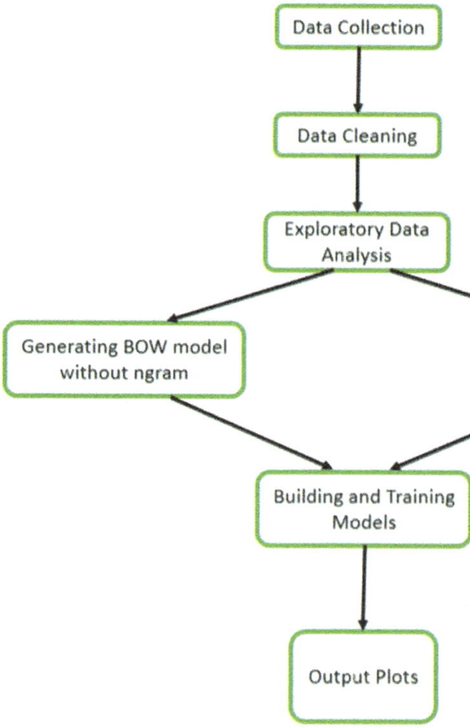

Fig. 1 Flow chart representing sequence of operations

4.2 Generating Bag of Words Models

The words collected in the corpus were converted into a Bag of Words model for creating the feature matrix for our training by the 'CountVectorizer'. One important thing to be taken under consideration is that our model needs to be able to recognise contextual references, for example the phrases 'showing positive change' and 'increasing positive cases' carry completely different meanings even though they both include the word positive. Let us take two sample texts from the dataset:

Text Sample 1: Anguished, I'll always recall our interactions: CM on **$$$###'s death.

Text Sample 2: "Can't bring students back from Kota until Centre revises lockdown rules" says CM.

Out of the two text samples above, Text 1 represents a news headline that is completely unrelated to Covid, whereas Text 2 represents a news headline that is Covid-relevant. It can be observed that both the categories of headlines contain the reference 'CM' even though it plays no role in determining whether the news is related to covid or not by itself. In fact, the entire dataset has innumerable recurrences of the word, here's another example:

Text Sample 3: Only 3 out of 529 media persons tested have been detected positive: Delhi CM.

Similar to this there are many other such words which have no contribution towards achieving our goal, which leads us to wonder if regardless of the high number occurrence of a particular word, its impact in determining our result is sufficient. Even though our training algorithm or in our case our Neural Network will take care of such occurrences, but if we assume that there is nearly 'n' occurrence of a particular word across the texts/headlines of both the category of our datasets (n occurrences for Real and n occurrences for Fake news headlines), then that word's presence has lower impact on our result then we desire it to. The solution to this problem can be obtained with the use of ngrams. With the use of ngrams, combinations of words for up to desired number of words can be generated which makes the training feature set more robust and at the same time provides a larger pool of words/phrases. In our model we have used ngrams with a range of (1–3) words.

4.3 Building and Training Model

In order to get an idea of how much of an impact the use of ngram has on a given text dataset, two separate feature matrices have been created for training the model for each of the two datasets (Covid_Relevance dataset and Covid_Fake_News dataset), one without ngram and in the second one ngram has been implemented, and after the models are trained, the two results have been compared. The final step before we proceed with the training involves splitting our training data into training and validation sets which have been performed in the ratio of 4:1. For training, a standard LSTM network has been considered, on which five different optimizers including CoCoB have been applied and at the end, the accuracies of the models trained by all of the optimizers have been plotted for making comparisons and drawing conclusions.

5 Experimental Setup

One of the most crucial steps of this work is data collection. Even though there is an abundance of COVID-related news data available, the sparsity of labeled and organized data makes the work difficult. Hence, multiple accessible sources have been considered for gathering the data suitable for this task. As for the RNN model that was used, along with the optimizers that have been considered, the specifications of the attributes as well as the learning parameters of the optimizers and the configuration of the system on which models have been trained, these have been discussed in further detail in the following subsections.

5.1 About the Dataset

The datasets used for the experimental analysis is a combination of data collected from the accessible data repositories such as Kaggle [15] and CORD-19 [16], which is the official data repository for COVID-related data. As Twitter is one of the largest sources of textual data, some of the data has been collected from Twitter with the help of Twitter APIs.

- The Covid_Fake_News dataset from Kaggle contains three files respectively for training, validation and testing with a combined total of over 10,000 entries. It contains tweets related to covid and a corresponding column indicating whether the tweets are Real or Fake.
- CORD-19—The Covid-19 Open Research Dataset is accumulated to allow easy access to data related to Covid to be considered as a standard reference for cross-checking data and facts. The CORD 19 dataset is the collection of papers from various legitimate sources such as PMC, CZI and WHO, with the legitimate supporting information.
- For the portion of data that has been gathered from Twitter, the tweepy [12] library has been used and all the guidelines and data withdrawal restrictions have been followed.
- The Covid_Relevance dataset is the combination of two datasets containing news articles along with some tweets. The two datasets involved are: News dataset taken from Kaggle as well as some tweets gathered from Twitter with total size of 45,814 which contains daily news unrelated to covid and Corona_NLP dataset retrieved from Kaggle with 41,157 rows which contains all the covid-related news/articles/tweets.

5.2 Deep Learning Models

The DL models used for this chapter comprise a standard LSTM model with a couple of dense layers. The LSTM or Long Short-Term Memory is a type of recurrent neural network that is capable of working around the long-term dependency problem with ease, which involves connecting the past data with the current task that is dependent on that information, a task that RNN is inherently designed to perform. Like all standard RNNs, LSTM also has a chain-like structure but the repeating module has a four-layered neural network structure instead of a single one as shown in Fig. 2., which makes it even more efficient. For a faster implementation of the LSTM network, CuDNNLSTM has been used.

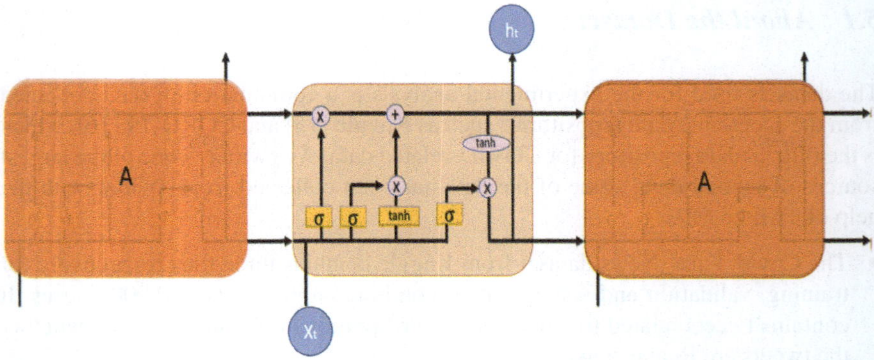

Fig. 2 Repeating modules in LSTM model structure, each containing four interacting layers

Table 1 Parameter settings of optimizers

Optimizers	Learning rate	Parameter-1	Parameter-2	Epsilon
Adam	0.001	beta_1 = 0.9	beta_2 = 0.999	1e-07
Adagrad	0.001	initial_accumulator_value = 0.1	NIL	1e-07
RMSProp	0.001	rho = 0.9	momentum = 0.0	1e-07
Adamax	0.001	beta_1 = 0.9	beta_2 = 0.999	1e-07

5.3 *Optimizers and Parameters*

For the optimizers that were considered, all the optimizers except CoCoB (no parameters are required) were set to their default parameters as provided by the Keras library, which are given in Table 1.

5.4 *System Configuration*

As per the system requirements of the latest stable release TensorFlow version (ver.2.5.0) which goes with python (ver. 3.6–3.9), GPU usage is not necessary but it definitely helps with the training especially considering the fact that we have used CuDNNLSTM for faster implementation of LSTM which runs exclusively on GPU. The system configurations of the system used for generating the results of this paper are:

CPU: Intel Core i5-10300H, ~2.5 GHz

GPU: NVIDIA GeForce RTX 3060

RAM: 8 GB

Operating System: Windows 10.0, 64-bit

Tensorflow Version: ver. 2.5.0

6 Result Analysis and Discussion

6.1 *Exploratory Data Analysis*

Before proceeding with conducting further operations on the data, it is wise to summarize the main characteristics of the dataset, find patterns, anomalies and get an overview of the data we're working with in order to get a better insight.

6.1.1 Polarity

The Polarity of words is essentially a float value ranging from $[-1, 1]$, that represents the sentiment embedded in a given text where values close to -1 represent negative sentiment, values closer to 1 represent positive emotions and values nearer to 0 are considered neutral. The value for Polarity of a text sample is calculated as the sum of polarity represented by all the individual words in a sentence divided by the total number of words present in the sentence. First, we shall consider the Covid_Relevance dataset. Given Fig. 3 represents two plots that represent the Sentiment Polarity of the news articles and tweets present in the dataset. One important thing to take note of is that the data has been arranged such that the first 44,940 text-data are unrelated to Covid and the rest are Covid-related, this has been done to make it easier for us to spot any patterns or trends of any significance.

In the scatter plot Fig. 3a, we can observe that the data relevant to covid are more polar after/around the mark of 44,940 are starting to appear more polarized as compared to the data before that, which leads us to the possible interpretation that the news which are related to Covid tends to trigger more intense and deep-seated sentiments from people as compared to other news in general.

Now, if we take a look at the histogram Fig. 3b, we can observe that most of the values are concentrated around the origin, a considerable amount is scattered between (-0.5 and 0.5) and a much smaller number of values are scattered towards the poles, which means most of the texts are neutral or semi-neutral and a relatively smaller proportion of the texts represent highly polar sentiments. Similarly, we can see the Polarity plots for Covid_Fake_News dataset as shown below in Fig. 4.

It can be observed that, unlike the previous case, here we can see that the polarity is rather well distributed all across the dataset on both Fake news and Real news categories, but yet again we can observe in the histogram that most of the values are concentrated around the neutral mark, i.e. zero, which indicates most of the texts represent neutral sentiments and relatively less portion of the texts represent highly polar sentiment in this dataset as well. Considering the fact that this dataset contains

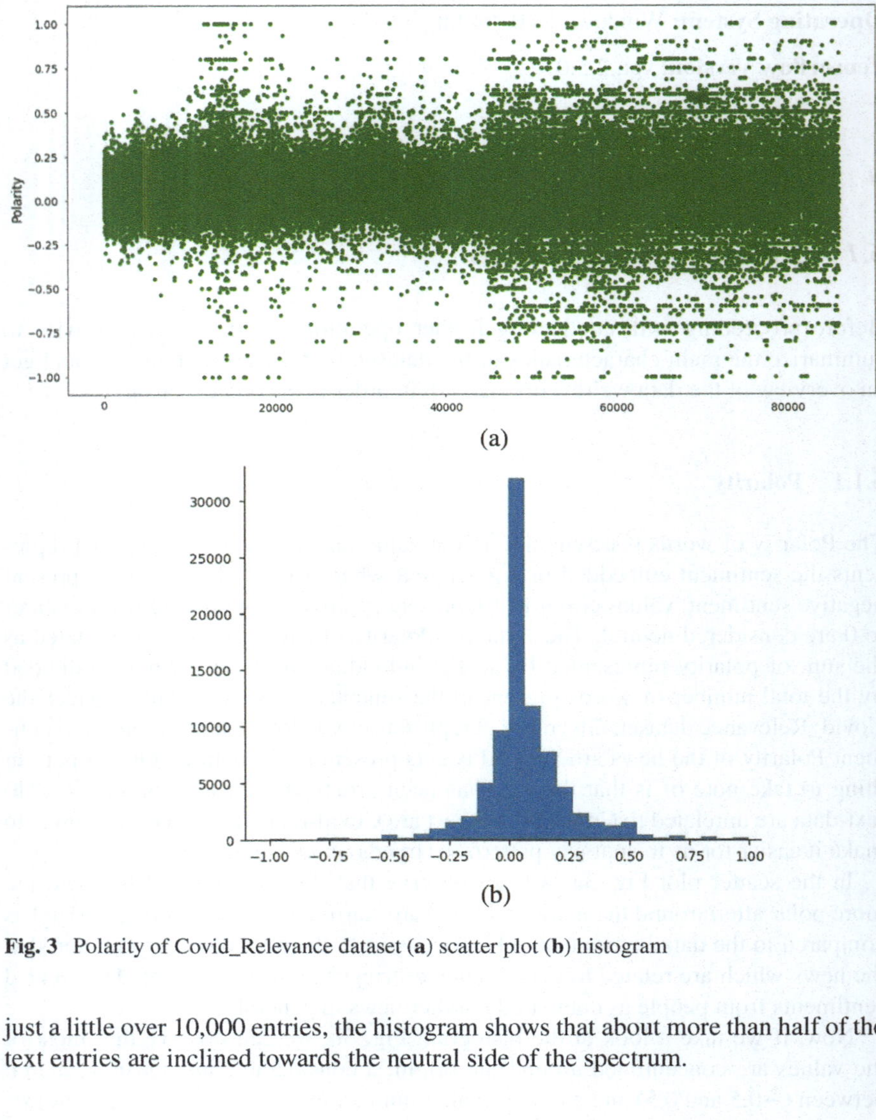

Fig. 3 Polarity of Covid_Relevance dataset (**a**) scatter plot (**b**) histogram

just a little over 10,000 entries, the histogram shows that about more than half of the text entries are inclined towards the neutral side of the spectrum.

6.1.2 Word Count

Given below in Fig. 5 is the representation of word-count per article/text-item of the dataset (after removing all the stop-words). For the Covid_Relevance dataset, it can be observed from the scatter plot that, on an average, the data related to covid has more words per article/tweet, and from the histogram, we can see that most of the articles/tweets in our dataset have somewhere around 10–25 words and very few articles/tweets exceed the 30 words mark. On an average, the word-count per text

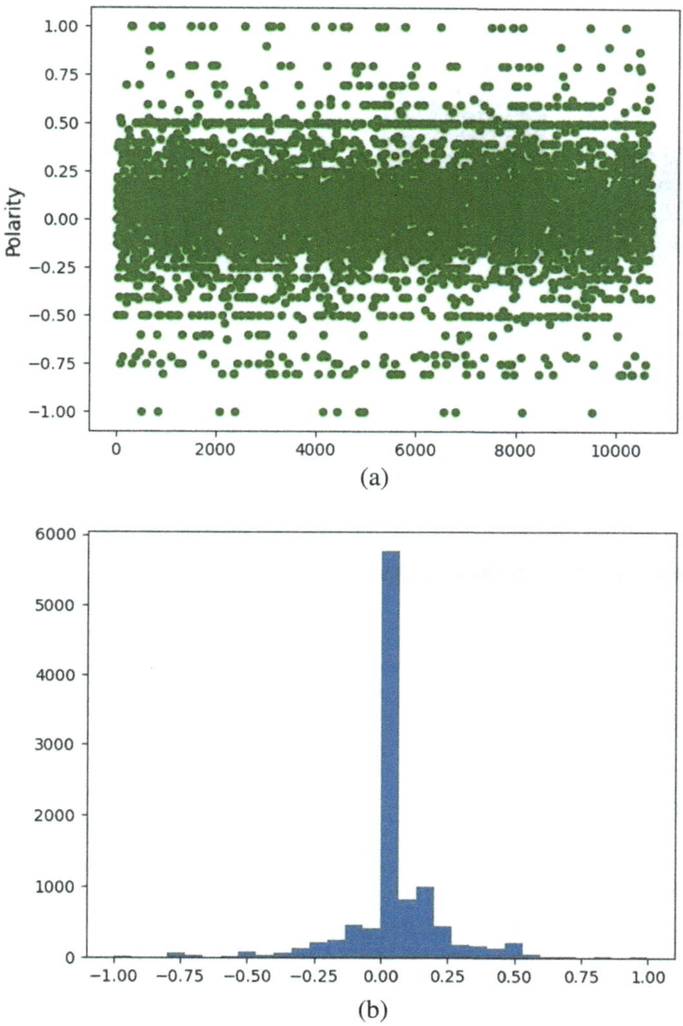

Fig. 4 Polarity of Covid_Fake_News dataset (**a**) scatter plot (**b**) histogram

entry in case of covid-related articles appears to be higher than the covid un-related articles.

In the scatter plot for the Covid_Fake_News dataset, we can observe that there are a few outliers that seem to have higher word count than the others, but besides that the overall word count per news article or tweet remains mostly below 30 words throughout the dataset as evident from the histogram.

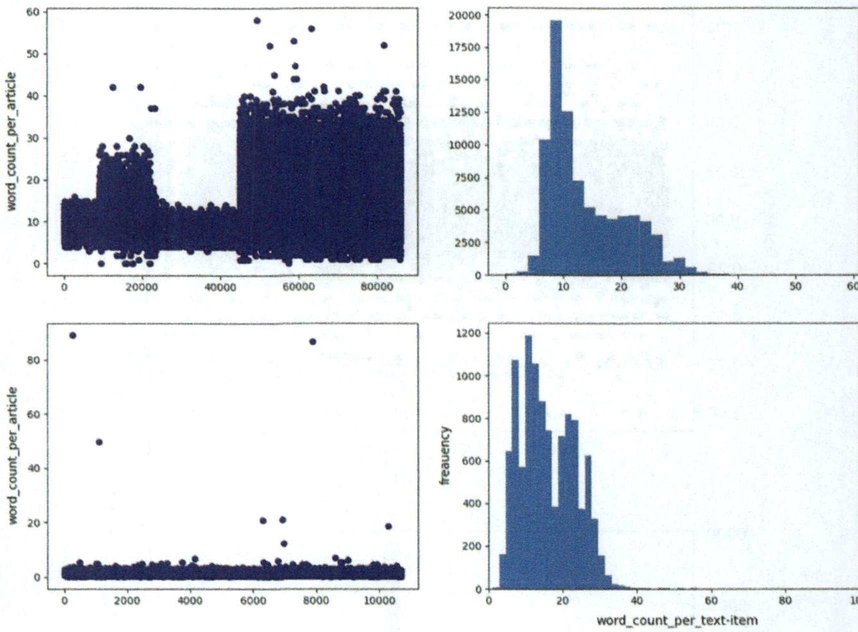

Fig. 5 Plots representing word count per sentence (**a**) scatter plot (Covid_Revance dataset) (**b**) histogram (Covid_Revance dataset) (**c**) scatter plot (Covid__Fake_News dataset) (**d**) histogram (Covid__Fake_News dataset)

6.1.3 Most Recurring Words

In Fig. 6. shown below, is the bar plot representation of the most recurrent words along with the number of times they appear all across the Covid_Relevance dataset in descending order. (**Note:** A magnified sub-plot of a segment of the plot is provided along with the original plot for obtaining a clearer view.)

6.1.4 Ngrams

For the ngrams to be implemented, a range of 1–3 words was considered since phrases of length more than 3 words are usually unlikely to have a relatively larger number of occurrences. Given below in Table. 2. are a few examples of some of the 1,2 and 3-word combinations that have been generated alongside a list of the words generated without applying ngram.

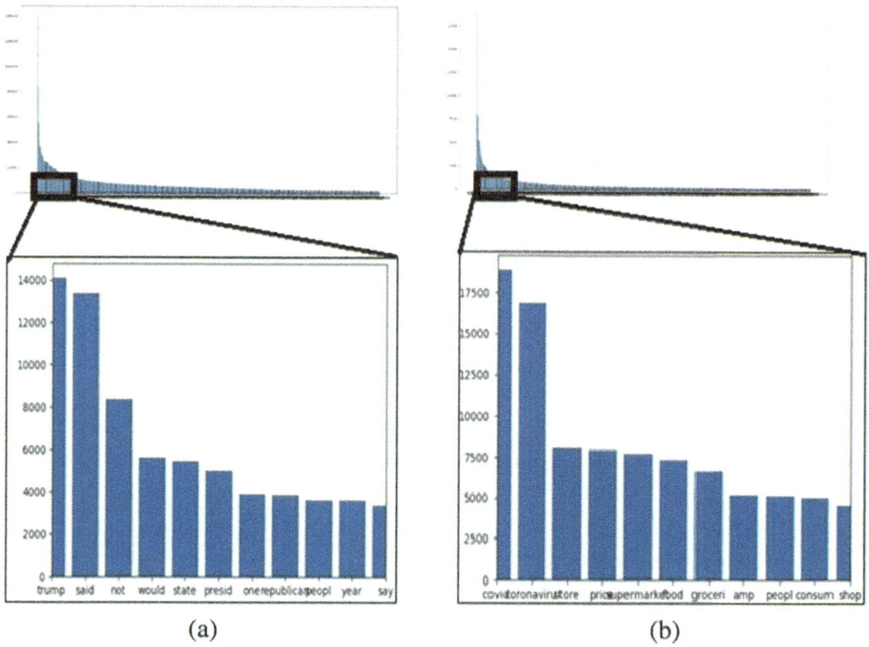

Fig. 6 Bar plots representing most recurrent words in the dataset (**a**) for Covid_Relevance dataset (**b**) for Covid_Fake_News dataset

Table 2 Examples of words generated with and without ngram

Without ngram	With ngram
Account	Accord
Accur	Account
Across	Acquir local unknown
Act	Across
Action	Act
Activ	Act cases
Actual	Act cases death
Ad	Action
Add	Activ case
Addit	Activ case death
Adult	Actual
Advis	Actual caus
Affect	Ad

6.2 Deep Learning Optimizer Analysis

For training our data, a feature matrix of 5000 most recurrent words for Covid_Relevance dataset and 1500 most recurrent words for Covid_Fake_News was taken and fitted into a standard LSTM network with a couple of added dense layers, with a dropout rate of 0.3. Each of the datasets were trained using five optimizers: CoCoB, Adam, Adagrad, Adamax and RMSprop and their results were recorded for comparison. The later four were set to their default learning rates and for loss and binary-cross-entropy function was used as the loss function.

6.2.1 Comparison Between the Optimizers

From the plots shown below Fig. 7., the accuracy and loss plots are shown for the models trained with all five optimizers mentioned previously. The Covid_Relevance

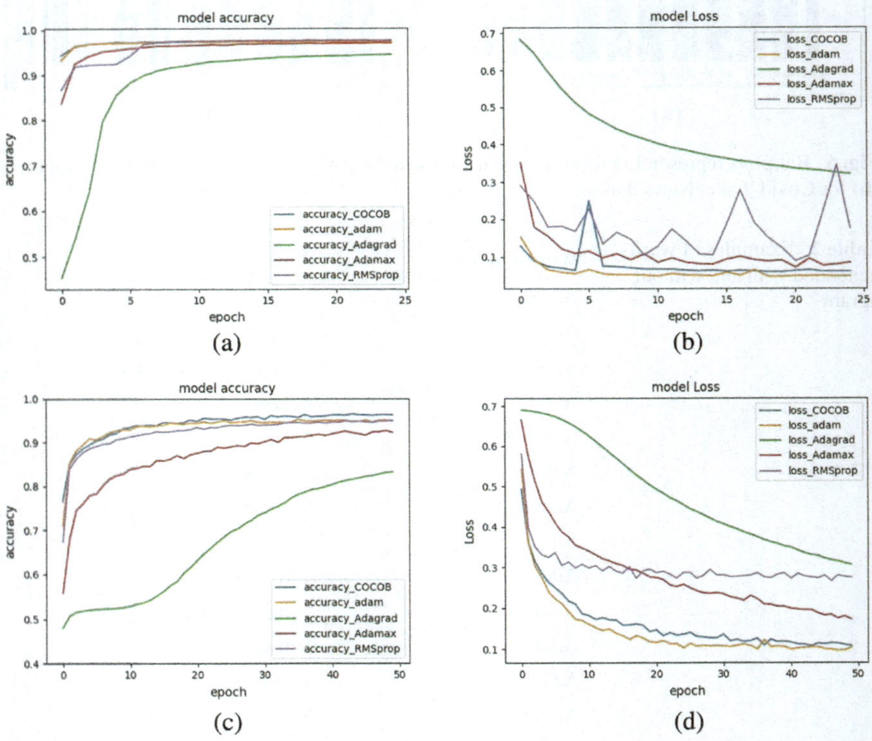

Fig. 7 Accuracy and loss plots (**a**) accuracy plot and (**b**) loss plot for Covid_Relevance dataset (**c**) accuracy plot and (**d**) loss plot for Covid_Fake_News dataset

feature matrix was trained for 25 epochs and the Covid_Fake_News dataset was trained for upto 50 epochs for obtaining better saturation.

From the graphs shown above, it is evident that CoCoB as an optimizer is clearly capable of keeping up with the other optimizers that are being considered and at times it even seems outperforming them. For the data currently being used for this training, only Adam is found to be able to keep up or at times outperform CoCoB by a very narrow margin and hence CoCob has proven to be an efficient optimizer that shows quite some promise for its future applications. If we take into consideration, the performance of each of the optimizers after the very first epoch into consideration, it can be observed that CoCoB outperforms all the other optimizers in the bunch by a considerable margin, with an accuracy of 93.8% and loss of 13.4% for the Covid_Relevance dataset and an accuracy of 76.6% and loss 49.1%.

6.2.2 Influence of Ngrams

In Fig. 8, the Covid_Relevance dataset upon training the feature matrix without ngram, after 25 epochs, came up with an accuracy of 96.8% and with ngram, it

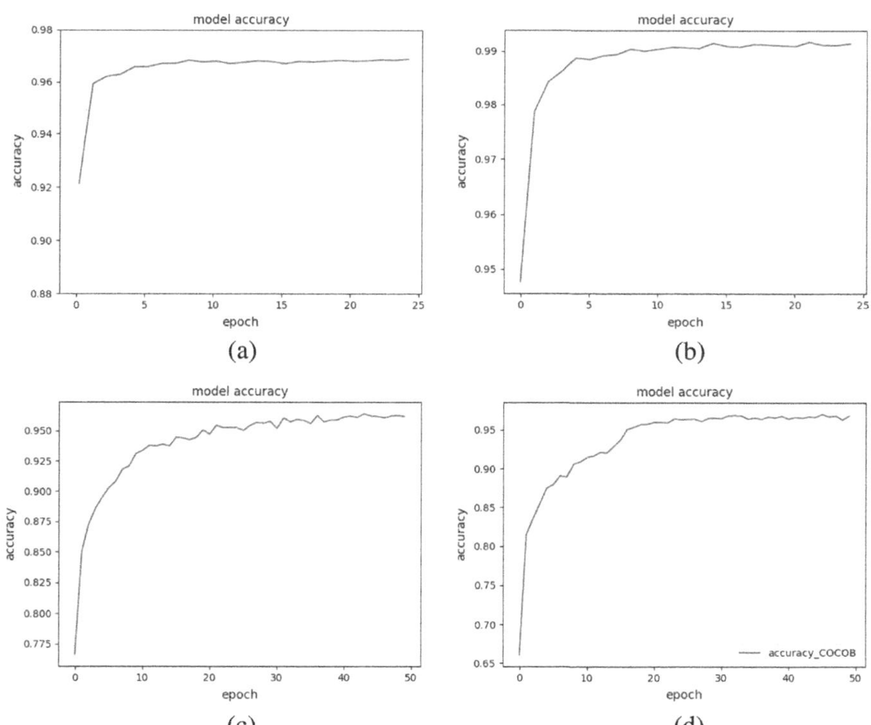

Fig. 8 Accuracy plots for data with and without ngram (**a**) without ngrams and (**b**) with ngrams for Covid_Relevance dataset (**c**) without ngrams and (**d**) with ngrams for Covid_Fake_News dataset

ended up with an accuracy of 99.1%, and for the Covid_Fake_News dataset, the model without ngram, after 50 epochs, showed a final accuracy of 95.46% on its best run and the one with ngram ended up with 96.84% accuracy.

After multiple training runs with and without the implementation of ngrams, it was observed that the implementation of ngram leads to increase in training accuracy ranges from 1.5% to upto 2.5%, which may not seem like a huge leap considering our accuracies obtained for these datasets are already quite high, but for smaller datasets with much smaller number of elements in the corpus to work with, it may end up making a considerable difference.

6.2.3 Confusion Matrix

In order to have a better look at the composition of the model's output, following are the confusion matrices for Covid_Relevance and Covid_Fake_News test data:

From the Confusion matrices in Fig. 9. we can observe that for both the datasets the number false negative values are more than that of false positive, which means that even though our model has a very high accuracy and will tend to misclassify our data very rarely, but the times it will end up misclassifying the data, the model has a higher tendency to falsely classify our **Covid Related** data as **Covid Unrelated** and **Real News** as **Fake News**.

Actual || Predicted	True	False
True	8062	165
False	109	8880

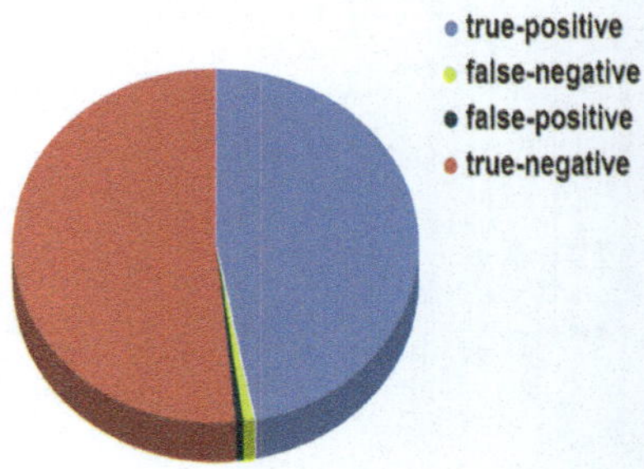

(a)

Actual || Predicted	True	False
True	986	107
False	69	978

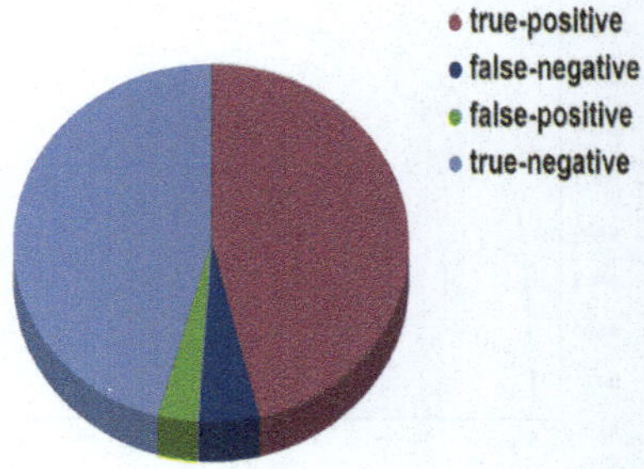

(b)

Fig. 9 Confusion matrix (**a**) Covid_Relevance dataset (**b**) Covid_Fake_News dataset

7 Conclusion and Future Direction

Performed comparative analysis of optimizers used in DL models in the context COVID-19 related fake news detection. The faster version of LSTM architecture CuDNNLSTM has been implemented along with several preprocessing steps such as cleaning, stemming and feature extraction operations on text data related/unrelated to covid and text data based on real and fake COVID-19 news. The performance of the newly developed CoCoB optimizer for DL models has been compared with four other widely used optimizers from the perspective of fake news detection. It was observed that Adagrad underperformed in comparison to all the other optimizers as expected, RMSProp and Adamax performed reasonably well, but adam and CoCoB showed the best performance and CoCoB even seemed to outperform adam. CoCoB also outperformed all the optimizers if we compare the results after the first epoch in both of the datasets.

Acknowledgements The work of Dr. Anupam Biswas is supported by the Science and Engineering Board (SERB), Department of Science and Technology (DST) of the Government of India under Grant No. Grant No.EEQ/2019/000657.

References

1. Dizikes, P., MIT News Office. (March, 2018). Massachusetts institute of technology news. https://news.mit.edu/2018/study-twitter-false-news-travels-faster-true-stories-0308.
2. World Health Organization News. (April, 2021). https://www.who.int/news-room/feature-stories/detail/fighting-misinformation-in-the-time-of-covid-19-one-click-at-a-time.
3. Vosoughi, S., Roy, D., & Aral, S. (2018). The spread of true and false news online. *Science, 359*(6380), 1146–1151.
4. Mahir, E. M., Akhter, S., & Huq, M. R. (2019, June). Detecting fake news using machine learning and deep learning algorithms. In *2019 7th International Conference on Smart Computing & Communications (ICSCC)* (pp. 1–5). IEEE.
5. Thota, A., Tilak, P., Ahluwalia, S., & Lohia, N. (2018). Fake news detection: A deep learning approach. *SMU Data Science Review, 1*(3), 10.
6. Girgis, S., Amer, E., & Gadallah, M. (2018, December). Deep learning algorithms for detecting fake news in online text. In *2018 13th International Conference on Computer Engineering and Systems (ICCES)* (pp. 93–97). IEEE.
7. Monti, F., Frasca, F., Eynard, D., Mannion, D., & Bronstein, M. M. (2019). Fake news detection on social media using geometric deep learning. *arXiv preprint* arXiv:1902.06673.
8. Hinton, G., Srivastava, N., & Swersky, K. (2012). Neural networks for machine learning lecture 6a overview of mini-batch gradient descent. *Cited on, 14*(8), 2.
9. Kingma, D. P., & Ba, J. (2014). Adam: A method for stochastic optimization. *arXiv preprint* arXiv:1412.6980.
10. Orabona, F., & Pál, D. (2016, December). Parameter-Free Convex Learning through Coin Betting. In *Workshop on Automatic Machine Learning* (pp. 75–82). PMLR.
11. Orabona, F., & Tommasi, T. (2017). Training deep networks without learning rates through coin betting. *Advances in Neural Information Processing Systems, 30*, 2160–2170.
12. Roesslein, J. (2020). Tweepy: Twitter for Python! https://Github.Com/Tweepy/Tweepy.

13. Van Rossum, G. (2020). *The Python Library Reference, release 3.8.2*. Python Software Foundation. https://docs.python.org/3/library/re.html
14. Bird, S., Klein, E., & Loper, E. (2009). *Natural language processing with Python: analyzing text with the natural language toolkit*. O'Reilly Media, Inc. https://www.nltk.org
15. Kaggle: https://www.kaggle.com. Accessed in July 2021.
16. US National Library for Medicine, CORD-19. https://www.ncbi.nlm.nih.gov, Accessed in July 2021.

Detecting Fake News on COVID-19 Vaccine from YouTube Videos Using Advanced Machine Learning Approaches

Wael M. S. Yafooz, Abdel-Hamid Mohamed Emara, and Mohamed Lahby

Abstract Fake news is considered a massive threat to many internet users, with the heavy usage of social media networks. Many news agencies develop their platforms to publish and share their news articles. Also, an ordinary user on the social media network has an account where the content can be posted and shared. Some users share fake news or rumors to achieve personal goals and benefits. Fake news is considered to be the most visible challenge on social media networks. It creates a threat to individuals and society while creating a negative impact. Many research works tackle this issue using propagation-based, content-based, and meta-data analysis approaches. This book chapter proposes a model to detect fake news about the COVID-19 vaccine on YouTube videos using a sentiment analysis approach through machine learn and deep learning approaches focusing on the Arabic language of middle-east people. The process started with building a dataset through the collected textual data using the comments that were later annotated into two classes. They are fake and real news. Two experiments have been conducted using machine learning classifiers and deep learning models. Through these experiments, the performance level of the model has reached 94% in terms of accuracy. In the deep learning approach, it has reached 99%.

Keywords Fake news · Machine learning · COVID-19 · YouTube video · Deep learning

W. M. S. Yafooz (✉)
Department of Computer Science, College of Computer Science and Engineering, Taibah University, Madinah, Saudi Arabia

A.-H. M. Emara
Computers and Systems Engineering Department, Faculty of Engineering, Al-Azhar University, Cairo, Egypt

M. Lahby
University Hassan II, Casablanca, Morocco

M. Lahby et al. (eds.), *Combating Fake News with Computational Intelligence Techniques*, Studies in Computational Intelligence 1001,
https://doi.org/10.1007/978-3-030-90087-8_21

1 Introduction

The digital transformation era has digitalized most of the daily human activities. One of such activities is news articles. People traditionally used TV channels, Radio, or news articles by the government sector. Hence, traditional media channels were considered a more credible source of information. Thus, in the early stage of the internet, towards the beginning of the nineteens, less prominence was given to online news articles. This was mainly due to the internet speed, technology advancements, and many newswires established on websites and social media accounts. The internet makes the news easily accessible while making it spread globally [1]. Also, re-shares through social media users make it more reachable. Substantial attention is given to online news articles during this time when social media platforms such as Twitter, Facebook, and YouTube are heavily used [2]. Unfortunately, many people are using social media platforms as a method of spreading fake news and rumours. The attractive headlines can attract several likes, re-shares in a shorter period even if the source is not verified or validated. Therefore, people can use different political, marketing, or entertainment reasons in achieving specific goals. This can be harmful, creating a serious impact on people, products, economies, countries, and societies [3].

In the literature, many research works have given attention to fake news and rumours. Fake news is falsified information aiming to mislead the readers with different intentions. Those intentions mainly are either financial or political [4]. The rumours are the information shared by someone without verifying the details. They can be true or false. Fake news is considered a threat found when using social media. It can cause damage and danger for many individuals and organizations. Thus, there are many studies conducted to detect fake news. These can be categorized into three approaches namely; content-based [5–11], context-based analysis [12–20], and propagation-based [23–27]. In the content-based approach, the researcher studies the content of the news articles and extracts the information. Later, the news is compared with a trusted resource composed of experts. This type is also known as linguistics feature extraction methods. While context-data analysis where the research concerns analysing the user profiles based on many factors and characteristics. The third approach propagation-based, the identified news propagates faster and deeper through social media networks is detected through robots [9] or user accounts created. This was done in 2016 the USA election.

During the Covid-19 pandemic, several fake news has been spread globally from discovering the symptoms, effect areas, and source of the pandemics. In addition, the Covid-19 vaccines began in Feb 2021 was the title of many rumours, making many people worried and scared about the vaccine. Therefore, this book chapter proposes a model to detect fake news on the covid-19 pandemic through YouTube videos by focusing on Arabic language using the sentiment analysis methods. The model starts to construct dataset which collected from user comments on YouTube videos. The pre-processing steps were started by removing the duplicated comments and cleaning the data. The annotation process was done by three Arabic speakers.

The annotated data were classified into two main predefined classes as fake and real news. There were two types of experiments that have been conducted utilizing machine learning classifiers and deep learning models. During the experiments, the N-grams and several feature extractions methods from the text have been utilized. The proposed model achieved a performance level of 94% in terms of accuracy and 99% using support vector machine and logistics regression classifiers in machine learning and deep learning respectively.

The rest of this book chapter is organized as follows: Sect. 2 gives an overview of the related studies. The methods of the proposed mode explain in Sect. 3. Section 4 presents the results and discussion of the conducted experiments. The book chapter concluded in Sect. 5.

2 Related Studies

The sections present the related studies focusing on methods of detecting fake news. These studies are categorized into three approaches. They are content-based, context-based analysis, and propagation-based analysis.

2.1 Content-Based Approach

Content-based is also known as knowledge-based or linguistics feature-based or fact-checking. It is the process of extracting the content of the news articles and confirming it through credible resources. This is known as fact-checking. It can be performed manually by domain experts. However, it is costly and time-consuming while the results are more accurate and generated automatically through the websites that are available to the public. Examples of such websites are PolitiFact [13], FullFact [28], and GossipCop [14]. The proposed model [5] identifies fake news automatically by proposing two datasets of seven domains. Then, the features from the news article content which contains misinformation were extracted where the model archive 78% of accuracy level using a linear SVM classifier. While Rashkin et al. [6] goes into more detail by identifying the difference between news types, it analysis writing style and lexical resources to compare the types with the authentic language. The types are divided into four categories called satire, hoaxes, propaganda, and trusted. They use political news from Politick and trained a model using LSTM. Similarly works in analysis writing style proposed by Potthast et al. [7], attempt to differentiate news from writing style in relation with hyper partisan for both right-wing and left-wing unmasking which is known as authorship verification. They used approximately 1627 news articles with help of annotations from experts from Buzz Feed. Their experimental results show that the model reached a 78% of accuracy level in terms of F-score in classifying the hyper partisan and mainstream while 81% accuracy level in the differentiation between satire and hyper partisan. Ahmed et al. [8] proposed a method using six classical machine learning classifiers where the text is represented by utilizing n-grams with features of 1000–50,000. The model accuracy reached 92% in linear SVM. Ghosh, & Shah [9] used deep learning models to identify fake news in two stages. In the first stage, they used to check the truthfulness of the information using information retrieval to secure the closed articles from knowledge-based. In the second stage, they detected the writing style where it always the author tone is the aggressive way to attract the readers. Some researchers also used the concept of polarizing content to detect fake news [10]. Yang et al. [11] introduced a method to detect the credibility of the news based on Sina Weibo, which is a social platform in China. They collected misinformation from Sina Weibo and extracted the features from the microblogs. However, it is hard to detect fake news using only content-based analysis.

2.2 Context-Based Approach

The context-based analysis approach is also known as meta-data or user profiles. In this approach, the researchers focus on the user meta-data, the source, the surrounding, and the relevant information related to the fake news. The two main components of fake news are the news creators and the content. In this approach, less attention is given to the aforementioned by the researchers. Instead, they attempt to identify and analyse the social media accounts through which false news such as bots and Spammers are unfurled. Shu et al. [12] studies the relationship between fake news and users, through user profile analysis using the factors such as location, age, image profile, and historical post on social media. The FakeNewsNet dataset collected from Politifact [13] and Gossipcop [14] is used in this. The filtering process is utilized in removing the bots' accounts. The model shows an accuracy level of 96%. Works in [15] propose a framework called tri-relationship, to discover the link between the users, publisher, and the news articles using the auxiliary information. They highlight if the publisher is considered an untrusted resource of information and if the user publishes or re-shares their news. The users are considered as low credible. Shu et al. [16] highlights that some of the social media accounts can be cyborg created by humans to be used by bots. Krishnan & Chen [17] proposes methods of statistical analysis of meta-data associated actors such as number of friends, number of followers, number of Tweets, and verify user or not on Twitter. Similarly, Saez-Trumper [18] develops a tool called fake tweet buster to analyse the user account. It can identify the account history while calculating the credit score of an account. Similarly, Atodiresei et al. [19] proposes tools for Twitter user accounts in storing information in a dataset while building score similarity measure-based user accounts with employer-named entity recognition methods. Also, Ghanem et al. [20] collected data from approximately 62 million fake Twitter accounts that spread misinformation based on the creation time and screen names. They study 33 attributes of the Twitter user accounts. Contrary, Shao et al. [21] studies the bot's account that circulated about 14 million messages, 400 k times to mislead the users during the USA election

2016. While Erşahin [22] detects the fake accounts using the numerical attributes with an Entropy Minimization Discretization method.

2.3 *Propagation Approach*

In the third approach, few researchers also concentrated on news propagation and how the news spread through propagation networks on the based-based theory which is known as diffusion cascades. In this approach, the news behaves differently. The path of the post or re-shares and information source can be used in detecting fake news. Ma et al. [23] develops a propagation tree kernels that represent a user through a node. The node contains the attributes such as textual information (meta-data and comments) and timestamp. The similarity between the nodes is computed using Jaccard similarity. Similarly, Malhotra et al. [24] develops Graphical Neural Networks representing users as nodes with 12 features extracted from Twitter platform. This also includes the extracted textual features from source tweets focusing on the retweet. The extracted feature was used through Bi-LSTM to predicate the source class with four classifications to detect fake news. Differently, Saad et al. [25] uses blockchain technology in following the propagation of fake news, news creators, and users who can be realizable. The transaction workflow is transparent in the process of writing and reading news articles or posts. Sivasankari & Vadivu, [26] builds social network graphs through tracing the path of propagation of fake news. The news obtained from websites, such as Politifact and compared through fact-checking. The graph networks between users make it easy to identify the fake news and the user who spread the news through the content similarity between users. Mishra [27] highlights the importance of focusing on how the user can influence another user using mutual attention way. The direct and indirect graphs are applied between pair users to detect fake news.

There are few datasets that has been used as a benchmark in literature such as LIAR [29], Fake News Challenge 2017 Dataset [30], BuzzFeed [31], Twitter 15 [23], Twitter 16 [32], FakeNewsNet [33], Pheme [34], CredBank [35] and BuzzFace [36]

3 Methods and Materials

This section presents the research methodology utilized in detecting fake news through the sentiment analysis approach. There are four main interrelated phases of the proposed model. They are data acquisition, annotation process, classification, and result evaluation as shown in Fig. 1.

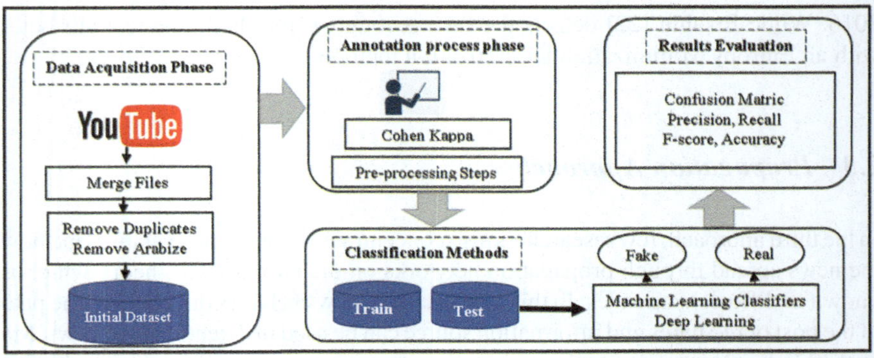

Fig. 1 Phases of the proposed model

3.1 Data Acquisition Phase

There are two main steps in the initial phase. They are data extraction and data cleaning. In the data extraction method, the data is collected through user comments on YouTube videos using Python 3.6 and YouTube API. Following the proposed criterion, the videos published between Feb. 2021 to August 2021, with more than 1 K likes and more than 2 K user comments were utilized. The user comments were downloaded into Ms excel file. After that, through the data cleaning steps as used in [37], all files were merged into one single file. In data cleaning, the duplicate data is removed and Arabize where the chat language and Arabic words by English character were removed. Out of the total 16,320 comments extracted from YouTube, the cleaned user comments lie at 8221. The output of these steps becomes the input for the next phase.

3.2 Annotation Process Phase

In the second phase, three Arabic native speakers were help in the annotating process. In this phase, the user comments indicating fake news were labelled '0', and the comments indicating real news were labelled '1'. The annotators were three in number because the utilized comments were numbered at 6112. The number of fake news is counted at 2943 whilst 3169 real news, the dataset are considered a balanced dataset as shown in Table 1. To assess the dataset validation, Cohen Kappa has been utilized as an equations in 1–5 from the confusion matrix as shown in Table 2. The results were achieved at a rate of 80% indicating accuracy based on literature.

$$\text{Kappa(K)} = \frac{Po - Pe}{1 - Pe} \tag{1}$$

Table 1 Dataset description

Items	Description	Max length	Minimum length	Average
Real	3169	402	3	22
Fake	2943			
Total	6112			

Table 2 Confusion matrix

		Judge 1	
		Yes	No
Judge 2	Yes	A	D
	No	C	B

Po = number of agreement of two Judges /total. The equation denoted which represent like the follows:

$$\mathrm{po} = \mathrm{A} + B/A + \mathrm{B} + \mathrm{C} + \mathrm{D}\frac{\mathrm{A} + \mathrm{B}}{\mathrm{A} + \mathrm{B} + \mathrm{C} + \mathrm{D}} \tag{2}$$

$$\mathrm{pe} = \mathrm{P(correct)} + \mathrm{P(incorrect)} \tag{3}$$

P(Yes) = number of Judge 1 said correct/yes divided by the total multiple the judge 2 said correct/total

$$\mathrm{P(Yes)} = \frac{\mathrm{A} + \mathrm{D}}{\mathrm{A} + \mathrm{B} + \mathrm{C} + \mathrm{D}} * \frac{\mathrm{A} + \mathrm{C}}{\mathrm{A} + \mathrm{B} + \mathrm{C} + \mathrm{D}} \tag{4}$$

P(No) = number of Judge 1 said incorrect/no divided by the total multiple the second rates said incorrect/total

$$\mathrm{P(No)} = \frac{\mathrm{C} + \mathrm{D}}{\mathrm{A} + \mathrm{B} + \mathrm{C} + \mathrm{D}} * \frac{\mathrm{B} + \mathrm{D}}{\mathrm{A} + \mathrm{B} + \mathrm{C} + \mathrm{D}} \tag{5}$$

3.3 Classification Methods

This section explains the two classification methods that were used in conducting the experiments. These methods were carried out using the classical machine learning classifiers and deep learning methods. In machine learning classifiers, six classifiers were utilized. They are *k-Nearest Neighbours(KNN) with parameter of n_neighbors = 4, Naive Bayes (NB)-GaussianNB, Decision Trees(DT) with parameter of max_depth = 10, Random Forest (RF), Support Vector Machine (SVM)* and

Logistics Regression (LR). In deep learning, the architecture used the artificial neural network (ANN), the model consist of five layers are input, three hidden layers, and the output layer. The input dimension contained 1500 features. The hidden layers consisted of 128, 64, 32 neurons with "Relu" as activation function and dropout with '30%' in second layer and '50%' in third layer in order to reduce the validation loss (regularization) whereas, the output layer consisted of one neuron with "sigmoid" function as activation functions. The optimizer is "Adam" and "SGD" and the loss function is "binary_crossentropy".

3.4 Results Evaluation

The last phase is to evaluate the performance of the proposed model. In this phase, confusion matrix has been used with precision, recall, f-score, and accuracy as shown in mathematical equations from (6–9)

$$Precision = \frac{True\ Positive}{True\ Positive + False\ Positive} \tag{6}$$

$$Recall = \frac{TP}{TP + FN} \tag{7}$$

$$F1\text{-}score = \frac{2 * (Precision * Recall)}{Precision + Recall} \tag{8}$$

$$Accuracy = \frac{True\ Positive + True\ Negative}{True\ Positive + True\ Negative + False\ Positive + False\ Negative} \tag{9}$$

4 Results Discussion

This section presents the results of the proposed model using the machine learning classifiers and deep learning models. The model performance evaluation in terms of the precision(P), recall(R), f-score, and accuracy (ACC), Also fivefold cross-validation (Val.) were performed. Two experiments were carried out using six machine learning classifiers (MLCs). The first experiment was conducted with 500 feature extractions and the second with 1000 features extraction. In both experiments, the n-gram utilizes with unigram, bigram, trigram, and quad-gram. The results of the experiments are as shown in Table 3 for unigram and bigram with

Table 3 Model Performance based unigram and bigram with 500 features

MLCs	Unigram						Bigram				
	Class	P (%)	R (%)	F-score (%)	Acc (%)	Val (%)	P (%)	R (%)	F-score (%)	Acc (%)	Val (%)
KNN	0	77	73	75	74	69	73	74	74%	73	72
	1	70	74	72			73	72	72		
SVM	0	96	93	94	**94**	95	93	96	94	94	**94**
	1	93	96	94			96	93	94		
NB	0	87	96	91	91	92	81	98	89	89	**89**
	1	96	87	92			98	83	90		
LR	0	93	96	94	**94**	93	93	96	94	94	94
	1	95	93	94			96	92	94		
DT	0	66	96	79	81	83	66	96	79	81	83
	1	97	73	83			97	73	83		
RF	0	91	93	92	92	92	91	93	92	92	92
	1	93	91	92			93	90	91		

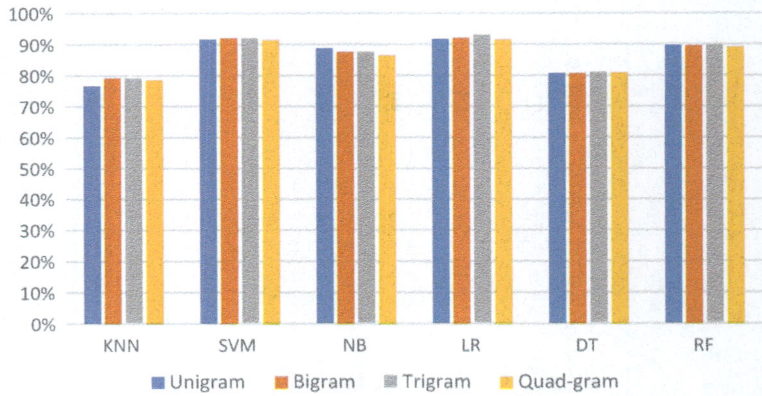

Fig. 2 Accuracy for the unigram to the quad-gram with 500 features

500 features. Figure 2 shows the comparison between the accuracy for the unigram, bigram, trigram, and quad-gram.

Experiments results show that the highest accuracy level recorded is 94% with SVM and LR in both experiments done through unigram and bigram with 500 features. The lowest accuracy recorded was with KNN. The highest accuracy reached 93% with trigrams and 92% with quad-grams. The result of the second experiment of 1000 features is shown in Table 4. Also, the highest accuracy reaches 94% for both classifiers SVM and LR. The lowest accuracy level is 72% and 73% for KNN with the unigram and bigram experiments. Figure 3 shows the comparison between accuracy for the four experiments using 1000 features extraction through bigram, unigram, trigram, and quad-gram.

In the deep learning experiment, the ANN architecture can be used in the aforementioned setting. The model performance in terms of accuracy reached 98% in the testing. The training process with 20 epochs reached 99% as shown in Fig. 4. The loss function in both validation and testing decreased as shown in Fig. 5 using Adam optimizer while the same setting also applied the 'SGD' optimizer, the results as shown in Figs. 6 and 7 for accuracy and loss function, respectively.

5 Conclusion

This book chapter proposes a model to detect Fake news on social media mainly on YouTube using sentiment analysis through classical MLC and DL. The dataset has been collected and annotated by an Arabic native speaker and also assessed using Kappa measures. The model using LR and VSM in N-grams reached an accuracy level of 94% while in ANN 99% with few epochs using Adam Optimizer. The lowest accuracy level was recorded in the model with KNN. The best performance for the

Table 4 Model performance based unigram and bigram with 500 features

MLCs	Uni-gram						Bi-gram				
	Class	P (%)	R (%)	F-score (%)	Acc (%)	Val (%)	P (%)	R (%)	F-score (%)	Acc (%)	Validation (%)
KNN	0	77	73	75	74	69	73	74	74	73	72
	1	70	74	72			73	72	72		
SVM	0	96	93	94	94	95	93	96	94	94	94
	1	93	96	94			96	93	94		
NB	0	87	96	91	91	92	81	98	89	89	89
	1	96	87	92			98	83	90		
LR	0	93	96	94	94	93	93	96	94	94	94
	1	95	93	94			96	92	94		
DT	0	66	96	79	81	83	66	96	79	81	83
	1	97	73	83			97	73	83		
RF	0	91	93	92	92	92	91	93	92	92	92
	1	93	91	92			93	90	91		

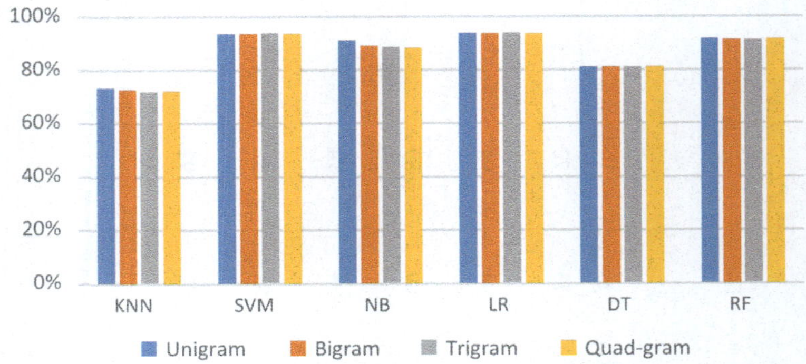

Fig. 3 Accuracy for the unigram to the quad-gram with 1000 features

Fig. 4 Accuracy using 'Adam' optimizer

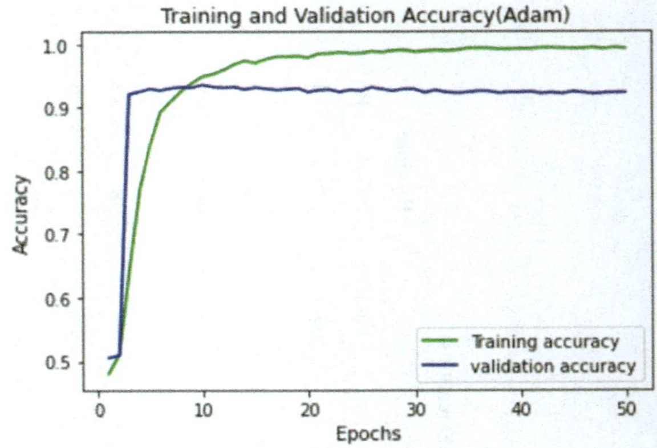

Fig. 5 Loss using 'Adam' optimizer

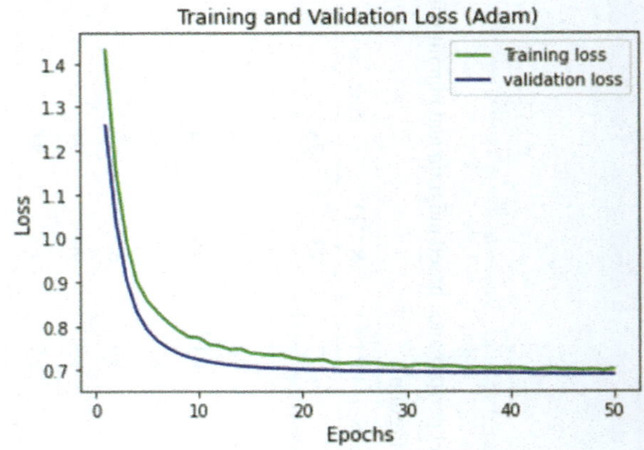

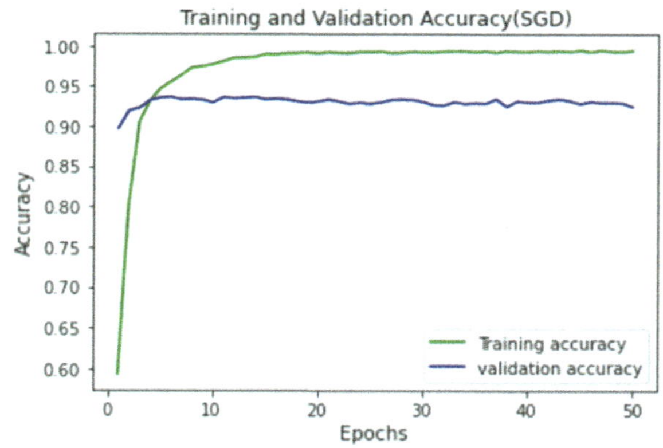

Fig. 6 Accuracy using 'SGD' optimizer

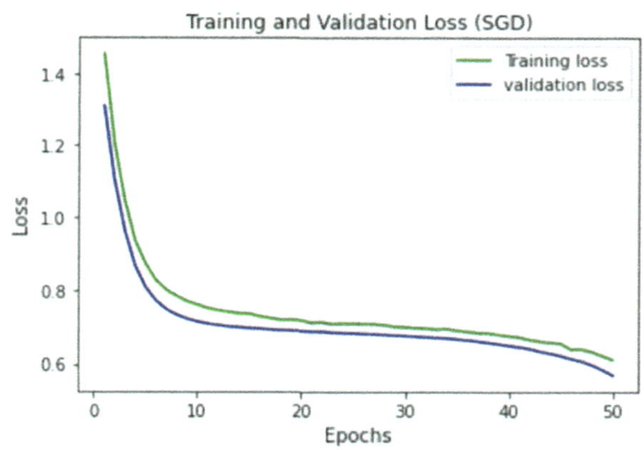

Fig. 7 Loss using 'SGD' optimizer

model has been recorded through ANN. The future work will focus on building big datasets and making them available for the public by applying deep learning models.

References

1. Yafooz, W. M., Abidin, S. Z., & Omar, N. (2011, November). Challenges and issues on online news management. In *2011 IEEE International Conference on Control System, Computing and Engineering* (pp. 482–487). IEEE.
2. Yafooz, W. M., Abidin, S. Z., Omar, N., & Hilles, S. (2016, September). Interactive Big Data Visualization Model Based on Hot Issues (Online News Articles). In *International Conference on Soft Computing in Data Science* (pp. 89–99). Springer, Singapore.
3. Allcott, H., & Gentzkow, M. (2017). Social media and fake news in the 2016 election. *Journal of Economic Perspectives, 31*(2), 211–236.
4. Aldwairi, M., & Alwahedi, A. (2018). Detecting fake news in social media networks. *Procedia Computer Science, 141*, 215–222.

5. Pérez-Rosas, V., Kleinberg, B., Lefevre, A., & Mihalcea, R. (2017). Automatic detection of fake news. arXiv preprint arXiv:1708.07104.
6. Rashkin, H., Choi, E., Jang, J. Y., Volkova, S., & Choi, Y. (2017, September). Truth of varying shades: Analyzing language in fake news and political fact-checking. In *Proceedings of the 2017 Conference on Empirical Methods in Natural Language Processing* (pp. 2931–2937).
7. Potthast, M., Kiesel, J., Reinartz, K., Bevendorff, J., & Stein, B. (2017). A stylometric inquiry into hyperpartisan and fake news. arXiv preprint arXiv:1702.05638.
8. Ahmed, H., Traore, I., & Saad, S. (2017, October). Detection of online fake news using n-gram analysis and machine learning techniques. In *International Conference on Intelligent, Secure, and Dependable Systems in Distributed and Cloud Environments* (pp. 127–138). Springer, Cham.
9. Ghosh, S., & Shah, C. (2018). Towards automatic fake news classification. *Proceedings of the Association for Information Science and Technology, 55*(1), 805–807.
10. Vicario, M. D., Quattrociocchi, W., Scala, A., & Zollo, F. (2019). Polarization and fake news: Early warning of potential misinformation targets. *ACM Transactions on the Web (TWEB), 13*(2), 1–22.
11. Yang, F., Liu, Y., Yu, X., & Yang, M. (2012, August). Automatic detection of rumor on sina weibo. In *Proceedings of the ACM SIGKDD Workshop on Mining data Semantics* (pp. 1–7).
12. Shu, K., Zhou, X., Wang, S., Zafarani, R., & Liu, H. (2019, August). The role of user profiles for fake news detection. In *Proceedings of the 2019 IEEE/ACM international conference on advances in social networks analysis and mining* (pp. 436–439).
13. PolitiFact. (2021). http://www.politifact.com, [Online] available , accessed 22, August 2021.
14. Gossipcop. (2021). https://www.gossipcop.com/, [Online] available , accessed 22, August 2021.
15. Shu, K., Wang, S., & Liu, H. (2019, January). Beyond news contents: The role of social context for fake news detection. In *Proceedings of the twelfth ACM international conference on web search and data mining* (pp. 312–320).
16. Shu, K., Bernard, H. R., & Liu, H. (2019). Studying fake news via network analysis: detection and mitigation. In *Emerging research challenges and opportunities in computational social network analysis and mining* (pp. 43–65). Springer, Cham.
17. Krishnan, S., & Chen, M. (2018, July). Identifying tweets with fake news. In *2018 IEEE International Conference on Information Reuse and Integration (IRI)* (pp. 460–464). IEEE.
18. Saez-Trumper, D. (2014, September). Fake tweet buster: a webtool to identify users promoting fake news ontwitter. In *Proceedings of the 25th ACM conference on Hypertext and social media* (pp. 316–317).
19. Atodiresei, C. S., Tănăselea, A., & Iftene, A. (2018). Identifying fake news and fake users on Twitter. *Procedia Computer Science, 126*, 451–461.
20. Ghanem, B., Ponzetto, S. P., & Rosso, P. (2020, October). FacTweet: profiling fake news twitter accounts. In *International Conference on Statistical Language and Speech Processing* (pp. 35–45). Springer, Cham.
21. Shao, C., Ciampaglia, G. L., Varol, O., Flammini, A., & Menczer, F. (2017). The spread of fake news by social bots. arXiv preprint arXiv:1707.07592, 96, 104.
22. Erşahin, B., Aktaş, Ö., Kılınç, D., & Akyol, C. (2017, October). Twitter fake account detection. In *2017 International Conference on Computer Science and Engineering (UBMK)* (pp. 388–392). IEEE.
23. Ma, J., Gao, W., & Wong, K. F. (2017). Detect rumors in microblog posts using propagation structure via kernel learning. *Association for Computational Linguistics.*
24. Malhotra, B., & Vishwakarma, D. K. (2020, September). Classification of propagation path and tweets for rumor detection using graphical convolutional networks and transformer based encodings. In *2020 IEEE Sixth International Conference on Multimedia Big Data (BigMM)* (pp. 183–190). IEEE.
25. Saad, M., Ahmad, A., & Mohaisen, A. (2019, June). Fighting fake news propagation with blockchains. In *2019 IEEE Conference on Communications and Network Security (CNS)* (pp. 1–4). IEEE.

26. Sivasankari, S., & Vadivu, G. (2021). Tracing the fake news propagation path using social network analysis. *Soft Computing* 1–9.
27. Mishra, R. (2020). Fake news detection using higher-order user to user mutual-attention progression in propagation paths. In *Proceedings of the IEEE/CVF Conference on Computer Vision and Pattern Recognition Workshops* (pp. 652–653).
28. Fullfact. (2021). https://fullfact.org/, [online], available , retrieved 26 , August , 2021
29. Chen, T., Li, X., Yin, H., & Zhang, J. (2018, June). Call attention to rumors: Deep attention based recurrent neural networks for early rumor detection. In *Pacific-Asia conference on knowledge discovery and data mining* (pp. 40–52). Springer, Cham.
30. Rubin, V. L., Chen, Y., & Conroy, N. K. (2015). Deception detection for news: Three types of fakes. *Proceedings of the Association for Information Science and Technology, 52*(1), 1–4.
31. Shu, K., Sliva, A., Wang, S., Tang, J., & Liu, H. (2017). Fake news detection on social media: A data mining perspective. *ACM SIGKDD explorations newsletter, 19*(1), 22–36.
32. Ma, J., Gao, W., Mitra, P., Kwon, S., Jansen, B. J., Wong, K. F., & Cha, M. (2016). Detecting rumors from microblogs with recurrent neural networks.
33. Shu, K., Mahudeswaran, D., Wang, S., Lee, D., & Liu, H. (2018). Fakenewsnet: A data repository with news content, social context and spatialtemporal information for studying fake news on social media. arXiv preprint arXiv:1809.01286.
34. Zubiaga, A., Liakata, M., Procter, R., Wong Sak Hoi, G., & Tolmie, P. (2016). Analysing how people orient to and spread rumours in social media by looking at conversational threads. *PloS one, 11*(3), e0150989.
35. Mitra, T., & Gilbert, E. (2015, April). Credbank: A large-scale social media corpus with associated credibility annotations. In *Ninth international AAAI conference on web and social media.*
36. Santia, G. C., & Williams, J. R. (2018, June). Buzzface: A news veracity dataset with facebook user commentary and egos. In *Twelfth International AAAI Conference on Web and Social Media.*
37. Yafooz, W. M., & Alsaeedi, A. (2021). Sentimental analysis on health-related information with improving model performance using machine learning.

The manufacturer's authorised representative in the EU is Springer Nature Customer Service Centre GmbH, Europaplatz 3, 69115 Heidelberg, Germany. If you have any concerns regarding our products, please contact ProductSafety@springernature.com

Printed and bound by CPI Group (UK) Ltd, Croydon, CR0 4YY
30/09/2026
02225291-0016